Progress in the Chemistry of Organic Natural Products

More information about this series at http://www.springer.com/series/10169

A. Douglas Kinghorn • Heinz Falk •
Simon Gibbons • Jun'ichi Kobayashi
Editors

Progress in the Chemistry of Organic Natural Products

Volume 101

With contributions by

S.-G. Liao · J.-M. Yue
F.-R. Chang · C.-C. Liaw · J.-R. Liou · T.-Y. Wu · Y.-C. Wu

Springer

Editors
A. Douglas Kinghorn
Div. Medicinal Chemistry & Pharmacognosy
The Ohio State University
College of Pharmacy
Columbus, OH, USA

Heinz Falk
Institute of Organic Chemistry
Johannes Kepler University
Linz, Austria

Simon Gibbons
Research Department of Pharmaceutical and Biological Chemistry
UCL School of Pharmacy
London, United Kingdom

Jun'ichi Kobayashi
Graduate School of Pharmaceutical Science
Hokkaido University
Sapporo, Japan

ISSN 2191-7043 ISSN 2192-4309 (electronic)
Progress in the Chemistry of Organic Natural Products
ISBN 978-3-319-22691-0 ISBN 978-3-319-22692-7 (eBook)
DOI 10.1007/978-3-319-22692-7

Library of Congress Control Number: 2015957989

Springer Cham Heidelberg New York Dordrecht London

Printed on acid-free paper

Springer International Publishing AG Switzerland is part of Springer Science+Business Media (www.springer.com)

Contents

Dimeric Sesquiterpenoids

Shang-Gao Liao and Jian-Min Yue

Contents

The original version of this chapter was revised as the affiliation of J.-M. Yue was incorrect. The erratum to this chapter is available at DOI 10.1007/978-3-319-22692-7_3.

S.-G. Liao
School of Pharmacy, Guizhou Medical University, Zhangjiang Hi-Tech Park, Shanghai 201203, People's Republic of China

Engineering Research Center for the Development and Application of Ethnic Medicine and TCM (Ministry of Education), School of Pharmacy, Guizhou Medical University, Guizhou 550004, People's Republic of China
e-mail: lshangg@163.com

J.-M. Yue (✉)
State Key Laboratory of Drug Research, Shanghai Institute of Materia Medica, Chinese Academy of Sciences, Zhangjiang Hi-Tech Park, Shanghai 201203, People's Republic of China
e-mail: jmyue@simm.ac.cn

A.D. Kinghorn, H. Falk, S. Gibbons, J. Kobayashi (eds.), *Progress in the Chemistry of Organic Natural Products, Vol. 101*, DOI 10.1007/978-3-319-22692-7_1

1 Introduction

It is widely accepted that a large number of proteins responsible for cellular function exist as dimers (hetero- or homo-) or need to be activated by dimerization before mediating certain signaling pathways [1, 2]. Simultaneously targeting both monomeric moieties of the dimeric proteins has shown potential in the development of various therapeutic agents [3–5]. As natural or synthetic dimeric molecules might be able to act on both moieties of a dimeric protein, dimeric sesquiterpenoids (DSs), which are generated biogenetically from coupling of two sesquiterpenoid molecules (either identical or different), are in essence potential biologically active molecules and have attracted great attention in recent years for their particular structures and biological activities. With a composition of at least 30 carbons, and their generation from sesquiterpenoids of a variety of structural types, and in showing variations of the connecting patterns of the two identical (for homo-DS) or different (for hetero-DS) sesquiterpenoid units, this makes the elucidation of DS structures and their synthetic construction quite challenging. Moreover, the biological effects of the DSs, particularly their potential anti-inflammatory, antimalarial, antitumor, antiviral, immunosuppressive, neurotropic, and potassium channel blocking activities, have rendered these molecules promising candidates for further drug development. A general trend observed is that some DSs are more potent than their monomeric precursors for many biological activities.

Two recent reviews have been written by Zhan et al. [6] and Lian and Yu [7], covering the isolation, structural determination, biological activities, biogenesis, and synthesis of natural DSs up to June 2010.

In this contribution, a general view of the classification and distribution of DSs (including those reported recently) will be provided. Strategies for the structural elucidation of DSs and their analogues will be presented. Chemical efforts toward the construction of DSs, particularly strategies for the convergence of the two sesquiterpenoid units, will be reviewed. Moreover, the biological activities of DSs will be discussed under each type of activity for the purposes of providing information regarding the structural features required by their target proteins.

2 Classification and Distribution

Based on coupling patterns and structural features of the two constitutional sesquiterpenoid units, DSs can be classified into disesquiterpenoids (type A) and pseudo-disesquiterpenoids (type B) [6]. In type A, two sesquiterpenoid units are connected directly by at least one C–C bond. In contrast, in type B, the two units are changed to two aza-sesquiterpenoid moieties, or are connected by an ester group, an O-/S-ether linkage, one or two amide groups, or a nitrogen/urea group. Dimerosesquiterpenoids [6], which originate from coupling of two merosesquiterpenoids that are biogenetically formed from direct carbon-carbon connection of a sesquiterpenoid and a nonsesquiterpenoid, are not included among the DS group.

Taking into consideration the structural types of monomeric sesquiterpenoids, disesquiterpenoids (type A) can also be classified into bisabolane, germacrane, guaiane, eremophilane, cadinane, eudesmane, lindenane, miscellaneous sesquiterpene, and compound disesquiterpenoids. Compound disesquiterpenoids refer to those DSs that are formed by coupling of two sesquiterpenoid units of different structural types. Pseudo-disesquiterpenoids (type B) may also have similar constitutions, but will not be discussed in this respect. It should be noted that since non-carbon-carbon-connected pseudo-disesquiterpenoids are more prone to metabolism to the constitutional monomers or their derivatives before reaching the target proteins, DSs of type A and dimeric aza-sesquiterpenoids seem to be of greater overall importance.

2.1 Disesquiterpenoid DSs

2.1.1 Bisabolane Disesquiterpenoids

Dimeric sesquiterpenoids of this type are generally present in either the Plantae (plant) or Animalia (sponge) kingdoms (Table 1). The genera *Perezia*, *Coreocarpus*, and *Baccharis* from the Asteraceae and the genus *Meiogyne* from the Annonaceae in the plant kingdom, as well as *Axinyssa* (Halichondriidae) and *Lipastrotethya* (Dictyonellidae) in the animal kingdom are reported to produce bisabolane disesquiterpenoids. However, one bisabolane disesquiterpenoid, disydonol C (**1**), has been reported to occur in a marine-derived fungus (*Aspergillus* sp.) (Fig. 1) [13].

Table 1 Bisabolane DSs

Compound	Origin	Family	Kingdom	Ref.
Bacchopetiolone	*Baccharis petiolata*	Asteraceae	Plant	[8]
6′,6-Bis-2-(1,5-dimethyl-1,4-benzoquinone	*Coreocarpus arizonicus*	Asteraceae	Plant	[9]
6′,6-Bis-2-(1,5-dimethyl-4-hexenyl-6-isovaleroxy)-3-hydroxy-5-methyl-1,4-benzoquinone	*Coreocarpus arizonicus*	Asteraceae	Plant	[9]
cis-Dimer A	*Axinyssa variabilis*	Halichondriidae	Animal (sponge)	[10]
cis-Dimer A	*Lipastrotethya ana*	Dictyonellidae	Animal	[10]
cis-Dimer B	*Axinyssa variabilis*	Halichondriidae	Animal	[10]
cis-Dimer B	*Lipastrotethya ana*	Dictyonellidae	Animal	[10]
Dicurcuphenols A–E	*Didiscus aceratus*	Heteroxyidae	Animal (sponge)	[11]
Diperezone/biperezone	*Coreocarpus arizonicus*	Asteraceae	Plant	[9]
Diperezone/biperezone	*Perezia alamani* var. *oolepis*	Asteraceae	Plant	[12]
Disydonol C (**1**)	*Aspergillus* sp.	Trichocomaceae	Fungi	[13]
1-*epi*-Meiogynin A	*Meiogyne cylindrocarpa*	Annonaceae	Plant	[14]
Meiogynin A	*Meiogyne cylindrocarpa*	Annonaceae	Plant	[14]
trans-Dimer C	*Lipastrotethya ana*	Dictyonellidae	Animal	[10]
trans-Dimer C	*Axinyssa variabilis*	Halichondriidae	Animal	[15]
trans-Dimer D	*Axinyssa variabilis*	Halichondriidae	Animal	[15]

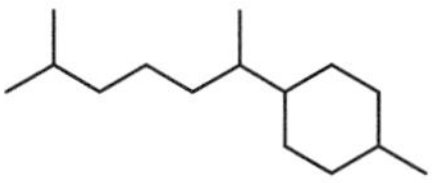

bisabolane

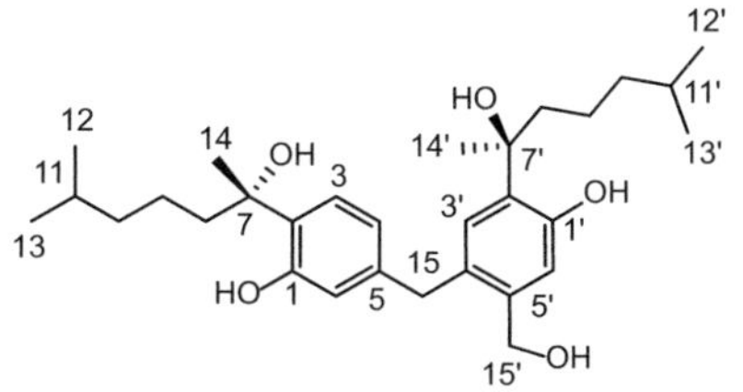

1 (disydonol C)

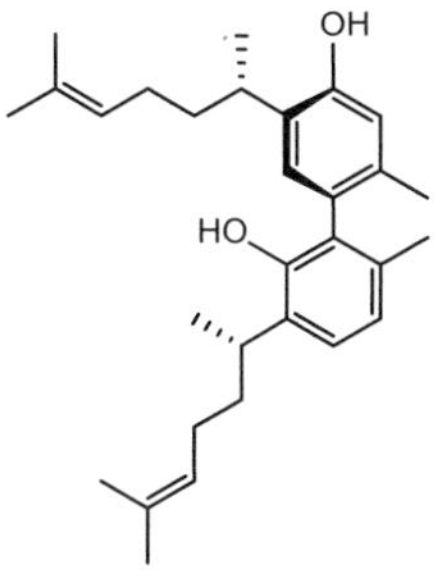

2 (dicurcuphenol B)

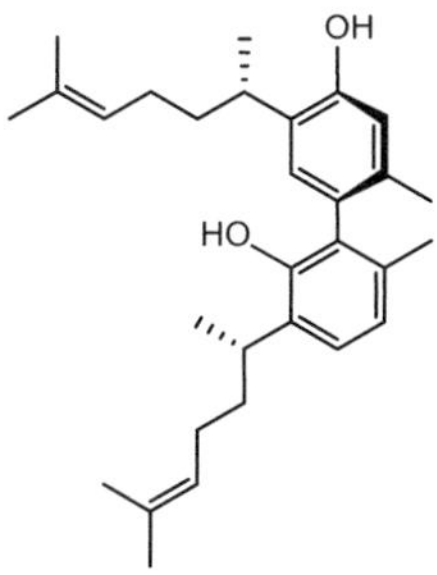

3 (dicurcuphenol C)

Fig. 1 Structures of bisabolane, disydonol C (**1**), and dicurcuphenols B (**2**) and C (**3**)

2.1.2 Germacrane Disesquiterpenoids

Dimeric sesquiterpenoids of this type are very rare. Up to the present, only a few species in the families Aristolochiaceae, Asteraceae, and Zingiberaceae have been reported to accumulate these compounds. Germacrane disesquiterpenoids are rigidly confined to the plant kingdom (Table 2, Fig. 2).

Table 2 Germacrane DSs

Compound	Origin	Family	Ref.
Artebarrolide	*Artemisia barrelieri*	Asteraceae	[16]
Difurocumenone	*Curcuma zedoaria*	Zingiberaceae	[17]
Elegain	*Gonospermum elegans*	Asteraceae	[18]
Helivypolide G	*Helianthus annuus*	Asteraceae	[19]
Mikagoyanolide (**4**)	*Mikania goyazensis*	Asteraceae	[20]
Mikagoyanolide (**4**)	*Tanacetopsis mucronata*	Asteraceae	[21]
Versicolactone	*Aristolochia versicolor*	Aristolochiaceae	[22]

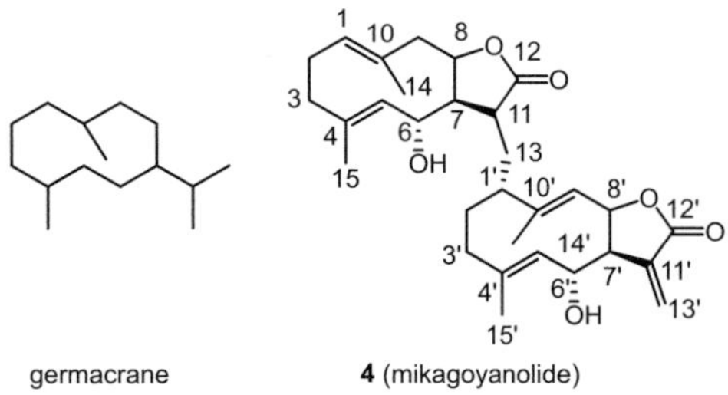

Fig. 2 Structures of germacrane and mikagoyanolide (**4**)

2.1.3 Guaiane, Pseudoguaiane, and Xanthane Disesquiterpenoids

Dimeric sesquiterpenoids of these classes may contain two guaiane, pseudoguaiane, secoguaiane, or xanthane units (Fig. 3), and are one of the largest aggregate groups of DSs. Asteraceae, which produces a considerable number of guaiane sesquiterpenoids, is also the major family producing guaiane disesquiterpenoids (Table 3). *Artemisia* is among the most important genera of this family, from a medicinal perspective. Several species in this genus have been reported to be guaiane disesquiterpenoid sources. In addition, *Daphne oleoides* of the Thymelaeaceae [33], *Salvia nubicola* of the Lamiaceae, *Joannesia princeps* of the Euphorbiaceae [41], and two species (i.e. *Xylopia aromatica* [58] and *Xylopia vielana* [71]) of the Annonaceae family were also reported to contain DSs. It is noteworthy that DSs of these types are limited to the plant kingdom.

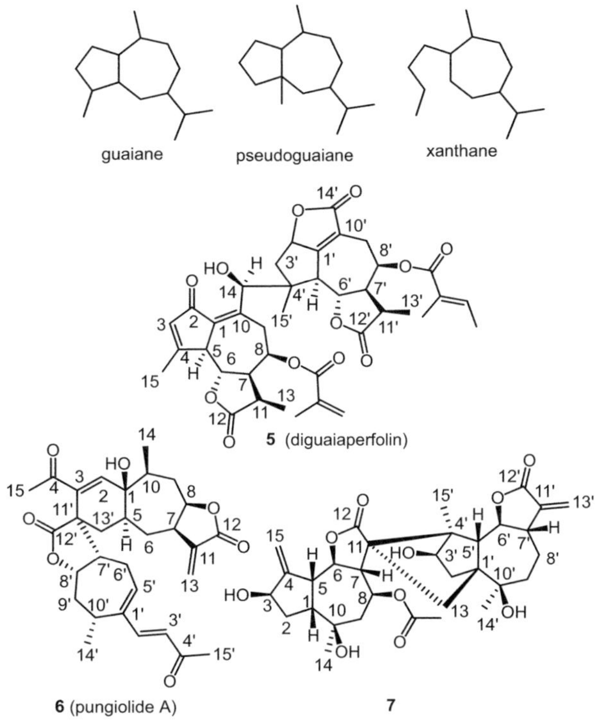

Fig. 3 Structures of guaiane, pseudoguaiane, xanthane, diguaiaperfolin (**5**), pungiolide A (**6**), and **7**

Table 3 Guaiane, pseudoguaiane, and xanthane DSs

Compound	Origin	Family	Refs.
Absinthin	*Artemisia absinthium*	Asteraceae	[23, 24]
Absinthin	*Artemisia caruifolia*	Asteraceae	[25]
Absintholide	*Artemisia caruifolia*	Asteraceae	[25]
Absintholide	*Artemisia absinthium*	Asteraceae	[26]
Achicollinolide	*Achillea collina*	Asteraceae	[27]
Achillinins B–C[a]	*Achillea millefolium*	Asteraceae	[28]
Ainsliadimer A	*Ainsliaea macrocephala*	Asteraceae	[29]
Ainsliadimer B	*Ainsliaea fulvioides*	Asteraceae	[30]
Anabsin	*Artemisia absinthium*	Asteraceae	[31]
Anabsin	*Artemisia caruifolia*	Asteraceae	[25]

(continued)

Table 3 (continued)

Compound	Origin	Family	Refs.
Anabsinthin	*Artemisia absinthium*	Asteraceae	[31]
Anabsinthin	*Artemisia anomala*	Asteraceae	[32]
Anabsinthin	*Artemisia caruifolia*	Asteraceae	[25]
Anabsinthin	*Daphne oleoides*	Thymelaeaceae	[33]
Artabsinolide B	*Artemisia caruifolia*	Asteraceae	[25]
Artanomadimers A–F	*Artemisia anomala*	Asteraceae	[34]
Artanomalide D	*Artemisia anomala*	Asteraceae	[32]
Artanomaloide A	*Artemisia anomala*	Asteraceae	[32, 35]
Artelein	*Artemisia leucodes*	Asteraceae	[36]
Artenolide	*Artemisia absinthium*	Asteraceae	[37]
Artesieversin	*Artemisia sieversiana*	Asteraceae	[38]
Artselenoide A	*Artemisia selengensis*	Asteraceae	[39, 40]
Artselenoide A	*Artemisia sylvatica*	Asteraceae	[40]
Assufulvenal	*Joannesia princeps*	Euphorbiaceae	[41]
Biennin C	*Hymenoxys biennis*	Asteraceae	[42]
2,12′-Bis-hamazulenyl	*Ajania fruticulosa*	Asteraceae	[43]
Bisnubenolide	*Salvia nubicola*	Lamiaceae	[44]
Bisnubidiol	*Salvia nubicola*	Lamiaceae	[45]
Bistataxacin	*Salvia nubicola*	Lamiaceae	[46]
Caruifolins B–D	*Artemisia caruifolia*	Asteraceae	[25]
Chrysanolide C	*Chrysanthemum indicum*	Asteraceae	[47]
Decathieleanolide	*Decachaeta thieleana*	Asteraceae	[48]
10-Desoxygochnatiolide A	*Gochnatia polymorpha*	Asteraceae	[49]
10-Desoxygochnatiolide A	*Gochnatia polymorpha* ssp. *floccosa*	Asteraceae	[49]
10-Desoxy-10β-*H*-gochnatiolide A	*Gochnatia hypoleuca*	Asteraceae	[49]
Dichrocepholides D–E	*Dichrocephala integrifolia*	Asteraceae	[50]

(continued)

Table 3 (continued)

Compound	Origin	Family	Refs.
Diguaiaperfolin (**5**)	*Eupatorium perfoliatum*	Asteraceae	[51]
Dihydroornativolide	*Geigeria ornativa*	Asteraceae	[52]
8*β*,8′*β*-Dihydroxy-10-desoxy-10*β*-*H*-gochnatiolide A	*Gochnatia hypoleuca*	Asteraceae	[49]
Distansolides A–B	*Achillea distans*	Asteraceae	[53]
10,11-Epiabsinthin	*Artemisia caruifolia*	Asteraceae	[25]
10′,11,11′-*epi*-Absinthin	*Artemisia sieversiana*	Asteraceae	[38]
10′,11′-*epi*-Absinthin	*Artemisia sieversiana*	Asteraceae	[38]
11-*epi*-Absinthin	*Artemisia sieversiana*	Asteraceae	[38]
11′-Epimaritimolide	*Ambrosia maritima*	Asteraceae	[35]
Gnapholide	*Pulicaria gnaphalodes*	Asteraceae	[54]
Gochnatiolide A	*Gochnatia polymorpha*	Asteraceae	[49]
Gochnatiolide A	*Gochnatia paniculata*	Asteraceae	[55]
Gochnatiolide B	*Gochnatia paniculata*	Asteraceae	[55]
Handelin (Yejuhua Lactone/Chrysanthelide)	*Handelia trichophylla*	Asteraceae	[56]
Helisplendidilactone	*Helichrysum splendidum*	Asteraceae	[57]
2*α*-Hydroxy-10-desoxy-1,10-dehydrogochnatiolide A	*Gochnatia polymorpha*	Asteraceae	[49]
2*α*-Hydroxy-10-desoxy-1,10-dehydro-11*α*,13,11′*α*,13′-tetrahydrogochnatiolide A	*Gochnatia polymorpha*	Asteraceae	[49]
8*β*-Hydroxy-10-desoxy-10*β*-*H*-gochnatiolide A	*Gochnatia hypoleuca*	Asteraceae	[49]
8-Hydroxy-10-desoxygochnatiolide A	*Gochnatia polymorpha*	Asteraceae	[49]
8-Hydroxy-10-desoxygochnatiolide A	*Gochnatia polymorpha* ssp. *floccosa*	Asteraceae	[49]
8′*β*-Hydroxy-10-desoxy-10*β*-*H*-gochnatiolide A	*Gochnatia hypoleuca*	Asteraceae	[49]
(11*α*,12*β*,13*α*,21*β*)-7-Hydroxy-16-oxo-17-isopropylidene-l*α*,5,5,9*β*,14*α*,20-hexamethyl-6-oxaheptacyclo[l0.9.1.$0^{2.10}.0^{4.7}.0^{12.24}.0^{13.19}$]docosa-2(l0),3,l9-triene	*Xylopia aromatica*	Annonaceae	[58]

(continued)

Table 3 (continued)

Compound	Origin	Family	Refs.
8α-Hydroxyxeranthemolide	*Anthemis austriaca*	Asteraceae	[59]
Isoabsinthin	*Artemisia absinthium*	Asteraceae	[60]
Leariifolianoids E[b]	*Inula leariifolia*	Asteraceae	[61]
Leariifolianoids F–G	*Inula leariifolia*	Asteraceae	[61]
Maritimolide	*Ambrosia maritima*	Asteraceae	[35]
Mexicanin F	*Helenium mexicanum*	Asteraceae	[62]
Microlenin	*Helenium microcephalum*	Asteraceae	[63, 64]
Microlenin acetate	*Helenium microcephalum*	Asteraceae	[65]
Millifolides A–B[c]	*Achillea millefolium*	Asteraceae	[66]
8-*O*-Acetylarteminolide	*Artemisia anomala*	Asteraceae	[32]
Ornativolide	*Geigeria ornativa*	Asteraceae	[52]
(5*S*,6*R*,7*R*,8*R*,11*R*)-2-Oxo-8-tigloyloxyguaia-1(10),3-dien-6,12-olide-14-carboxylic acid	*Eupatorium perfoliatum*	Asteraceae	[67]
Pungiolide C[d]	*Xanthium strumarium*	Asteraceae	[68]
Pungiolides A–B	*Xanthium pungens*	Asteraceae	[69]
Pungiolides D–E[d]	*Xanthium sibiricum*	Asteraceae	[70]
Seemarin	*Daphne oleoides*	Thymelaea-ceae	[33]
Vielanin A–C	*Xylopia vielana*	Annonaceae	[71]
Vielanins D–E	*Xylopia vielana*	Annonaceae	[72]
Xeranthemolide	*Anthemis austriaca*	Asteraceae	[59]
Unnamed (**7**)	*Chrysanthemum indicum*	Asteraceae	[73]

[a]1,10-Secoguaiaolide-guaiaolide
[b]Guaiane-pseudoguaiane
[c]1,10-Secoguaianolide-1,10-secoguaianolide
[d]Xanthane-xanthane

2.1.4 Eremophilane Disesquiterpenoids

Eremophilane disesquiterpenoids generally occur in the plant kingdom (Table 4) as dimeric furanoeremophilanes or dimeric eremophilenolides (Fig. 4) [6]. Most of the eremophilane DSs found to date have been isolated from species in the Asteraceae, with *Ligularia* being the most prevalent genus in this regard.

Table 4 Eremophilane DSs

Compound	Origin	Family	Ref.
Adenostin A	*Cacalia adenostyloides*	Asteraceae	[74]
Adenostin A	*Ligularia virgaurea*	Asteraceae	[75]
Adenostin B	*Cacalia adenostyloides*	Asteraceae	[74]
14-Angeloyloxy-12-(cacalohastin-14-yl) cacalohastine	*Senecio canescens*	Asteraceae	[76]
Bi-3*β*-angeloyloxy-8*β*-hydroeremophil-7(11)-en-12,8*α*(14b,6a)-diolide	*Ligularia lapathifolia*	Asteraceae	[77]
Bieremoligularolide	*Ligularia muliensis*	Asteraceae	[78]
Biliguhodgsonolide	*Ligularia hodgsonii*	Asteraceae	[79]
Biligulaplenolide	*Ligularia platyglossa*	Asteraceae	[80]
α,α′-Bis-3β-angeloyloxyfuranoeremophilane	*Farfugium japonicum*	Asteraceae	[81]
9*β*,9′*α*-Bis-1,8-dihydroligularenolide[a]	*Bedfordia salicina*	Asteraceae	[82]
9*β*,9′*β*-Bis-1,8-dihydroligularenolide[a]	*Bedfordia salicina*	Asteraceae	[82]
12-(Dehydrocacalohastin-14-yl)cacalohastin (**8**)	*Senecio canescens*	Asteraceae	[78]
12-(Dehydrocacalohastin-14-yl)cacalohastin (**8**)	*Senecio crispus*	Asteraceae	[83]
(1*S**,5*S**,10a*R**)-1-[(8′*S**,8a′*R**)-8′,8a′-Dimethyl-4′-oxo-1′,4′,6′, 7′,8′,8a′-hexahydronaphthalen-2′-yl]-4-hydroxy-1,4,5,10*α*-tetramethyl-1,2,3,4,5,6,7,9,10,10a-decahydroanthracen-9-one	*Eremophila mitchelli*	Scrophulariaceae	[84]
(4a*R*,4a*R*,5*S*,5*S*,9a*R*,9a*R*)-4,4,4a,4a,5,5,6,6,7,7,8,8-Dodecahydro-3,3,4a,4a,5,5-hexamethyl-2*H*,2*H*-9a,9a-binaphtho[2,3-*b*]furan-2,2-dione[a]	*Senecio tsoongianus*	Asteraceae	[85]
8*β*-[Eremophila-3′,7′(11′)-dien-12′,8′*α*;15′,6′*α*-diolide]-eremophil-3,7(11)-dien-12,8*α*;15,6*α*-diolide	*Ligularia atroviolacea*	Asteraceae	[86]
8*β*-[Eremophila-3′,7′(11′)-diene-12′,8′*α*;14′,6′*α*-diolide]eremophila-3,7(11)-diene-12,8*α*;14,6*α*-diolide	*Ligularia atroviolacea*	Asteraceae	[87]
Fischelactone	*Ligularia fischeri*	Asteraceae	[88]
Fischelactone B (**9**)	*Ligularia fischeri*	Asteraceae	[89]
(5*S*)-5,6,7,7a,7b,12b-Hexahydro-3,4,5,11,12b-pentamethyl-10-[(3*E*)-pent-3-en-1-yl]-furo[3″,2″: 6′,7′]naphtho[1′,8′: 4,5,6]pyrano[3,2-*b*]benzofuran-9-ol	*Ligularia virgaurea*	Asteraceae	[90]
(4a*R*,5*S*,9a*S*)-4a,5,6,7,8,9a-Hexahydro-3,4a,5-trimethyl-9a-[(4a*R*,5*S*,9a*R*)-	*Senecio tsoongianus*	Asteraceae	[85]

(continued)

Table 4 (continued)

Compound	Origin	Family	Ref.
4,4a,5,6,7,8-hexahydro-3,4a,5-trimethyl-2-oxonaphtho[2,3-*b*]furan-9a(2*H*)-yl]naphtho-[2,3-*b*]furan-2(4*H*)-one (**10**)			
Ligulamulienins A–B	*Ligularia muliensis*	Asteraceae	[91]
Ligularin A	*Ligularia virgaurea* ssp. *oligocephala*	Asteraceae	[92]
Ligulolide B	*Ligularia virgaurea* ssp. *oligocephala*	Asteraceae	[93]
Ligulolide D	*Ligularia virgaurea* ssp. *oligocephala*	Asteraceae	[92]
2-{[(5*S*)-5,6,7,8-Tetrahydro-9-hydroxy-3,5-dimethylnaphtho[2,3-*b*]furan-4-yl]methyl}-3,5-dimethyl-6-[(3*E*)-pent-3-en-1-yl]-1-benzofuran-4,7-dione	*Ligularia virgaurea*	Asteraceae	[90]
Tetrahydromitchelladione	*Eremophila mitchelli*	Scrophulariaceae	[84, 94]
Virgaurin A	*Ligularia virgaurea*	Asteraceae	[75, 90, 95]
Virgaurin A	*Ligularia virgaurea*	Asteraceae	[90, 96]
Virgaurin B	*Ligularia virgaurea*	Asteraceae	[75, 97, 98]
Virgaurin C	*Ligularia virgaurea*	Asteraceae	[75, 98]
Virgaurols A–B	*Ligularia virgaurea*	Asteraceae	[95]

[a] 15(→6)-Furanoeremophilane

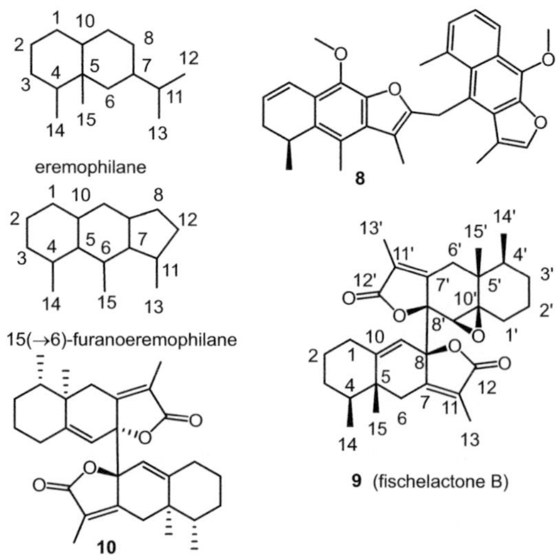

Fig. 4 Structures of eremophilane and its disesquiterpenoids **8**, fischelactone B (**9**), and **10**

2.1.5 Cadinane Disesquiterpenoids

Dimeric sesquiterpenoids of this class are very rare (Table 5). Species of the genus *Gossypium* in the family Malvaceae seem to be a good source of cadinane disesquiterpenoids (Fig. 5). In addition, *Curcuma parviflora* from the Zingiberaceae family also has been reported to produce a number of cadinane disesquiterpenoids [110, 112, 113].

Table 5 Cadinane DSs

Compound	Origin	Family	Refs.
Aquatidial	*Pachira aquatica*	Malvaceae	[99]
Bicalamenene	*Dysoxylum alliaceum*	Meliaceae	[100, 101]
(5*S*,5′*R*,8*S*,8′*S*)-5,5′-diisopropyl-3,3′,8,8′-tetramethyl-5,5′,6,6′,7,7′,8,8′-octahydro-[1,2′-binaphthalene]-1′,2-diol	*Ocotea corymbosa*	Lauraceae	[102]
8-Bis(7-hydroxycalamenene)	*Heritiera ornithocephala*	Malvaceae	[103]
8-Bis(7-hydroxycalamenene)	*Siparuna macrotepala*	Monimiaceae	[104]
(+)-7,7′-Bis[(5*R*,7*R*,9*R*,10*S*)-2-oxocadinan-3,6(11)-dien-12,7-olide]	*Eupatorium adenophorum*	Asteraceae	[105]
Dicadalenol	*Heterotheca inuloides*	Asteraceae	[106]
6,6′-Dimethoxygossypol	*Gossypium barbadense*	Malvaceae	[107]
6,6′-Dimethoxygossypol	*Gossypium hirsutum*	Malvaceae	[107]
Gossypol (**11**)	*Gossypium barbadense*	Malvaceae	[108]
Gossypol (**11**)	*Gossypium hirsutum*	Malvaceae	[108, 109]
6′-Methoxygossypol	*Gossypium barbadense*	Malvaceae	[107]
6′-Methoxygossypol	*Gossypium hirsutum*	Malvaceae	[107]
Parviflorene A	*Curcuma parviflora*	Zingiberaceae	[110]
Parviflorene J	*Curcuma parviflora*	Zingiberaceae	[111]
Parviflorenes B–F	*Curcuma parviflora*	Zingiberaceae	[111, 112]
Parviflorenes G–I	*Curcuma parviflora*	Zingiberaceae	[111, 113]

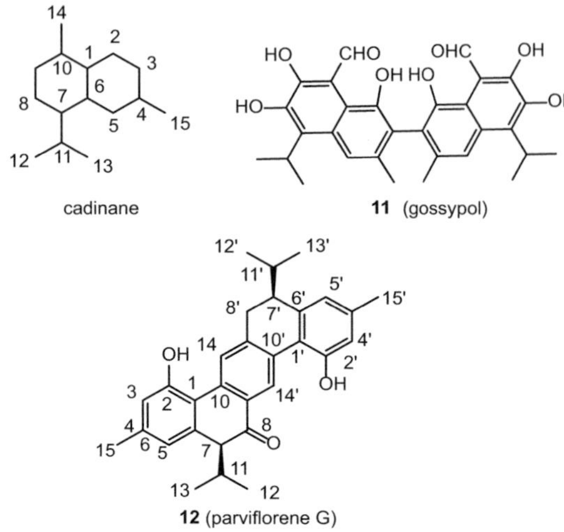

Fig. 5 Structures of cadinane and its disesquiterpenoids gossypol (**11**) and parviflorene G (**12**)

2.1.6 Eudesmane Disesquiterpenoids

Eudesmane disesquiterpenoids are generally present either as dimers of two eudesmane units (e.g. **13** and **14**) or as those of both an eudesmane and a secoeudesmane sesquiterpenoid component (e.g., **15**) (Fig. 6). Currently, only eleven eudesmane DSs have been reported (Table 6). Again, the Asteraceae is the main plant family that produces this type of DSs.

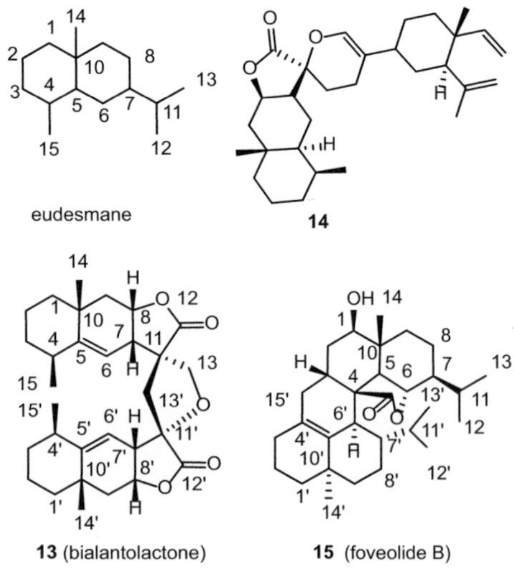

Fig. 6 Structures of eudesmane and its disesquiterpenoids bialantolactone (**13**), **14**, and foveolide B (**15**)

Table 6 Eudesmane DSs

Compound	Origin	Family	Ref.
Bialantolactone (**13**)	*Inula helenium*	Asteraceae	[114]
Biatractylolide	*Atractylodes macrocephala*	Asteraceae	[115]
Biatractylolide	*Trattinickia rhoifolia*	Asteraceae	[116]
Biepiasterolide	*Atractylodes macrocephala*	Asteraceae	[117]
Biepiasterolide	*Trattinickia rhoifolia*	Asteraceae	[118]
Bilindestenolide	*Lindera strychnifolia*	Lauraceae	[119]
3aa,3′,4′,5,6.7,8,8a,9,9a-Decahydro-5*β*-8*αβ*-5′-(4*aβ*-methyl-8-methyliden-2*β*-naphthyl)spiro[naphtha[2,3-*b*]furan-3,2′-2′*H*-pyran]-2(3*H*)-one	*Helenium autumnale*	Asteraceae	[120]
4a,5′-Ethenyl-4*β*-methyl-3*β*-[(1-methylethenyl) cyclohex-1*β*-yl]-3a,3′,4′,5,6,7,8,8a.9,9a-decahydro-5*β*,8*αβ*-dimethylspiro[naphtha[2,3-*b*]furan-3,2′-2′*H*-pyran]-2(3*H*)-one (**14**)	*Helenium autumnale*	Asteraceae	[120]
5′,4a-Ethenyl-4*β*-methyl-3*β*-[(1-methylethenyl) cyclohex-1*β*-yl]-3a,3′,4,4′,4a,5,6.7,8,8a,9,9a-dodecahydro-8a*β*-methyl-15-methylidenspiro[naphtha[2,3-*b*]furan-3,2′-2′*H*-pyran]-2(3*H*)-one	*Helenium autumnale*	Asteraceae	[120]
Foveolide B (**15**)	*Ficus foveolata*	Moraceae	[121]
Fruticolide	*Ferreyranthus fruticosus*	Asteraceae	[122]
Hydroxy-bis-dihydroencelin	*Montanoa speciosa*	Asteraceae	[123]
Muscicolides A–B	*Frullania muscicola*	Frullaniaceae	[124]

2.1.7 Lindenane Disesquiterpenoids

Lindenane disesquiterpenoids are one of the largest classes of DSs. However, up to the present, these DSs have been found only in the plant kingdom and are confined to the family Chloranthaceae (see Table 7 in the current chapter and Table 2 of the review by Zhan et al. [6]). *Chloranthus* and *Sarcandra* are the only two genera that have been reported to produce this type of DS (Fig. 7, Plate 1).

Table 7 Lindenane DSs isolated in recent years (2010–2013)

Compound	Origin	Family	Ref.
Chloramultilide A	*Chloranthus serratus*	Chloranthaceae	[125]
Chloramultilide C	*Chloranthus elatior*	Chloranthaceae	[126]
Chloramultilide C	*Chloranthus multistachys*	Chloranthaceae	[127]
Chloramultilide D	*Chloranthus multistachys*	Chloranthaceae	[127]
Chloramultiol G (**16**)	*Chloranthus multistachys*	Chloranthaceae	[128]
Chloramultiols A–F	*Chloranthus multistachys*	Chloranthaceae	[127]

(continued)

Table 7 (continued)

Compound	Origin	Family	Ref.
Cycloshizukaol A	*Chloranthus fortunei*	Chloranthaceae	[129]
Cycloshizukaol A	*Chloranthus multistachys*	Chloranthaceae	[127]
Henriol A	*Chloranthus serratus*	Chloranthaceae	[125]
Henriol D	*Chloranthus fortunei*	Chloranthaceae	[129]
Multistalides A–B	*Chloranthus multistachys*	Chloranthaceae	[130]
Sarcandrolides A–E	*Sarcandra glabra*	Chloranthaceae	[131]
Sarcandrolides F–J	*Sarcandra glabra*	Chloranthaceae	[132]
Sarcanolides A–B	*Sarcandra hainanensis*	Chloranthaceae	[133]
Shizukaol B	*Chloranthus spicatus*	Chloranthaceae	[134]
Shizukaol B	*Chloranthus fortunei*	Chloranthaceae	[129]
Shizukaol B	*Chloranthus japonicus*	Chloranthaceae	[135]
Shizukaol C	*Chloranthus spicatus*	Chloranthaceae	[134]
Shizukaol C	*Chloranthus multistachys*	Chloranthaceae	[127]
Shizukaol C	*Chloranthus fortunei*	Chloranthaceae	[129]
Shizukaol D	*Chloranthus fortunei*	Chloranthaceae	[129]
Shizukaol D	*Chloranthus multistachys*	Chloranthaceae	[127]
Shizukaol H	*Chloranthus spicatus*	Chloranthaceae	[134]
Shizukaols B, D	*Chloranthus serratus*	Chloranthaceae	[125]
Shizukaols E, G, M, O	*Chloranthus fortunei*	Chloranthaceae	[129]
Spicachlorantin B	*Chloranthus multistachys*	Chloranthaceae	[127]
Spicachlorantins A, C	*Chloranthus serratus*	Chloranthaceae	[125]
Spicachlorantins G–J	*Chloranthus spicatus*	Chloranthaceae	[136]
Unnamed (**17**–**18**)	*Chloranthus serratus*	Chloranthaceae	[125]

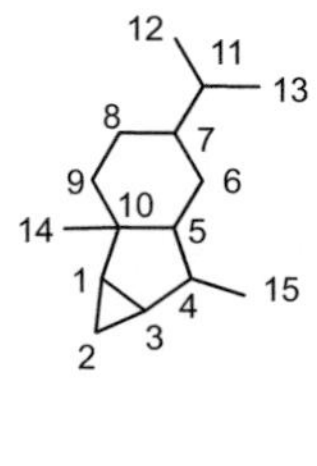

lindenane

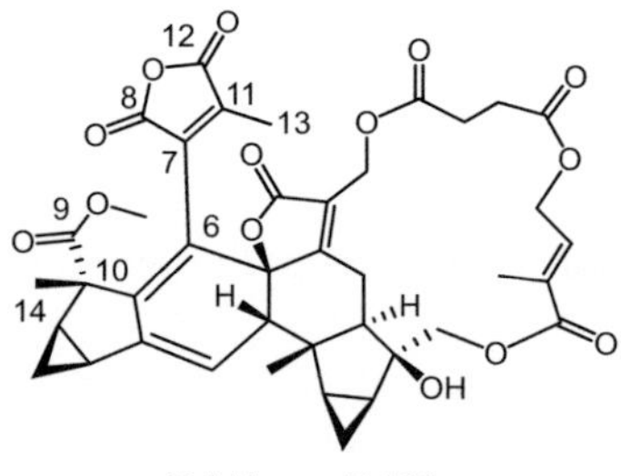

16 (chloramultiol G)

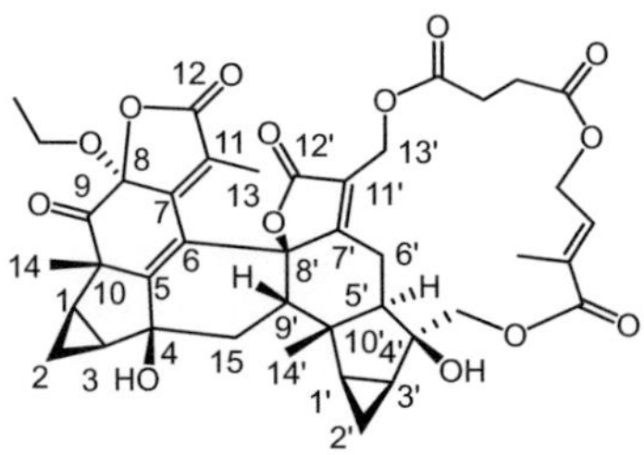

17

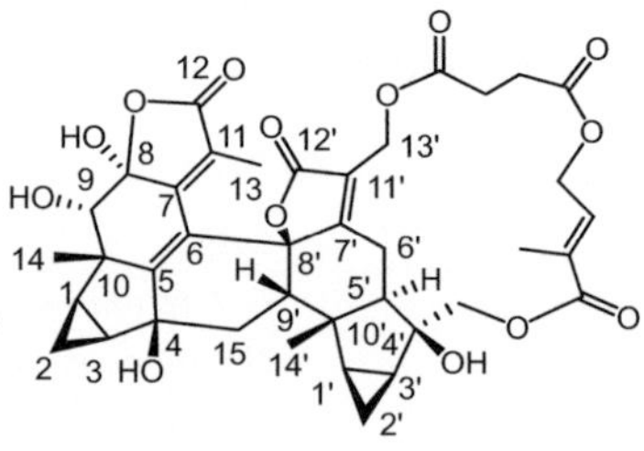

18

Fig. 7 Structures of lindenane and its disesquiterpenoids **16** and **17**, and the 8,9-*seco*-lindenane disesquiterpenoid chloramultiol G (**18**)

Plate 1 Lindenane disesquiterpenoids

2.1.8 Cuparane, Cyclolaurane, and Herbertane Disesquiterpenoids

The cuparane, cyclolaurane, and herbertane skeletons are related structurally. Since these types of monomeric sesquiterpenoids are not common, their dimerization has been reported only occasionally in several species of the families Herbertiaceae, Lejeuneaceae, Mastigophoraceae, Rhodomelaceae, and Scrophulariaceae (Table 8, Fig. 8).

Table 8 Cuparane, cyclolaurane, and herbertane DSs

Compound	Structural type	Origin	Family	Refs.
Aquaticenol	Herbertane (Isocuparane)	*Lejeunea aquatica*	Lejeuneaceae	[137]
Aquaticol (**19**)	Cuparane	*Veronica anagallis-aquatica*	Scrophulariaceae	[138, 139]
Laurebiphenyl	Cyclolaurane	*Laurencia nidifica*	Rhodomelaceae	[140]
Mastigophorenes A–B	Herbertane (Isocuparane)	*Herbertus sakuraii*	Herbertaceae	[141]
Mastigophorenes A–B	Herbertane (Isocuparane)	*Mastigophora diclados*	Mastigophoraceae	[142, 143]
Mastigophorenes C–D	Herbertane (Isocuparane)	*Mastigophora diclados*	Mastigophoraceae	[143, 144]
Unnamed (**20**)	Cyclolaurane	*Laurencia microcladia*	Rhodomelaceae	[145]
Unnamed (**21**)	Cyclolaurane	*Laurencia microcladia*	Rhodomelaceae	[146]

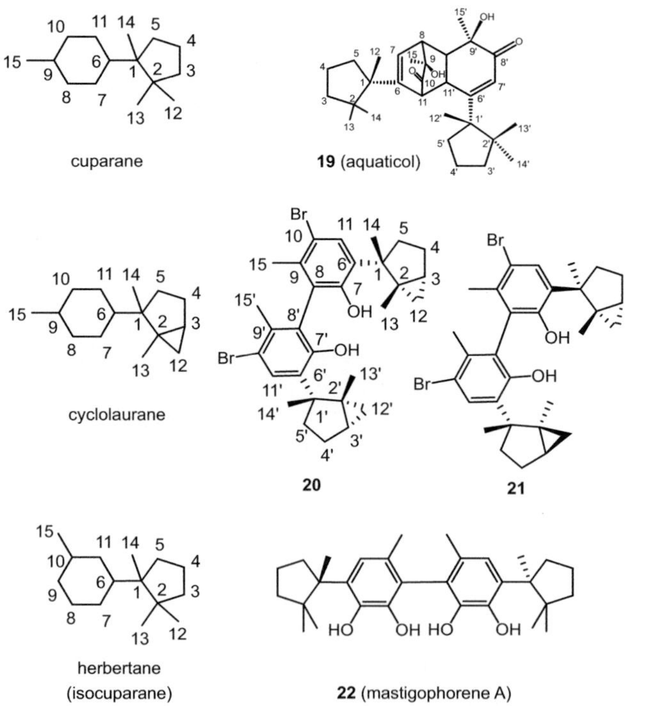

Fig. 8 Structures of cuparane, cyclolaurane, and herbertane, and their corresponding disesquiterpenoids aquaticol (**19**), **20**, **21**, and mastigophorene A (**22**)

2.1.9 Miscellaneous Disesquiterpenoids

Except for the structural classes mentioned above, DSs of other types can also be found in a number of fungi, protozoa, and other plants (Table 9). Representative DSs and their corresponding sesquiterpenes of each class are shown in Figs. 9, 10, 11, and 12.

Table 9 Miscellaneous DSs

Compound	Structural type	Origin	Family	Refs.
Alertenone (**24**)	Quadrane	*Alertigorgia* sp.	Agaricaceae[a]	[147]
Aurisins A, K	Aristolane	*Neonothopanus nambi*, PW1 and PW2	Marasmiaceae[a]	[148]
Bisacutifolones A–C	Pinguisane	*Porella acutifolia* subsp. *tosana*	Hepaticae[b]	[149]
Bis-cineradienone (**25**)	Cinerane	*Cineraria fruticulorum*	Asteraceae[b]	[150]
Bitaylorione (**27**)	1,10-Seco-aromadendrane	*Mylia taylorii*	Jungermanniaceae[b]	[151]
Bovistol (**31**)	Illudane	*Bovista* sp. 96042	Agaricaceae (basidiomycete)[a]	[152]
Cinnamacrin C (**29**)	Drimane	*Cinnamosma macrocarpa*	Canellaceae[b]	[153]
Distentoside	Dinorpterosin	*Dennstaedtia distenta*	Dennstaedtiaceae[b]	[154]
Methylmona-chosorin B	Dinorpterosin	*Dennstaedtia distenta*	Dennstaedtiaceae[b]	[154]
14-*O*-Methyl monachosorin A (Monomethyl monachosorin A)	Dinorpterosin	*Dennstaedtia distenta*	Dennstaedtiaceae[b]	[147]
14-*O*-Methyl monachosorin A (Monomethyl monachosorin A)	Dinorpterosin	*Monachosorum henryi*/*Monachosorum flagellare*/*Monachosorum maximowiczii*	Pteridaceae[b]	[147]
14′-*O*-Methylmona-chosorin A	Dinorpterosin	*Monachosorum flagellare*	Pteridaceae[b]	[147]
Monachosorins A–C	Dinorpterosin	*Monachosorum arakii*	Pteridaceae[b]	[155, 156]
Monachosorins A–C	Dinorpterosin	*Dennstaedtia distenta*	Dennstaedtiaceae[b]	[155]
Monachosorins A–C	Dinorpterosin	*Monachosorum henryi*/*Monachosorum flagellare*/*Monachosorum maximowiczii*	Pteridaceae[b]	[155]
Myltayloriones A–B	1,10-Seco-aromadendrane	*Mylia taylorii*	Jungermanniaceae[b]	[151]
Officinalic acid	Drimane	*Fomes officinalis*	Fomitopsidaceae[a]	[157, 158]
Vannusals A–B	Hemivannusane	*Euplotes vannus*	Euplotidae[c]	[159]
Unnamed (**32–33**)	Norilludalane	*Stereum ostrea* BCC 22955	Stereaceae[a]	[160]

[a]Fungi
[b]Plant
[c]Protozoa

Fig. 9 Structures of pterosin, quadrane, cinerane, and pinguisane, and their corresponding disesquiterpenoids **23–26**

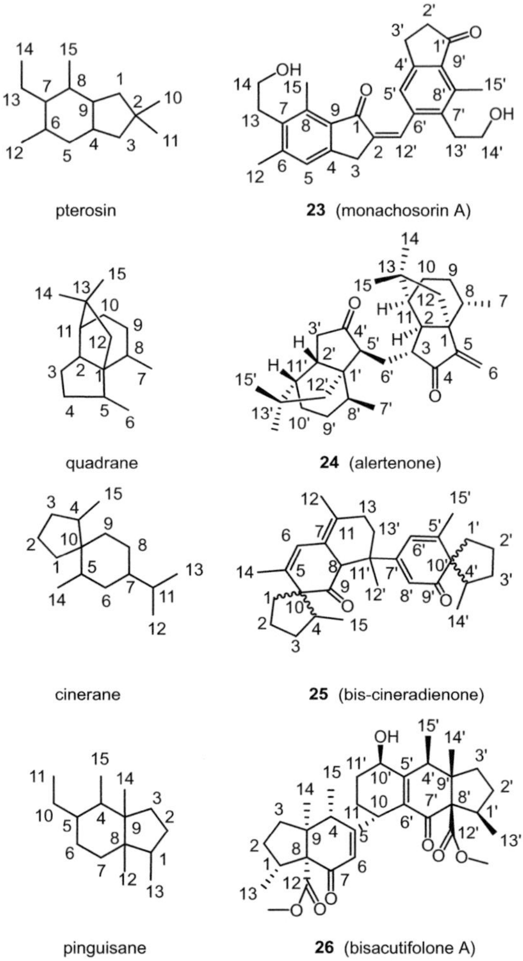

Fig. 10 Structures of aromadendrane, aristolane, and drimane, and their corresponding disesquiterpenoids **27–29**

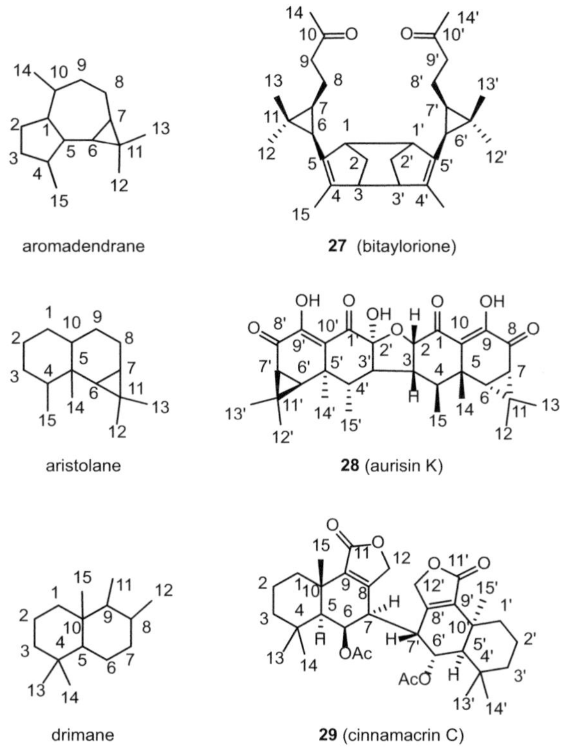

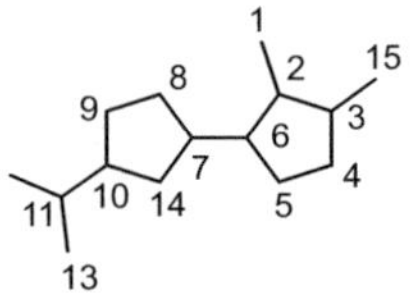

hemivannusane

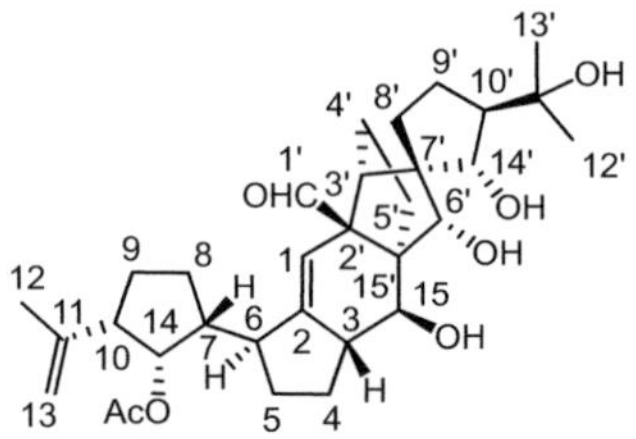

30 (vannusal B)

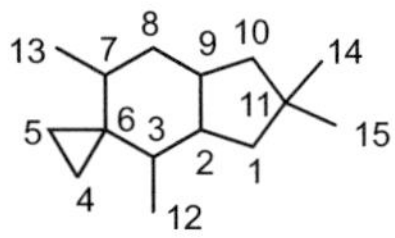

illudane

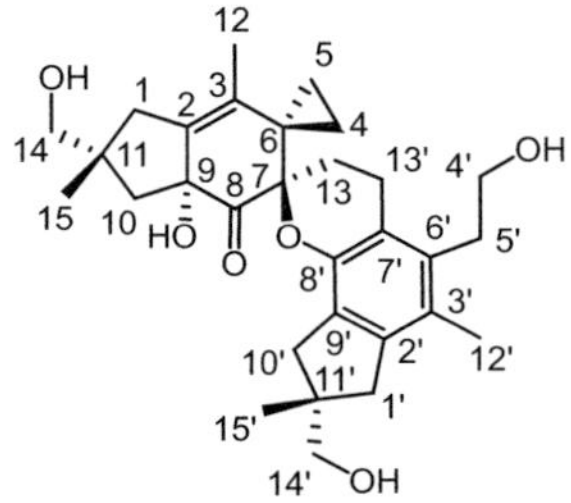

31 (bovistol)

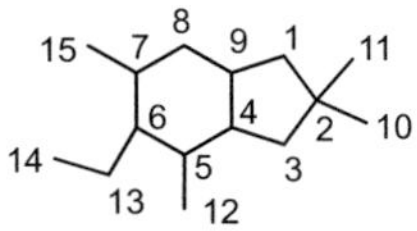

illudalane

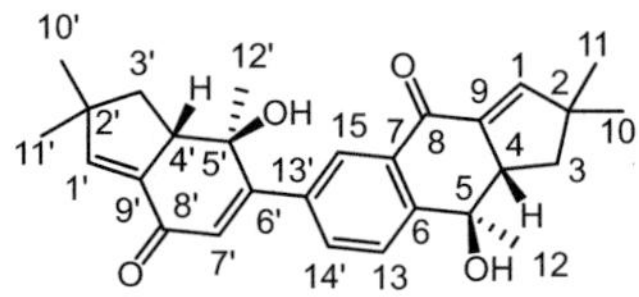

31

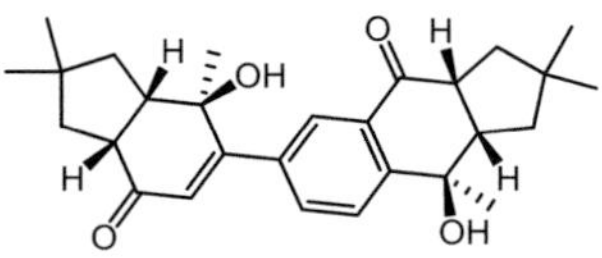

32

Fig. 11 Structures of hemivanusane, illudane, cinerane, and illudalane, and their corresponding disesquiterpenoids **32** and **33**

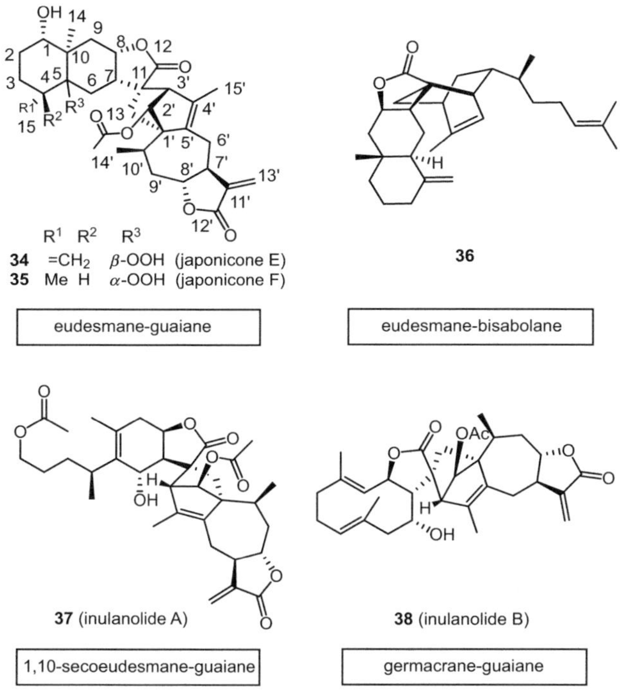

Fig. 12 Sesquiterpenoid structural types and their representative compound disesquiterpenoids **34–38**

2.1.10 Compound Disesquiterpenoids

Earlier investigations showed that the co-occurrence of different structural types of sesquiterpenoids might also lead to their dimerization (Table 10 and Figs. 12 and 13). Eudesmane–guaiane or 1,10-secoeudesmane–guaiane are the most typical coupling patterns thus far detected, although germacrane-guaiane, eudesmane-bisabolane, elemane–eudesmane, myliane-1,10-secoaromadendrane, and xanthane–guaiane couplings may also occur in various species of the family Asteraceae. A guaiane-aromadendrane coupling has been reported in *Chiloscyphus subporosus* from the Lophocoleaceae family [172].

Table 10 Compound DSs

Compound	Structural type	Origin	Family	Refs.
7*β*-(1*S*,5)-Dimethyl-4-hexenyl-3a′,4,4a′,5,6,7,8,8a,9a-decahydro-5,8′a*β*-dimethyl-5′-methylenspiro[bicyclo[2.2.2]oct-5-en-2,3′(2′*H*)-naphtho[2,3-*b*]furan]-2′-one (**36**)	Eudesmane-bisabolane	*Helenium autumnale*	Asteraceae	[120]
7*β*-(1*S*,5)-Dimethyl-4-hexenyl-3a′*α*,4,4a′*α*,5′,6′,7′,8′,8a′,9′,9a′*α*-octahydro-5, 5′*β*,8a′*β*-trimethyl-5′-methylenspiro[bicyclo[2.2.2]oct-5-en-2,3′(2′*H*)-naphtho[2,3-*b*]furan]-2′-one	Eudesmane-bisabolane	*Helenium autumnale*	Asteraceae	[120]
Inulanolide B (**38**)	Germacrane-guaiane	*Inula britannica* var. *chinensis*	Asteraceae	[161]
Inulanolide D	1,10-Secoeudesmane–guaiane	*Inula britannica* var. *chinensis*	Asteraceae	[161, 162]
Inulanolides A, C	1,10-Secoeudesmane–guaiane	*Inula britannica* var. *chinensis*	Asteraceae	[161]
Inulanolides A, C	1,10-Secoeudesmane–guaiane	*Inula japonica*	Asteraceae	[162]
Japonicone D	1,10-Secoeudesmane–guaiane	*Inula japonica*	Asteraceae	[163]
Japonicones A–C	Eudesmane–guaiane	*Inula japonica*	Asteraceae	[163]
Japonicones E–G	Eudesmane–guaiane	*Inula japonica*	Asteraceae	[164]
Japonicones H–L	1,10-Secoeudesmane–guaiane	*Inula japonica*	Asteraceae	[164]
Japonicones M–P	1,10-Secoeudesmane–guaiane	*Inula japonica*	Asteraceae	[165]
Japonicones Q, S	1,10-Secoeudesmane–guaiane	*Inula japonica*	Asteraceae	[162]
Japonicones R, T	Eudesmane–guaiane	*Inula japonica*	Asteraceae	[162]
Lappadilactone	Eudesmane–guaiane	*Saussurea lappa*	Asteraceae	[166]
Lineariifolianoid H	Germacrane-guaiane	*Inula lineariifolia*	Asteraceae	[61]
Lineariifolianoids A–D	Xanthane–guaiane	*Inula lineariifolia*	Asteraceae	[167]
Macrophyllidimer C	Elemane–eudesmane	*Inula macrophylla*	Asteraceae	[168]

(continued)

Table 10 (continued)

Compound	Structural type	Origin	Family	Refs.
Macrophyllidimers A and B	Elemane–eudesmane	*Inula macrophylla*	Asteraceae	[169]
Myltayloriones A–B	Myliane–1,10-secoaromadendrane	*Mylia taylorii*	Jungermanniaceae	[151]
Neojaponicone A	1,10-Secoeudesmane–guaiane	*Inula japonica*	Asteraceae	[165]
Rudbeckiolide	Eudesmane–guaiane	*Rudbeckia laciniata*	Asteraceae	[170]
Serratustones A and B	Elemane–eudesmane	*Chloranthus serratus*	Chloranthaceae	[171]
Unnamed (**39**)	Guaiane–aromadendrane	*Chiloscyphus subporosus*	Lophocoleaceae	[172]

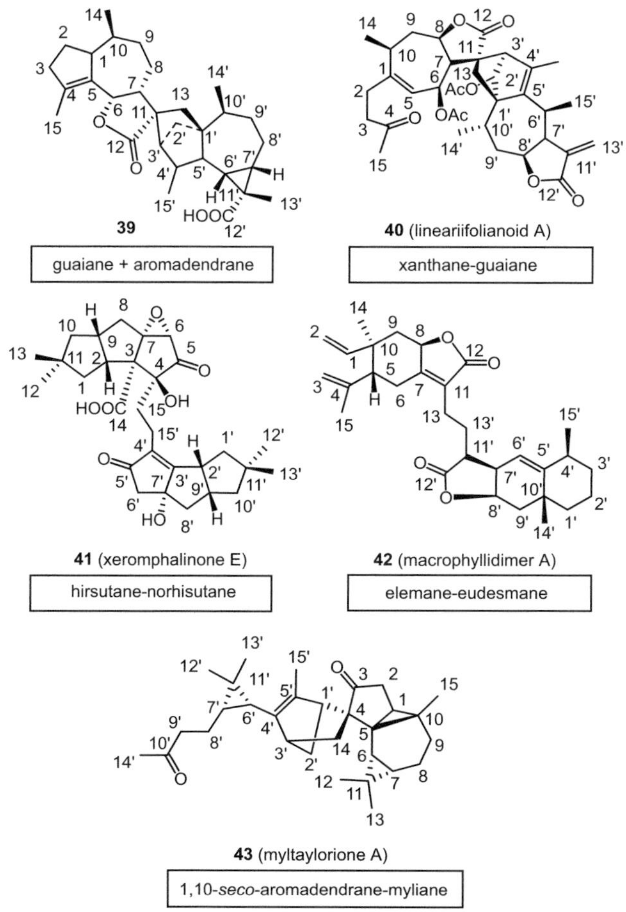

Fig. 13 Sesquiterpenoid structural types and their representative compound disesquiterpenoids **39–43**

2.2 *Pseudo-disesquiterpenoids*

2.2.1 Dimeric Aza-sesquiterpenoids

Aza-sesquiterpenoids or aminated sesquiterpenoids refer to sesquiterpenoids that incorporate at least one nitrogen atom into their molecules during enzyme-mediated cyclizations and *Wagner–Meerwein* rearrangements for construction of the carbocyclic skeletons or the subsequent functionalization process. It has been suggested that molecules possessing a nitrogen atom or atoms can no longer be simply considered a sesquiterpenoid and should be classified as an aza-sesquiterpenoid or a sesquiterpenoid alkaloid [173]. For this reason, dimeric aza-sesquiterpenoids or sesquiterpenoid alkaloid dimers [6] have been moved to the pseudo-disesquiterpenoid section herein rather than being included in the disesquiterpenoid group as adopted in the review by Zhan and associates [6]. Most of these DSs feature a thiaspirane (e.g., **44**) or a thiaspirane sulfoxide (e.g., **45**) structure formed by convergence of two C_{15} 3-furylquinolizidine (or 3-furyl-hemiaminoquinolizidine) units through a carbon–carbon bond and a carbon–sulfur bond (Fig. 14 and the review by Zhan et al. [6]).

Aza-sesquiterpenoids or sesquiterpenoid alkaloids can be found in the genera *Nuphar* and *Dendrobium* of the family Orchidaceae, and in some species of the Celastraceae and Hippocrateaceae [174]. However, up to the present, only the aquatic macrophytes of the genus *Nuphar* have been reported to produce dimeric aza-sesquiterpenoids [6]. These aza-sesquiterpenoids are also termed *Nuphar* alkaloids.

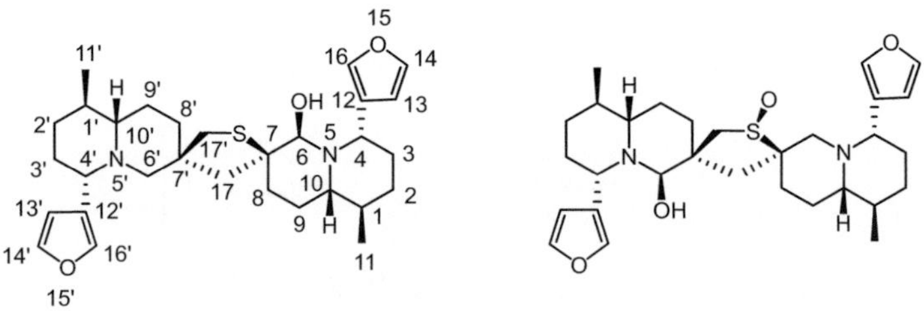

44 (6-hydroxythiobinupharidine) **45** (nupharpumilamine)

Fig. 14 Structures of 6-hydroxythiobinupharidine (**44**) and the sulfoxide nupharpumilamine (**45**)

2.2.2 Miscellaneous Pseudo-disesquiterpenoids

Pseudo-disesquiterpenoids other than dimeric aza-sesquiterpenoids can be formed by the connection of two sesquiterpenoid moieties via an ester group (e.g. **46–50**), an O-/S-ether linkage (e.g. **51**–**52**), a nitrogen atom (**53**), a urea group (**54**), two separate amide bonds connected by a piperidine ring (**55**), one or two isocitric acid esters (**56–58**), and two N–N-coupled indole units (**59–60**) (Table 11 and Figs. 15–19). A number of sesquiterpenoid structural types have been determined in this class of DSs. In the ester-connecting pseudo-disesquiterpenoids, their occurrence has been confined mainly to the plant family Asteraceae. A thujopsane-ester-

Table 11 Pseudo-disesquiterpenoid DSs

Compound	Connection type	Origin	Family	Refs.
Amarantholidosides VI–VII	Farnesane-*O*-farnesane	*Amaranthus retroflexus*	Amaranthaceae	[179]
Arrivacins A–B	Guaiane-ester-guaiane	*Ambrosia psilostachya*	Asteraceae	[180]
Artemilinin A	Eudesmane–*O*-guaiane	*Artemisia argyi*	Asteraceae	[181]
N,N′-Bis[(6*R*,7*S*)-7,8-dihydro-*α*-bisabolen-7-yl] urea	Bisabolane-(NH-CO-NH)-bisabolane	*Halichondria* sp.	Halichondriidae[a]	[182]
N,N-11-Bis[(1*Z*,4*Z*)-7*α* *H*-germacra-1(10),4-dienyl] urea	Germacrane-(NH-CO-NH)-germacrane	*Axinyssa* n. sp.	Halichondriidae[a]	[183]
Bisparthenolidine	Germacrane-*N*-germacrane	*Michelia rajaniana*	Magnoliaceae	[184]
Bisparthenolidine	Germacrane-*N*-germacrane	*Paramichelia baillonii*	Magnoliaceae	[185]
Capsicodendrin	Drimane-di-*O*-drimane	*Capsicodendron dinisii*	Canellaceae	[186]
Chinensiol	Himachalane-*O*-himachalane	*Juniperus chinensis* var. *tsukusiensis*	Cupressaceae	[187]
Cinnafragrin D	Drimane-di-*O*-drimane	*Cinnamosma macrocarpa*	Canellaceae	[153]
Cinnafragrins A–B	Drimane-di-*O*-drimane	*Cinnamosma fragrans*	Canellaceae	[188]
Cinnafragrolide	Drimane-di-*O*-drimane	*Cinnamosma fragrans*	Canellaceae	[188]
Conyaegyptin	Farnesane-xyloside-farnesane	*Conyza aegyptiaca*	Asteraceae	[189]
3-Costoyloxydehydroleucodin	Guaiane-ester-eudesmane	*Podachaenium eminens*	Asteraceae	[190]
Costunolact-12*β*-ol dimer	Germacrane-*O*-germacrane	*Magnolia virginiana*	Magnoliaceae	[191]
Cryptoporic acid D	Cyclic-(*O*-isocitric acid ester-drimane)$_2$-	*Cryptoporus volvatus* infected by *Paecilomyces variotii*	Marasmiaceae[b]	[192]

(continued)

Table 11 (continued)

Compound	Connection type	Origin	Family	Refs.
Cryptoporic acids C, E	Drimane-*O*-isocitric acid ester-drimane	*Cryptoporus volvatus*	Polyporaceae[b] (Agaricomycetes)	[193, 194]
Cryptoporic acids D, E, J, K	Cyclic-(*O*-isocitric acid ester-drimane)$_2$-	*Marasmius cladophyllus* F070624009	Marasmiaceae[b]	[195]
Cryptoporic acids D, F–G	Cyclic-(*O*-isocitric acid ester-drimane)$_2$-	*Cryptoporus volvatus*	Polyporaceae[b] (Agaricomycetes)	[193, 194]
Dicurcuphenol ether F	Bisabolane-*O*-bisabolane	*Didiscus aceratus*	Heteroxyidae[a]	[11]
3*β*-*O*-(1,2-Didehydro-3-oxo-costoyloxy)-4*β*,10*β*-dihydroxy-guaia-1(2),11(13)-dien-6*α*,12-olide	Guaiane-ester-eudesmane	*Warionia saharae*	Asteraceae	[185]
3*β*-*O*-(1,2-Didehydro-3-oxo-costoyloxy)-4*β*,10*β*-dihydroxy-guaia-1(2),11(13)-dien-6*β*,12-olide	Guaiane-ester-eudesmane	*Warionia saharae*	Asteraceae	[185]
1*α*,3*β*-Di-(3,4-dihydroxyphenyl)-2*α*,4*β*-dibazzanenyl cyclobutane dicarboxylate	Guaiane-glucoside-guaiane	*Bazzania pompeana*	Lepidoziaceae	[196]
1*α*,10*α*-Dihydrolactucin 8-*O*-isohypoglabrate	Guaiane-ester-guaiane	*Hypochoeris oligocephala*	Asteraceae	[197]
Disydonols A–B	Bisabolane-*O*-bisabolane	*Aspergillus* sp.	Trichocomaceae[b]	[13]
Dithiofurodysinin disulfide (**52**)	Furodysane-S–S-furodysane	*Ceratosoma brevicaudatum*	Chromodorididae[a]	[177]
Dixiamycins A–B (**59**–**60**)	Drimane-indolo-N–N-indolo-drimane	*Streptomyces* sp. SCSIO 02999	Streptomycetaceae[c]	[198]
ent-Cryptomeridiol-4-yl-hinokiate (**47**)	Thujopsane-ester-eudesmane	*Chamaecyparis obtusa*	Cupressaceae	[175]
Halichonadin A	Eudesmane-(NH–CO–NH)–eudesmane	*Halichondria* sp.	Halichondriidae[a]	[199]
Halichonadin E	Aromadendrane–(NH–CO–NH)-eudesmane	*Halichondria* sp.	Halichondriidae[a]	[200]
Halichonadin H	Eudesmane-(NH–CO–CH(OH)–CO–NH)-eudesmane	*Halichondria* sp.	Halichondriidae[a]	[201]
Halichonadins G, I	Eudesmane-(NH–CO–NH)-eudesmane	*Halichondria* sp.	Halichondriidae[a]	[201]
Halichonadins K, L	Eudesmane-diamide-eudesmane	*Halichondria* sp.	Halichondriidae[a]	[202]
Haploporic acid A	Cyclic-(*O*-isocitric acid ester-drimane)$_2$-	*Haploporus odorus*	Polyporaceae[b] (Basidiomycete)	[203]
3-Hydroxyambrosin damsinate (**46**)	Guaiane-ester-guaiane	*Ambrosia hispida*	Asteraceae	[204]
15-Hydroxycryptoporic acid H	Cyclic-(*O*-isocitric acid ester-drimane)$_2$-	*Marasmius cladophyllus* F070624009	Marasmiaceae[b]	[195]

(continued)

Table 11 (continued)

Compound	Connection type	Origin	Family	Refs.
14-Hydroxyhypocretenolide-β-D-glucopyranoside-4′,14′-hydroxyhypocretenoate	Guaiane-glucoside-guaiane	*Leontodon hispidus*	Asteraceae	[205]
8α-Hypoglabroyloxy-jaquinelin	Guaiane-ester-α-copane	*Hypochaeris glabra*	Asteraceae	[206]
Lactucains A–C	Guaiane-*O*-guaiane	*Lactuca indica*	Asteraceae	[207]
Lactucin-8-*O*-hypoglabrate	Guaiane-ester-α-copane	*Hypochaeris glabra*	Asteraceae	[206]
Ligulamulienins A–B	Furanoeremophilane-*O*-12-noreremophilane	*Ligularia muliensis*	Asteraceae	[91]
Ligumacrophyllal	8,9-Secoeremo-philane-*O*-8,9-secoeremophilane	*Ligularia macrophylla*	Asteraceae	[208]
Neocreolophin	Norhirsutane-di-*O*-norhirsutane	*Creolophus cirrhatus*	Hericiaceae[b] (Basidiomycete)	[178]
Picriosides A–B	Guaiane-ester-guaiane	*Picris hieracioides* var. *japonica*	Asteraceae	[209]
Podachaenin	Guaiane-ester-eudesmane	*Podachaenium eminens*	Asteraceae	[210]
Roseolide A	Cyclic-(*O*-isocitric acid ester-drimane)$_2$-	*Roseoformes subflexibilis*	Polyporaceae[b] (Basidiomycete)	[211]
Vernodalidimers A–B	Elemane-isobutanoic acid ester-elemane	*Vernonia anthelmintica*	Asteraceae	[212]
Virgaurin D	Furanoeremophilane-*O*-furano-eremophilane	*Ligularia virgaurea*	Asteraceae	[75]
Virgaurols C–D	Eremophilane-ester-eremophilane	*Ligularia virgaurea*	Asteraceae	[213]
Xeromphalinone F (**50**)	Hirsutane-ester-norhirsutane	*Xeromphalina* sp.	Mycenaceae[b]	[176]

[a]Animal
[b]Fungi
[c]Bacteria

eudesmane pseudo-disesquiterpenoid, *ent*-cryptomeridiol-4-yl-hinokiiate (**47**), was reported to be present in *Chamaecyparis obtusa* of the family Cupressaceae [175]. In turn, a hirsutane-ester-norhirsutane type pseudo-disesquiterpenoid, xeromphalinone F (**50**), was purified from a Basidiomycota fungus, *Xeromphalina* sp. [176]. Ether-linked pseudo-disesquiterpenoids are not particularly rare, but their occurrence is limited to only certain species in the families Amaranthaceae, Asteraceae, Canellaceae, Cupressaceae, and Magnoliaceae. In addition, a furodysane disulfide pseudo-disesquiterpenoid (**52**) and a norhirsutane pseudo-disesquiterpenoid (neocreolophin) have been isolated from the Australian nudibranch *Ceratosoma brevicaudatum* [177] and the mycelial cultures of *Creolophus*

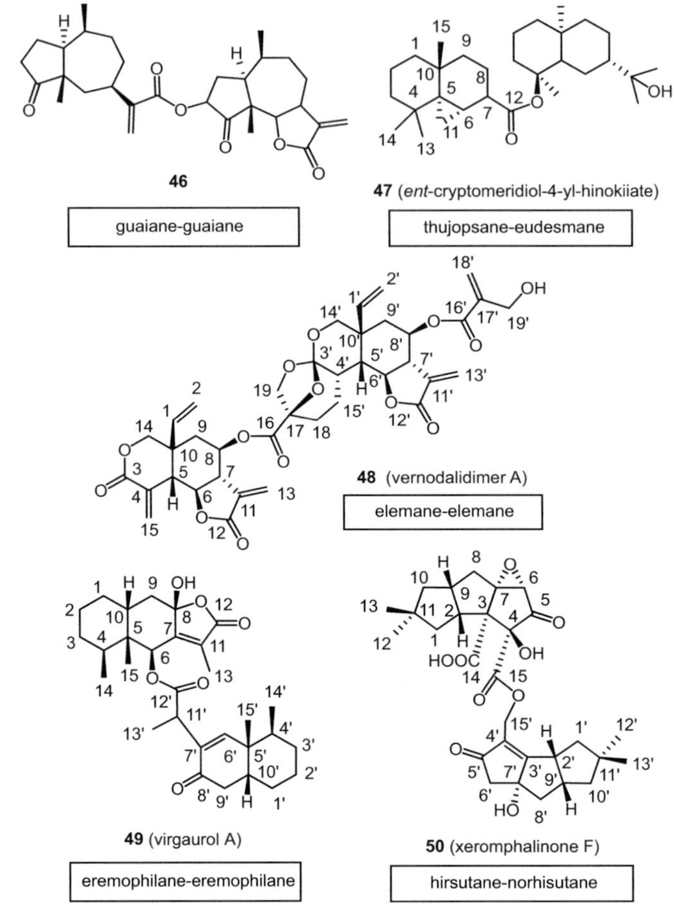

Fig. 15 Typical structures of pseudo-disesquiterpenoids with two moieties connected by an ester group (**46–50**)

cirrhatus (a basidiomycete) [178], respectively. However, neocreolophin was shown ultimately to be an artifact formed from dimerization of the natural product creolophin E. Several urea- or diamide-connected pseudo-disesquiterpenoids were reported, but their occurrence is confined rigidly to the marine sponges *Halichondria* sp. and *Axinyssa* n. sp. (Halichondriidae) (Table 11). The former is important as all these pseudo-disesquiterpenoids except one were reported to be present in the species. Fungi from the Polyporaceae and Marasmiaceae families are reported to produce a number of isocitric acid-connecting pseudo-disesquiterpenoids, which are generally linked by two separate isocitric acid units esterified at one end and etherified at the other (Fig. 18). Dixiamycins A–B (**59–60**), two N–N-coupled atropo-diastereomers with a dimeric indolo-sesquiterpene skeleton and a stereogenic N–N axis between sp^3-hybridized nitrogen atoms, were isolated from a marine-derived actinomycete, *Streptomyces* sp. [198].

Fig. 16 Structures of lactucain C (**51**) and dithiofurodysinin disulfide (**52**)

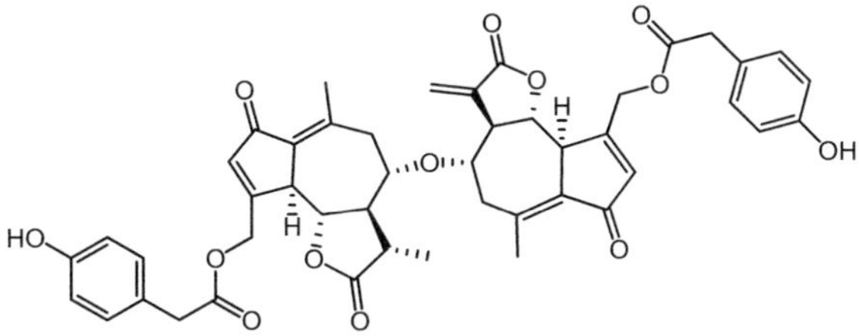

51 (lactucain C)

guaiane-guaiane

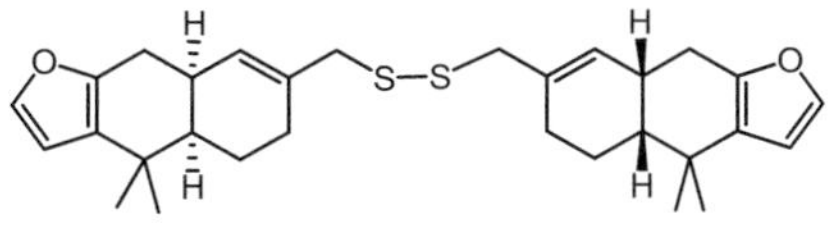

52 (dithiofurodysinin disulfide)

furodysane-S-S-furodysane

Fig. 17 Structures of bisparthenolidine (**53**), halichonadin A (**54**), and halichonadin K (**55**)

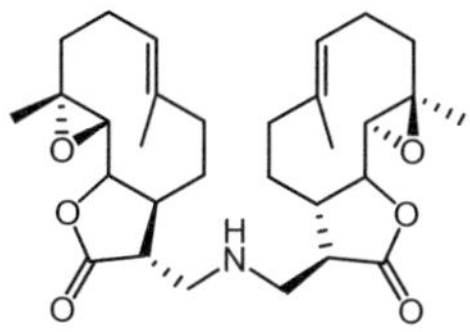

53 (bisparthenolidine)

germacrane-germacrane

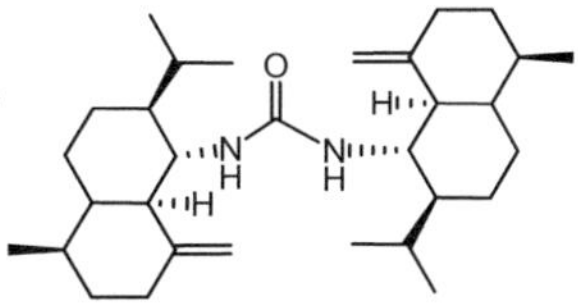

54 (halichonadin A)

eudesmane-eudesmane

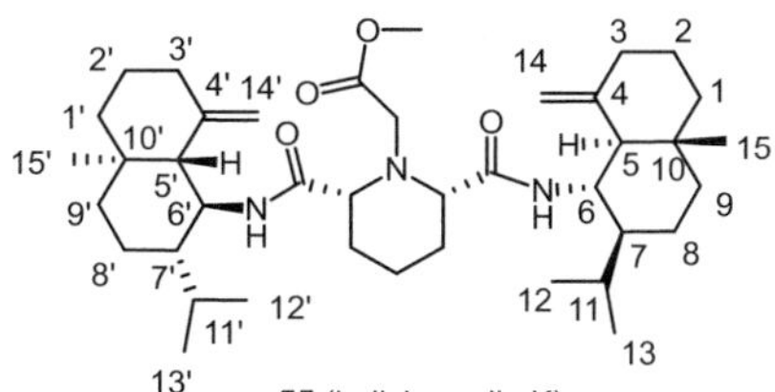

55 (halichonadin K)

eudesmane-eudesmane

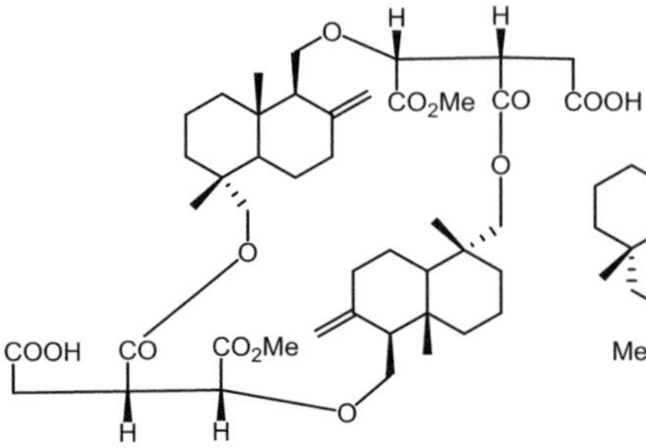

56 (cryptoporic acid D)

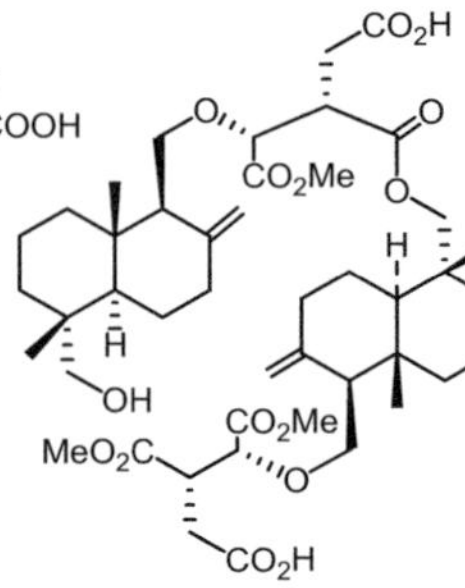

57 (cryptoporic acid E)

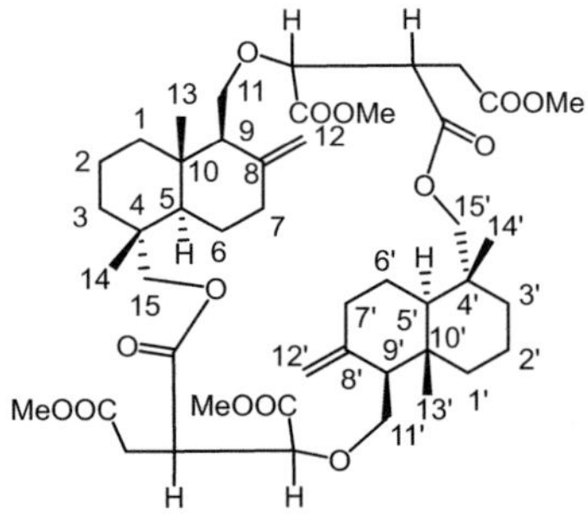

58 (cryptoporic acid J)

drimane-drimane

Fig. 18 Structures of three drimane pseudodisesquiterpenoids crytoporic acids D (**56**), E (**57**), and J (**58**)

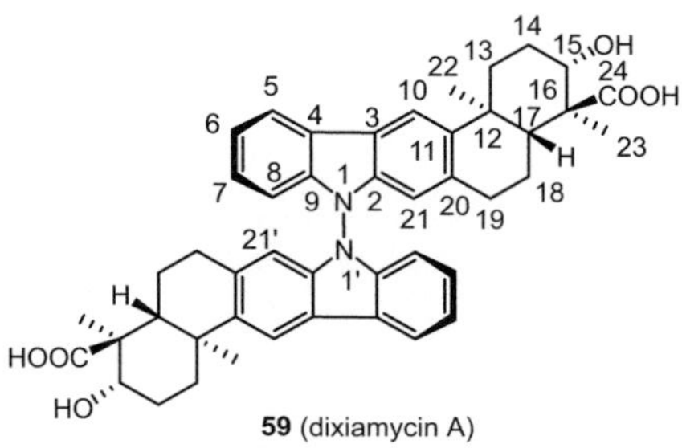

59 (dixiamycin A)

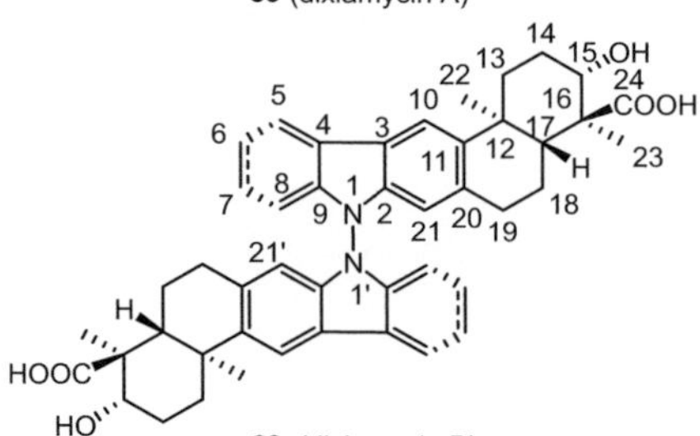

60 (dixiamycin B)

Fig. 19 Structures of dixiamycins A (**59**) and B (**60**)

3 Structural Elucidation

3.1 General

With 30 skeletal carbons, a wide variety of sesquiterpenoid scaffolds, and various convergent patterns, DSs unsurprisingly exhibit a large number of structural variations. Structural elucidation of DSs is therefore not always an easy task. The most immediate obstacle is the determination of the molecular formula for a DS. The earlier use of electron ionization (EI) may not give a molecular ion but rather fragment ions instead. However, the application of soft-ionization techniques e.g. [electrospray ionization (ESI), atmospheric pressure chemical ionization (APCI), chemical ionization (CI), and fast atom bombardment (FAB)] in the last several decades has allowed the reliable determination of DS molecular weights. Another challenge is that the same (for homodimers) or similar (for heterodimers) structures of the two sesquiterpenoid units of a DS usually give coincident spectroscopic signals, making the structural elucidation process quite a demanding task. The availability of various additional spectroscopic techniques (e.g. high-field 1D- and 2D-NMR techniques, high-resolution mass spectrometry, circular dichroism, and X-ray crystallography) and computational methods have made the structural determination of complex DSs a feasible task when either milligram or submilligram quantities are available. As a result, a considerable number of DSs have been structurally determined in recent years.

Chemical methods, either by chemical transformation, derivatization or synthesis, have led to great success in establishing the structures of diverse DSs. The application of these chemical methods leads to either the proof of structures of DSs, or, in some cases, their structural revision.

3.2 Mass Spectrometry

The dimeric feature of DSs renders to a given substance of this type a molecular weight close to the molecular weight adduct of its two constitutional sesquiterpenoids. However, determination of a dimeric structure needs an ambiguous molecular weight (and/or its molecular formula). Electron impact (EI) may be used occasionally for this purpose. This ionization technique can provide information not only on the molecular weights of DSs (in the case of HREIMS, the chemical formulas from exact *m*/*z* values of their molecular ion), but also on their structural fragments. The EIMS fragmentation patterns of halichonadin A (**54**) (*m*/*z*, 468, M^+) and its reduction product also supported the proposed structure (Fig. 20) [199].

Fragment ions generated from linear cleavage, RDA fragmentation and dehydration in the EI mass spectrum also have been very useful for structural confirmation, as in the case of ligulamulienin A (**61**, Fig. 21) [91].

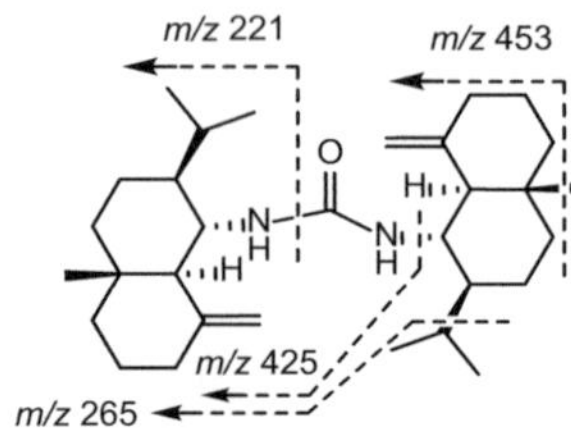

Fig. 20 EIMS (M^+, *m*/*z* 468) fragmentation pattern of halichonadin A (**54**)

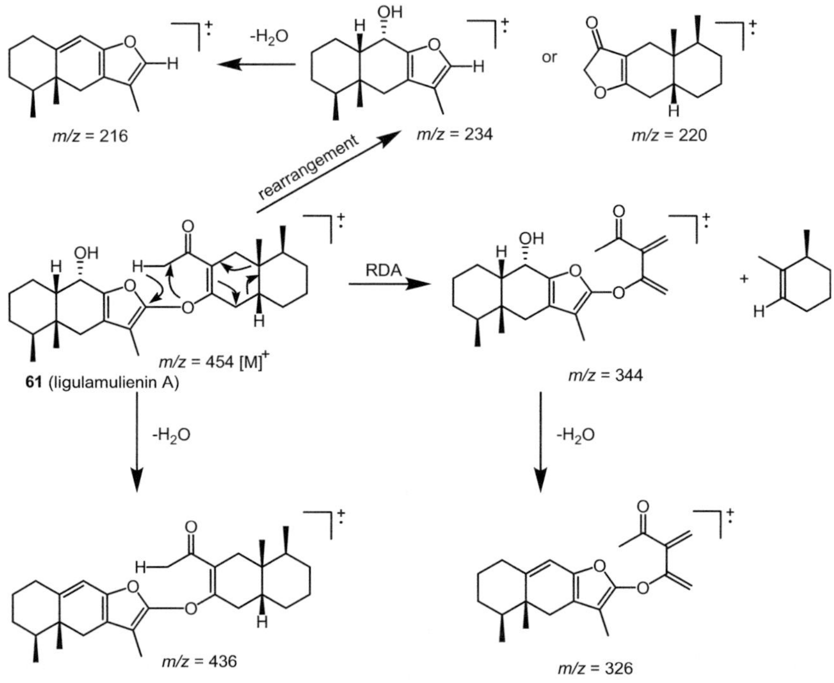

Fig. 21 Major EI-MS fragmentions of ligulamulienin A (**61**)

While the elemental formula of arrivacin A (**62**) ($C_{30}H_{42}O_7$) was suggestive of a triterpenoid, analysis of the NMR spectra and MS fragmentation data suggested that this compound is actually a DS. The desorption chemical ionization (DCI) fragment ions (Fig. 22) also suggested the presence of an ester link between the two C_{15} moieties [180].

In recent years, soft-ionization techniques (ESI, APCI [202], CI [180], FD [213], and FAB [113, 198, 203]) have been used widely to measure the molecular weights of DSs. In fact, high-resolution mass spectrometric techniques have become a routine tool for the determination of their molecular formulas. High-resolution MS data of various pseudo-molecular cations (e.g. $(M+H)^+$ and/or $(M+Na)^+$ in the positive-ion mode [121, 132, 148, 162, 171, 181, 202]) or anions (e.g.,

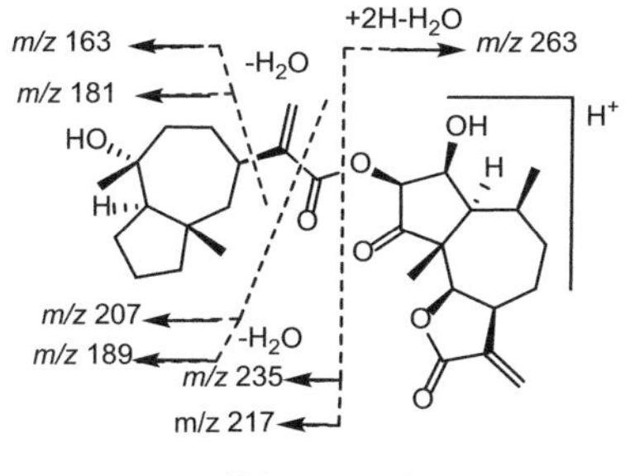

Fig. 22 DCI mass fragmentation pattern of the $[M+H]^+$ of arrivacin A (**62**)

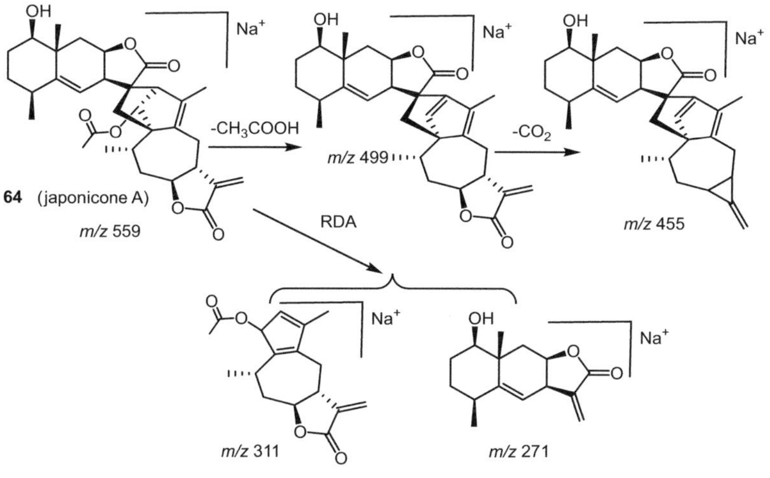

Fig. 23 Proposed fragmentation patterns for japonicone A (**63**)

$(M-H)^-$ in the negative-ion mode [47]) have been used to determine the molecular formulas of DSs.

Moreover, the ESI-MS fragmentation patterns revealed may be very useful for the identification and characterization of DSs. A rapid and effective HPLC/(+)ESI-MS^n method has been developed to analyze structurally related DSs in the species of *Inula japonica* [214]. All the standards investigated exhibited strong $(M+Na)^+$ ions but low-abundance $(M+H)^+$ ions in the positive-ion mode. The fragmentation patterns for DSs of the eudesmane–guaiane (Fig. 23), 1,10-secoeudesm-5(10)-ene-guaiane (Fig. 24), 1,10-secoeudesm-5-ene-guaiane, and xanthane–guaiane structural types have been proposed.

The typical fragmentation of the $[M+Na]^+$ (*m/z* 559) peak of japonicone A (**63**) (Fig. 23, eudesmane–guaiane type) includes the loss of an acetic acid unit from C-2′ and *RDA* fragmentation between the two constitutional sesquiterpenoid units.

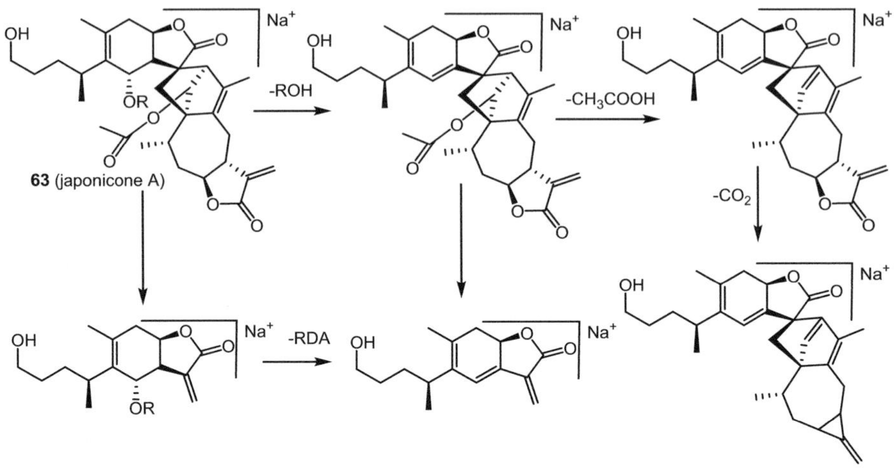

Fig. 24 Typical fragmentation pattern of 1,10-secoeudesm-5(10)-ene-guaiane type disesquiterpenoids

3.3 Nuclear Magnetic Resonance Spectroscopy

The basic structural feature of DSs is that they all possess sesquiterpenoid scaffolds. Therefore, the NMR structural elucidation techniques employed for sesquiterpenoids are also applicable to DSs. A NMR structural determination of a DS usually begins with the elucidation of each sesquiterpenoid unit and ends with the coupling pattern of the two substructures.

Since DSs generally co-occur in the producing organism with their sesquiterpenoid precursors or precursor analogues, so the NMR signals corresponding to each sesquiterpenoid unit should therefore be present. For symmetrical homodimers, NMR signals corresponding to one half of the DS are to be observed. However, for heterodimers and asymmetric homodimers, different NMR signals corresponding to each half of the DS would be observed. Identical signals may also occur. 2D-NMR analysis (particularly HMBC for planar connectivities and NOESY/ROESY for relative configuration determination) then provides information for the coupling patterns of the two substructures. The use of these methods can be found in papers that have dealt with structures of DSs and have been published in the past decade.

In complex structures where NMR experimental data can not provide ambiguous structural information, computational NMR spectroscopy (particularly Density Functional Theory-NMR, DFT-NMR) [215], which has seen a marked increase in accuracy, affordability, and application over the past decade [216], is a very useful tool for predicting or verifying possible structures for DSs. For example, vannusals are a class of DSs with an unusual hexacyclic backbone isolated from a tropical population (Sil 21) of the morphospecies *Euplotes vannus* [159]. Due to the uncertain configuration of the hexacyclic backbone, the original structure (**64**)

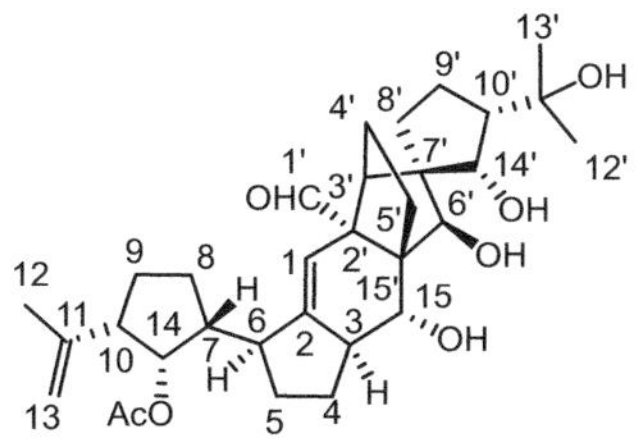

64

Fig. 25 The original structure (**64**) assigned to vannusal B by NMR spectroscopy

(Fig. 25) assigned to vannusal B by NMR spectroscopy has been revised to **30** (Fig. 10) by total synthesis [217–219]. However, further work showed that DFT-NMR calculations were useful for the proposal of a correct structure for vannusal B [220].

If feasible, direct derivatization with chiral auxiliary reagents might be used for assigning the absolute configurations for DSs by ^{1}H NMR spectroscopy (e.g. the modified Mosher method) [112, 149, 163]. However, care should be taken when using this procedure [167].

3.4 Single-Crystal X-Ray Diffraction

Over the past few years, single-crystal X-ray diffraction has become accepted widely as one of the most reliable techniques for determining the relative or absolute configurations of organic compounds. Indeed, XRD has played a pivotal role in determining the relative or absolute configurations of DSs by crystals in the native state [29, 30, 61, 62, 73, 79, 80, 105, 112, 116, 149, 167, 169, 171, 172, 187, 221–223] or their derivatives [224, 225] as well as cocrystals with solvent stabilizer [202].

It should be noted for DSs that contain only light atoms (C, H, O and N), the anomalous scattering under MoK_{α} radiation is rather small, therefore only the relative configuration can be determined in these cases. Whereas, when CuK_{α} radiation is used, the anomalous scattering effect can be observed strongly enough for the establishment of the absolute configuration of DSs. As an example, the relative and absolute configurations of japonicone A were suggested to be as shown in structure **65** by the use of XRD and the modified Mosher method (Fig. 26) [163]. However, X-ray crystallographic analysis of japonicone A as an anomalous dispersion with copper radiation [Flack parameter, 0.10 (14)] led to the exactly opposite absolute configuration (**63**), the same as leariifolianoids A–D [167]. Based on this observation and from a biosynthesis standpoint, japonicones [163–165] and inulanolides [161] may possess similar stereochemistry.

Fig. 26 Structures of originally assigned (**65**) and revised japonicone A (**63**), inulanolide A (**37**), and japonicone M (**66**)

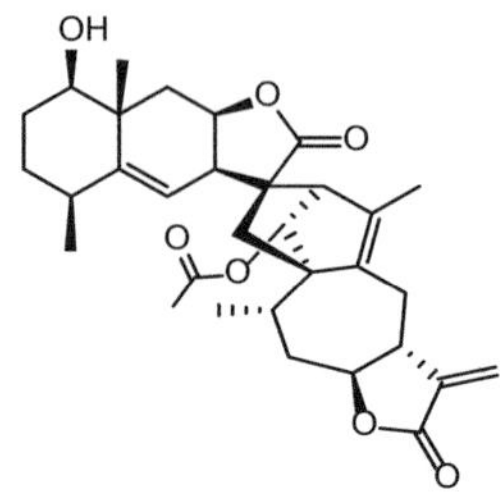

63 (japonicone A)

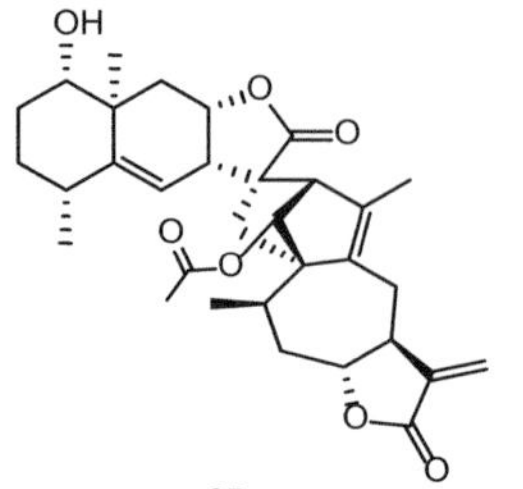

65

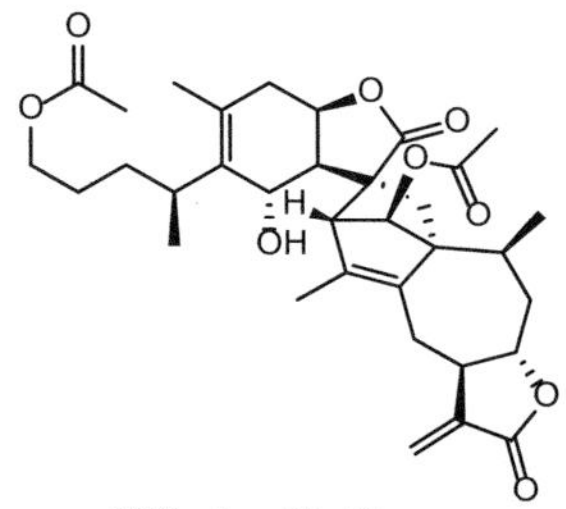

37 (inulanolide A)

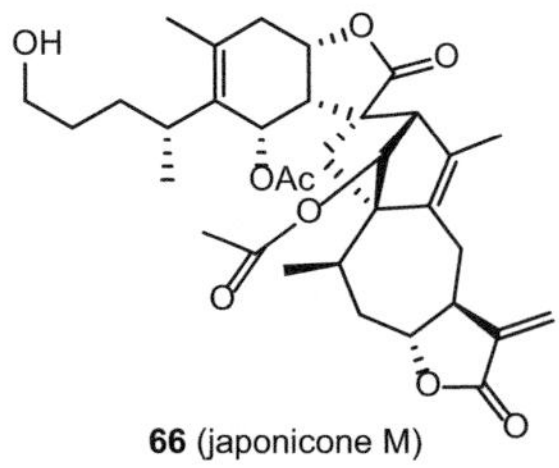

66 (japonicone M)

3.5 CD and ECD Calculations

Circular dichroism (CD) is one of the chiroptical methods that can be used to determine the absolute configurations of DSs. Configurations of DSs may be established by applying empirical rules (e.g. the octant rule) [17, 79] or by comparing their CD spectra to those of structural analogues [113, 136, 181, 226, 227].

For DSs that possess two identical or similar chromophores (this is generally the case due to their dimeric feature) and a single prevailing conformation, exciton chirality CD (ECCD), a non-empirical chiroptical method based on the chirality (positive or negative) of the CD couplet of two chromophores, has proved to be a convenient and reliable method for directly establishing the absolute configuration of DSs [125, 130, 132, 133, 171, 221, 223, 228]. This method can also be used for DSs that carry only one chromophore if a suitable chromophore can be introduced to a desirable site of the molecule [149, 223].

Electronic circular dichroism (ECD), the circular dichroism resulting from an electronic transition, is now one of the most often used CD methods. With the development of quantum chemistry, ECD can be predicted theoretically by quantum chemical calculation. In the past several years, time-dependent density functional theory (TDDFT) calculations of ECD spectra have become one of the most efficient tools for assigning absolute configurations of organic molecules [229]. Therefore, the absolute configurations of DSs can be easily established simply by comparing the measured CD spectrum with the TDDFT calculated CD spectrum of the assumed configuration [114, 133, 165, 171, 198, 229, 230].

NMR analysis showed that japonicones M–P [165] should have a similar stereochemistry to inulanolides A–D [161]. However, ECD calculations showed different configurations (e.g. **66**) [165]. These contradictory results may be able to be resolved by the XRD method [167].

3.6 Chemical Methods

Earlier structural determination of DSs relied heavily on chemical methods. The structures and stereochemistry of microhelenins A, B, and C, and microlenin acetate were solved in large part by physical methods as well as chemical transformations and correlations [65].

A notable example is the structural elucidation of vannusals A–B by extensive chemical synthesis [217–219], through which the original structures of vannusals A–B have been revised.

A chloranthalactone A photodimer was originally assigned the structure **67** [231], but this was revised as **68**, a head-to-head and *anti* dimer of chloranthalactone A, by NOE and photodimerization observations (Fig. 27).

The absolute stereochemistry of halichonadin L (**69**) was concluded via transformation of halichonadin K (**55**) to **69** (Scheme 1) [202], and the structure of arrivacin B (**70**) was confirmed by dehydration of arrivacin A (**62**) [180] (Scheme 2).

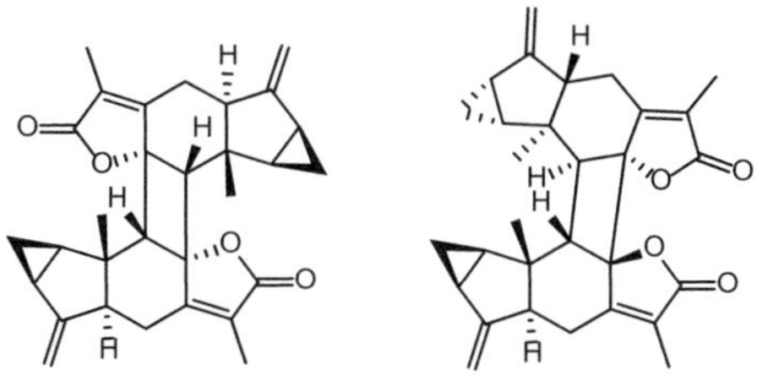

67 **68** (chloranthalactone A photodimer)

Fig. 27 Structure assigned to chloranthalactone A photodimer (**68**)

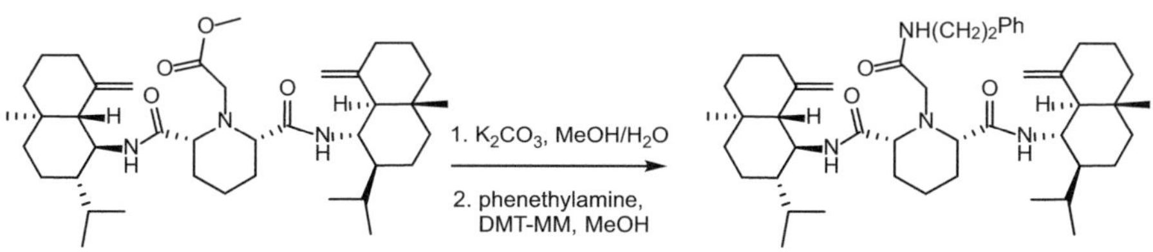

55 (halichonadin K) **69** (halichonadin L)

Scheme 1 Transformation of halichonadin K (**55**) to halichonadin L (**69**)

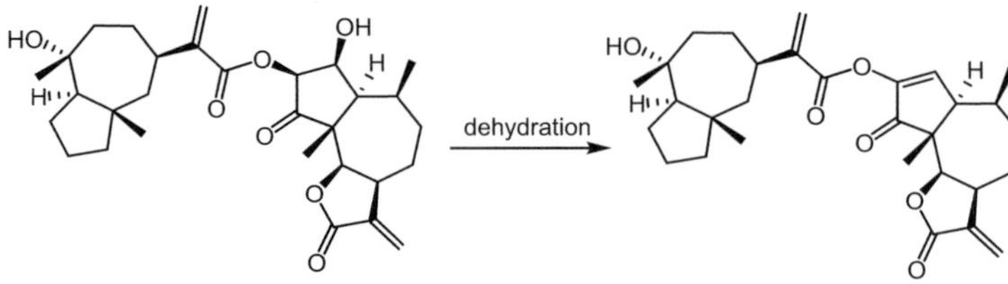

62 (arrivacin A) **70** (arrivacin B)

Scheme 2 Dehydration of arrivacin A (**62**) to arrivacin B (**70**)

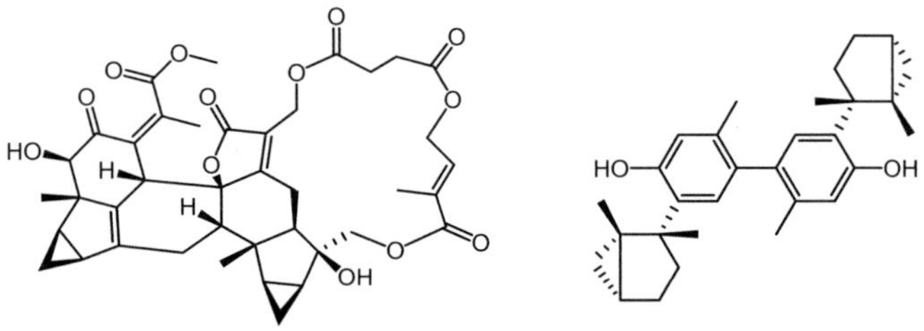

71 (shizukaol B) **72** (laurebiphenyl)

Fig. 28 Structures of shizukaol B (**71**) and laurebiphenyl (**72**)

During the structural confirmation of shizukaol B (**71**) (Fig. 28), alkaline hydrolysis of shizukaol B yielded succinic and γ-hydroxytiglic acids, which were identified as their methyl esters by GC-MS [213].

The structure of laurebiphenyl (**72**) was confirmed by transformation of debromolaurinterol into laurebiphenyl via oxidative coupling with manganese dioxide [140].

In the structural elucidation of DSs that possess polycyclic fused systems with densely functionalized quaternary carbons, the use of NMR spectroscopic methods alone may fail to provide a correct structure. Gochnatiolide B (**74**) was initially isolated by Bohlmann and co-workers from *Gochnatia* spp. in the 1980s and was suggested to have the structure of **73** mainly by 1D-NMR spectroscopic analysis (Fig. 29) [49, 55]. However, biomimetic syntheses of gochnatiolide B and its analogues gochnatiolides A and C indicated that this DS should have the structure **74** [232]. This synthesis effort also established the absolute configurations of gochnatiolides A and C.

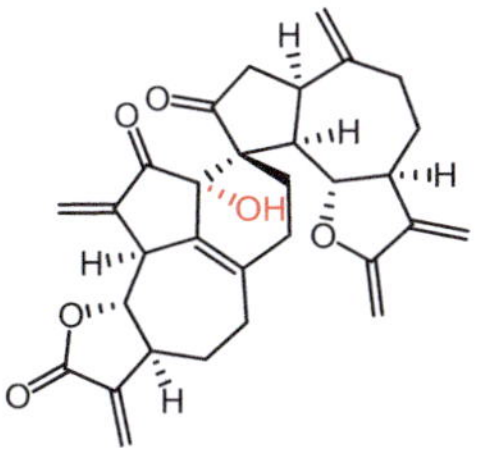

73

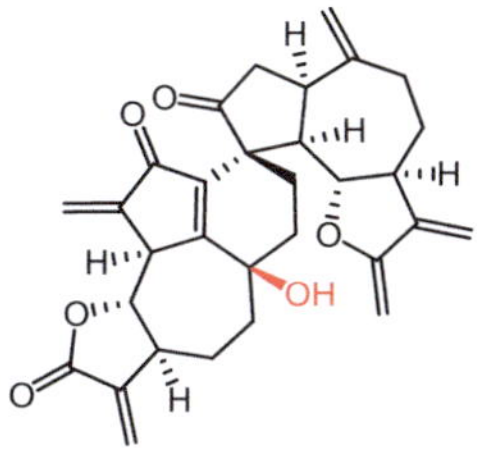

74 (gochnatiolide B)

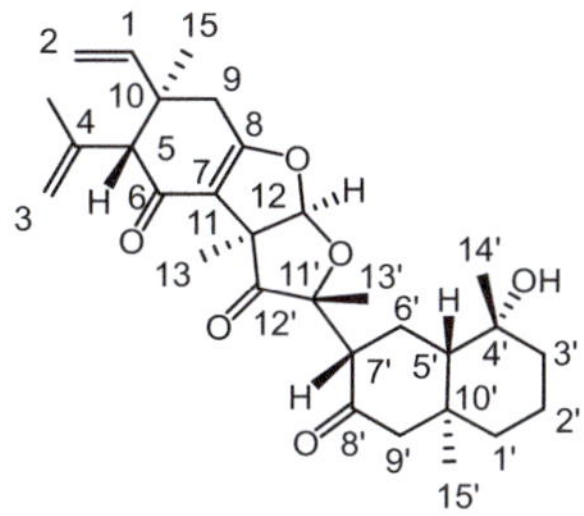

75 (serratustone A)

Fig. 29 Structures of gochnatiolide B (**74**) and serratustone A (**75**)

3.7 Structural Elucidation of Serratustone A

Serratustone A (**75**) is an elemane–eudesmane type DS isolated from *Chloranthus serratus* [171]. Its HR-ESI(+)MS showed a sodiated molecular ion $(M + Na)^+$ at *m*/*z* 519.2704 (calcd 519.2723), consistent with a molecular formula of $C_{30}H_{40}O_6$ and incorporating 11° of unsaturation. The IR spectrum showed the presence of hydroxy (3430 cm^{-1}), carbonyl (1765 and 1689 cm^{-1}), and vinyl (1657 and 1632 cm^{-1}) groups. The NMR data (with DEPT and HSQC) revealed resonances corresponding to a monosubstituted terminal double bond (δ_H 5.73 ppm, H-1; 4.99 and 4.94 ppm, H_2-2; δ_C 145.7 ppm, C-1; 113.3 ppm, C-2), a disubstituted double bond (δ_H 5.02 and 4.80 ppm, H_2-3; δ_C 115.2 ppm, C-3; 141.4 ppm, C-4), a tetrasubstituted double bond (δ_C 114.9 ppm, C-7; 175.0 ppm, oxygenated, C-8), and three carbonyls (δ_C 192.5 ppm, C-6; 208.9 ppm, C-8′; 209.9 ppm, C-12′). Other characteristic signals included an olefinic methyl (δ_C 25.0 ppm, C-14), five tertiary methyls (δ_C 16.5, 19.0, 23.2, 25.6, and 30.7 ppm; C-13, C-15′, C-13′, C-15, and C-14′), an sp^3 oxygenated methine (δ_C 115.1 ppm, C-12), and two sp^3 oxygenated quaternary carbons (δ_C 71.4 ppm C-4′; 83.2 ppm, C-11′). In addition, the only remaining proton that was distinguished by the HSQC data was supportive of a hydroxy group, in agreement with the IR spectrum. These observations accounted for six out of the eleven degrees of unsaturation, requiring **75** to be pentacyclic, and suggested a likely DS skeleton for this compound. While lindenane-type dimeric sesquiterpenoids are common secondary metabolites from this plant family (see Sect. 2.1.7), the NMR properties of compound **75** were apparently distinct from

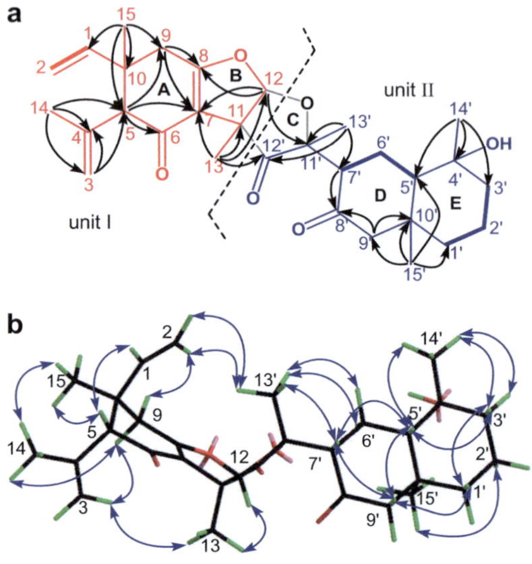

Fig. 30 (**a**) ^{1}H$-^1$H COSY (*thick solid line*) and selected HMBC correlations (*pointing arrows*) for **75**. (**b**) Key ROESY correlations (*double sided arrows*) for **75**

previously reported data, implying an unusual framework and/or a novel dimerization pattern.

Analysis of the ^{1}H$-^1$H COSY data (Fig. 30a) for **75** revealed three structural fragments (C-1 to C-2, C-1′ to C-3′, and C-5′ to C-7′), which only provided limited information on the overall structure of this compound. Fortunately, the acquisition of excellent HMBC data (Fig. 30a) enabled the connectivities of these fragments to be determined in addition to the isolated methines, methylenes and methyls across those deprotonated carbons and heteroatoms, and this led to the planar structure of **75**, as shown. Thus, the HMBC correlations from H_3-15 to C-1, C-5, C-9 and C-10 suggested the attachment of the monosubstituted vinyl, CH-5, CH_2-9 and Me-15 to C-10; the cross-peaks of H_3-14 to C-3, C-4 and C-5 were used to link the isopropenyl to C-5; the correlations from H-5 to C-6, C-7 and C-9, and from H_2-9 to C-7 and C-8 indicated a cyclohexenone motif (ring A), as drawn; and the cross-peaks of H_3-13 to C-7, C-11 and C-12, together with H-12 to C-7 and C-8 suggested the presence of a dihydrofuran (ring B) functionality fused with ring A and with Me-13 linked to C-11. The structural unit I of **75** was thereby (Fig. 30a, in red) characterized, representing an elemane-type sesquiterpenoid. Similarly, structural unit II (Fig. 30a, in blue) of a eudesmane-type sesquiterpenoid was also constructed on observing the HMBC correlations between H_3-14′/C-3′, C-4′ and C-5′; H_3-15′/C-1′, C-5′, C-9′ and C-10′; H-7′ and H-9′/C-8′; and H_3-13′/C-7′, C-11′ and C-12′. Finally, the connection of the two units (I and II) via a 3-keto-tetrahydrofuran ring (ring C) was aided by the key HMBC correlations of H_3-13/C-12′ and H-12/C-11′ (Fig. 30a), and supported by the deshielded C-12 (δ_C 115.1 ppm) and C-11′ (δ_C 83.2 ppm) carbon signals, indicating that a new carbon framework for a sesquiterpenoid dimer was formed by a C-11$-$C-12′ linkage.

The relative configuration of **75** was determined on the basis of a ROESY experiment (Fig. 30b). The H-9α proton showed correlations with both H-3b and H_3-14, suggestive of a quasi 1,3-di*axial* relationship for H-9α and 5-isopropenyl (C-3 to C-14 via C-4), and these functionalities were assigned arbitrarily in an α-orientation, with H-9β and H-5 therefore quasi *equatorial* and β-oriented. Consequently, the cross-peak of H_3-14/H_3-15 indicated an equatorially positioned Me-15 and hence an *axial* and β-configured vinyl group at C-10. The ROESY correlation of H_2-2/H_3-13′ suggested a co-planar relationship for the C-1–C-2 fragment and Me-13′, while the further observation of correlations of H-12/H_3-13 and H_3-13/H-3b confirmed the *cis* junction of the heterocyclic rings B and C, and supported H-12, Me-13 and 5-isopropenyl as being co-facial. As for structural unit II, the ROESY correlation pairs of H-5′/H-7′, H-7′/H-9′β, and H-9′β/H-1′β indicated that H-5′, H-7′, H-9′β, and H-1′β are all axially located, and these were randomly assigned with a β-configuration, with thereby Me-15′ being α-oriented. In addition, Me-14′ was assumed to be equatorially located and thus β-configured on the basis of the cross-peaks of H_3-14′/H_2-3′ and H-5′. Finally, Me-13′ and H-7′ were determined tentatively (due to the rotational nature of the C-7′–C-11′ bond) to have a *cis* relationship based on the observation of ROESY correlations of H_3-13′ with H-7′ and H_2-6′, together with consideration of steric hindrance resulting from the bulky substituents on both sides. The relative configuration of **75** was secured eventually by single crystal X-ray diffraction (Fig. 31), with the results in good accordance with that acquired in solution for **1** from the ROESY data.

The absolute configuration of **75** was determined with the CD exciton chirality method, and was secured by calculation of electronic circular dichroism (ECD) spectra using density functional theory (DFT). The UV spectrum exhibited a strong absorption band at λ_{max} 264 nm attributable to the conjugated α,β-unsaturated ketone (application of Woodward's rule gave ca. λ_{max} 267 nm). The first negative Cotton effect at λ_{max} 292 ($\Delta\varepsilon - 22.0$) and the second positive Cotton effect at λ_{max} 265 ($\Delta\varepsilon + 24.5$) in the CD spectrum (Fig. 32), arising from the exciton coupling between the two chromophores of α,β-unsaturated ketone and the adjacent Δ^3

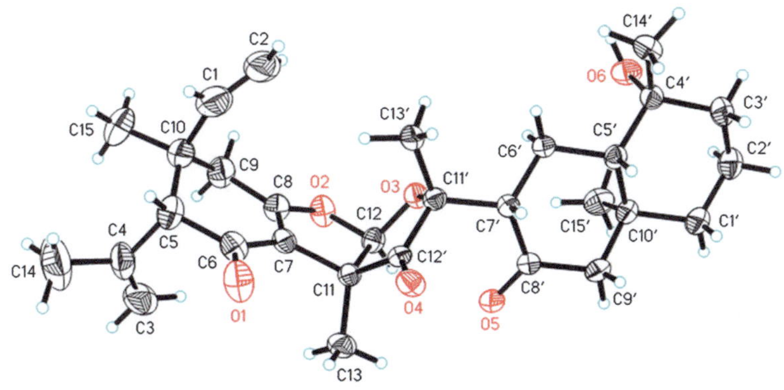

Fig. 31 Single-crystal X-ray structure of **75**

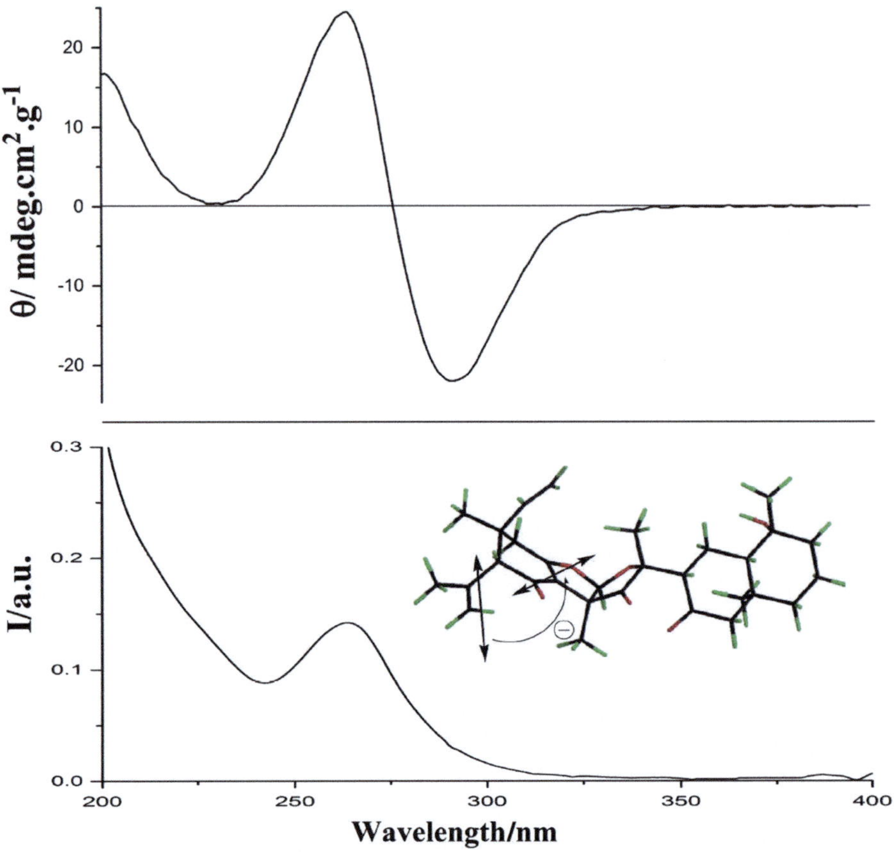

Fig. 32 CD and UV spectra (in MeOH) and the exciton chirality of **75**

double bond, indicated a negative chirality for **75**. The absolute configuration of **75** was therefore assigned as drawn.

Finally, the calculation of electronic circular dichroism (ECD) using time-dependent density functional theory (TDDFT) was applied to secure the absolute configuration of **75**. The calculated ECD curves of **75** and its enantiomers in both the gas phase and methanol are illustrated in Fig. 33. The calculated ECD of **75** matched very well with the experimental data, while the calculated ECD curves of their enantiomers were opposite to the experimental ones, confirming the absolute structures of compound **75** as assigned by the experimental CD method.

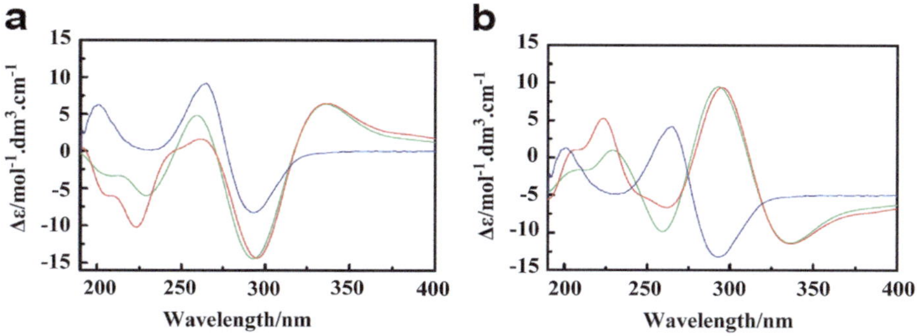

Fig. 33 Calculated ECD spectra of **75** (**a**) and its enantiomers of **75** (**b**); experimental ECD (*blue*); calculated ECD in gas phase (*green*) and in MeOH (*red*)

4 Biological Activity

Sesquiterpenoids have been reported biologically to be one of the most important groups of secondary metabolite structures from organisms [233–235]. The dimeric and sesquiterpenoid structural features of DSs not only render them "target-related" molecular properties (see Introduction), but also these compounds exhibit numerous different types of biological activities. An appealing aspect is that some DSs have shown more "drug-like" and "biologically friendly" properties than their monomeric precursors. DSs might be able to act simultaneously on both moieties of a dimeric protein in eliciting biological activities.

4.1 Cytotoxic and Antitumor Activity

Potential antitumor activity is an important biological activity, as DSs of most classes were reported to be cytotoxic for cancer cells.

It has been reported that several proteins responsible for cell proliferation and differentiation exist as hetero or homodimers or become activated through dimerization [1]. Many of these proteins have become popular targets for the development of antitumor agents. For example, proteins of the Bcl-2 family are able to form heterodimers with apoptotic proteins (e.g. Bak). Bcl-xL and Bcl-2, two antiapoptotic proteins that are involved in cancer development and resistance to treatment, can be molecular targets in anticancer therapy. Agents that disrupt the heterodimerization through down-regulating Bcl-xL expression or activity can be used to modulate cell death, leading to cell apoptosis as a possible mechanism in cancer treatment or prevention [14].

Artemisinin is a highly oxygenated guaiane-type sesquiterpenoid that contains a unique 1,2,4-trioxane ring structure. It has been reported that artemisinin and its analogues are promising cancer chemotherapeutic drug candidates by selectively causing apoptosis and metastasis in a number of cancer cell lines [236, 237]. However, the known medium potency and short plasma half-lives of these compounds in vivo have required the development of more potent and target-selective analogues of artemisinin. Artemisinin dimers, which possess the potential to target simultaneously two sites of a dimeric cancer-related protein, have been shown to be significantly more active than their monomers in inhibiting the growth of cancer cell lines [238, 239]. For example, a diphenyl phosphate dimer 838 was at least 200-fold more potent in inhibiting the growth of HCT116 and 1205Lu cells and 37-fold more potent in inhibiting the growth of HeLa cells, as compared to artesunate, the most potent monomer used in the assay. The endoperoxide bridge seems to be critical in mediating all biological activities of artemisinin derivatives [238].

Detailed discussion of artemisinin-derived dimers as potential anticancer agents can be found in some recent reviews [237, 240, 241].

Bioassay-guided purification of the AcOEt bark extract of *Meiogyne* cf. *M. cylindrocarpa* (Annonaceae), a highly potent Bcl-xL and Bak modulator, led to the isolation of meiogynin A (**76**) and its 1-epimer, 1-*epi*-meiogynin A (**77**), which are DSs of the bisabolane type [14]. Evaluation of the binding affinity of **76** to Bcl-xL and Bak showed this compound to be an antagonist to the Bcl-xL/Bak association with a *K*i value of 10.8 ± 3.1 μ*M*. Interestingly, 1-*epi*-meiogynin A (**77**) did not show any obvious affinity for this biological target (*K*i > 100 μ*M*), which suggests a configurational preference. Compound **76** also showed cytotoxic activity when tested against the KB cancer cell line, with an IC_{50} value of 4.0 μ*M*.

Three new phenolic bisabolane DSs, disydonols A–C (**78**, **79**, and **1**) were isolated from the fermentation broth of *Aspergillus* sp., a marine-derived fungus isolated from the sponge *Xestospongia testudinaria* collected from the South China Sea. In vitro cytotoxicity assays against the HepG-2 and Caski human tumor cell lines showed that the pseudo-disesquiterpenoids disydonols A (**78**) and B (**79**) exhibited moderate cytotoxicity, with IC_{50} values in the range of 2.9–12.4 $\mu g/cm^3$. However, the disesquiterpenoid disydonol C (**1**) was found to be non-cytotoxic ($IC_{50} > 100\ \mu g/cm^3$) against these two tumor cell lines. The reduction of cytotoxicity was attributed to the mesomeric effect evident in disydonol C (**1**).

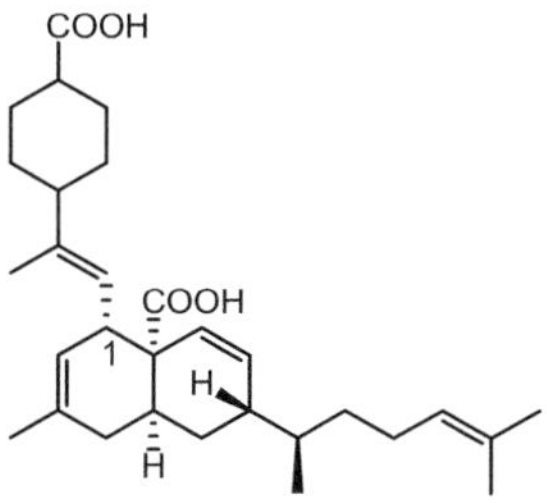

76 (meiogynin A)

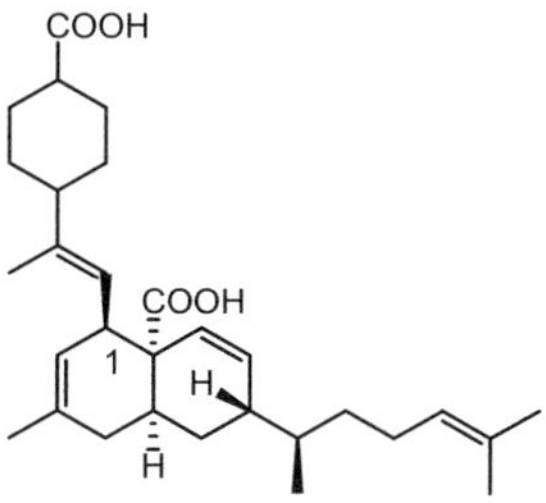

77 (1-*epi*-meiogynin A)

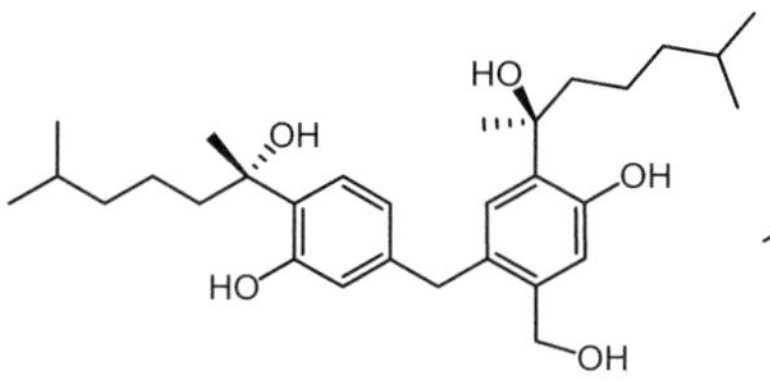

1 (disydonol C)

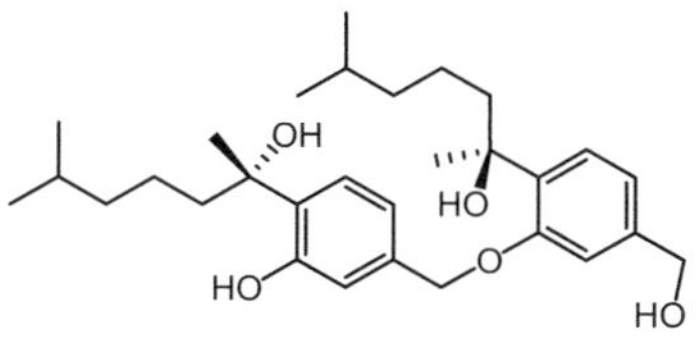

78 (disydonol A)

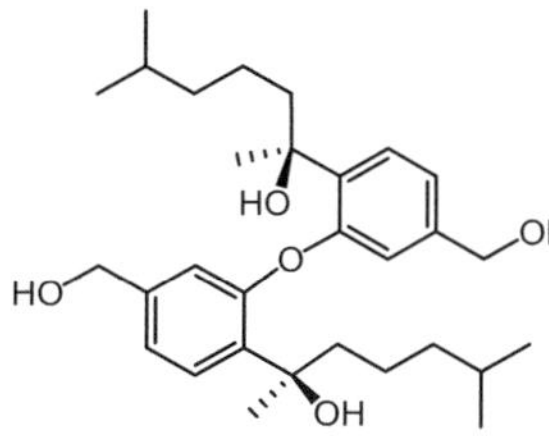

79 (disydonol B)

Microlenin (**80**) is a guaiane-type DS isolated from the plant *Helenium microcephalum* grown in Texas [63, 64]. Compound **80** showed potent antineoplastic activity against Ehrlich ascites carcinoma growth in mice at 5 and 10 mg/kg/day with 99.9 and 90.4% inhibition, more potent than the co-occurring monomeric sesquiterpene lactones [242]. Metabolic studies demonstrated that DNA synthesis (DNA polymerase, purine synthesis, and dihydrofolate reductase) and protein synthesis (polypeptide synthesis initiation) were inhibited significantly by microlenin (**80**).

Artanomalide D (**81**), 8-*O*-acetylarteminolide (**82**), and artanomaloide (**83**) are three guaiane DSs isolated from the aerial parts of *Artemisia anomala* [32]. Cytotoxicity evaluation of the DSs and monomers showed that artanomalide D (**81**) resulted in potent growth inhibition, with IC_{50} values of 1.9, 3.0, 8.5, and 1.8 μ*M*, respectively, against HCT-8, Bel-7402, BGC-823, and A2780 cancer cells.

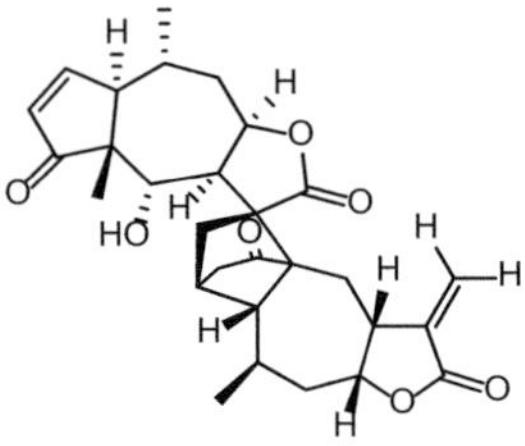

80 (microlenin)

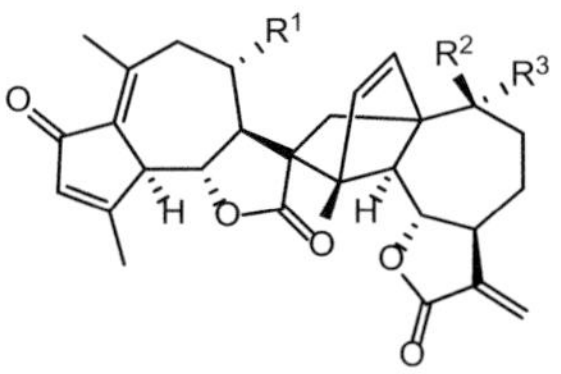

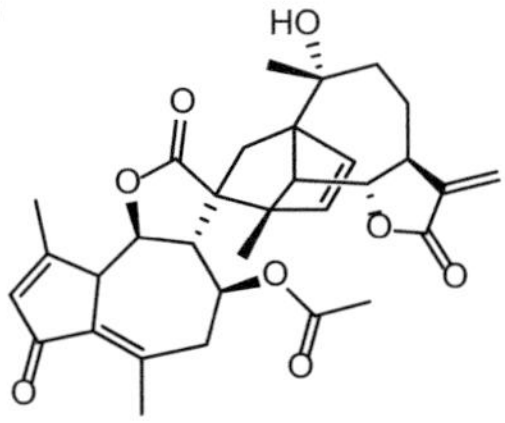

	R^1	R^2	R^3
81 (artanomalide D)	OH	OAc	Me
82 (8-*O*-acetylarteminolide)	OAc	Me	OH

83 (artanomaloide)

Among artanomadimers A–F, six guaiane DSs also obtained from the aerial part of *Artemisia anomala* [34], artanomadimers A (**84**) and F (**85**) displayed cytotoxicities against the BGC-823 cell line with IC_{50} values of 2.7 and 6.3 μ*M*, respectively, while the other isolates did not show significant cytotoxicity.

Pungiolides D (**86**) and E (**87**) are two xanthane DSs isolated from the aerial parts of *Xanthium sibiricum* [70]. Cytotoxicity evaluation against the SNU387 and A-549 human cancer cell lines showed these two DSs to possess cell growth inhibitory activity against SNU387 cells (IC_{50} values of 14.6 and 11.7 μ*M*, respectively) but they exhibited no discernible cytotoxicity against A-549 cells ($IC_{50} > 40$ μ*M*).

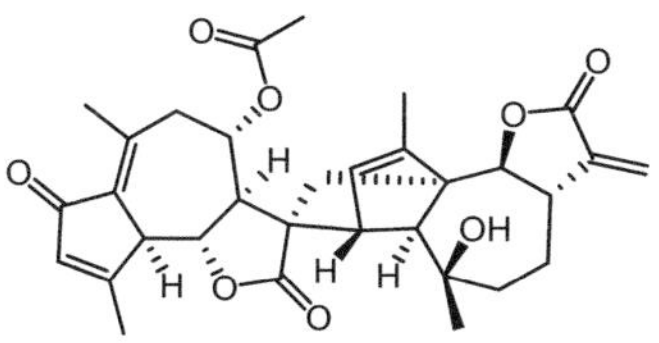

84 (artanomadimerA)

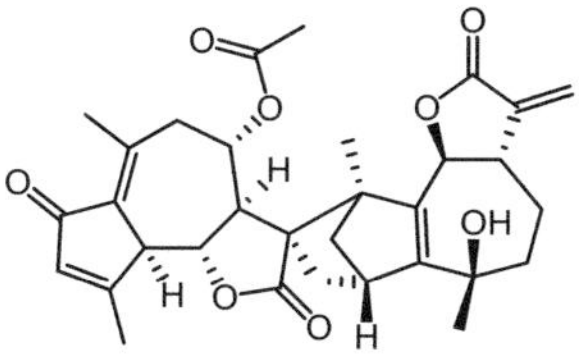

85 (artanomadimer F)

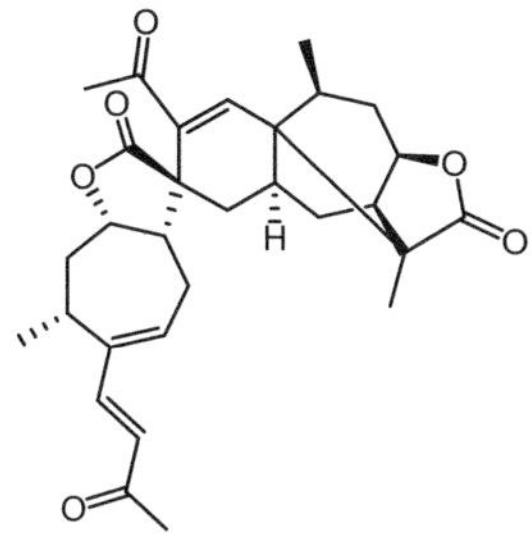

86 (pungiolide D)

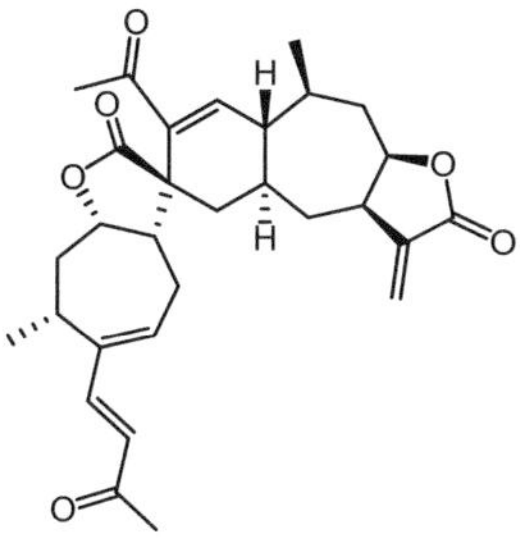

87 (pungiolide E)

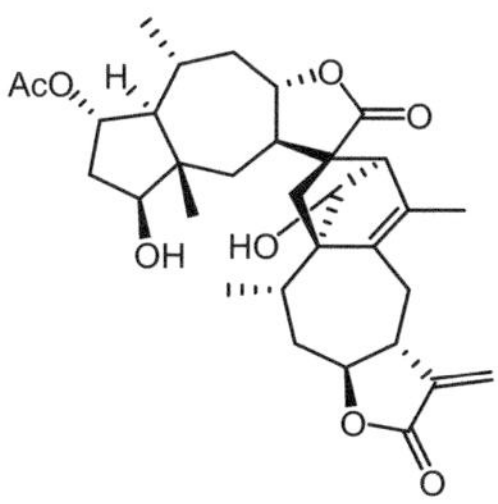

88 (leariifolianoid E)

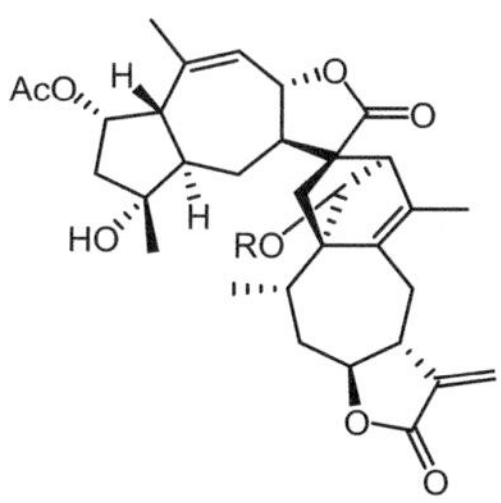

89 R = Ac (lineariifolianoid F)
90 R = H (lineariifolianoid G)

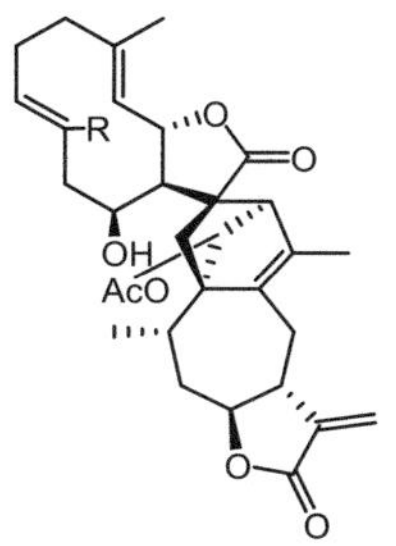

91 R = CH_2OH (lineariifolianoid H)

Lineariifolianoids E (**88**, guaiane-pseudoguaiane type), F–G (**89**, **90**, guaiane-guaiane type), and H (**91**, germacrane-guaiane type) were isolated from the aerial parts of *Inula lineariifolia* [61]. Cytotoxicity evaluation against two human breast cancer cell lines (MCF-7 and MDA-MB-231) and one normal breast cell line

(MCF-10A) showed that the four DSs all had selective cytotoxicity for cancer cells (IC_{50} values in the range of 1.6–16.5 μ*M*). The DSs were more potent than their co-occurring monomers, which was attributed to the enhanced cellular penetration as a result of dimerization. The α,β-unsaturated cyclopentenone and α-methylene-γ-lactone groups were suggested as being essential for the enhanced activity of **88** as the most active isolate. Further studies also showed that **88** gave significant, dose-dependent cytotoxic effects, and that cell cycle arrest and apoptosis were potential mechanisms of action.

Parviflorene A (**92**) is a cadinane-type DS isolated from an extract of a tropical zingiberaceous plant, *Curcuma parviflora* [110]. Cytotoxicity evaluation revealed the activity of **92** against vincristine (VCR)-resistant P388 cells with IC_{50} values of 3.2 and 3.0 μg/cm^3, respectively, in the presence and absence of 12.5 ng/cm^3 of VCR. However, the IC_{50} value against a sensitive P388 strain was 3.2 μg/cm^3, which indicated that **92** had no reversal effect on multidrug resistance. This compound also exhibited cytotoxicity against B16 melanoma cells (IC_{50} 4.1 μg/cm^3).

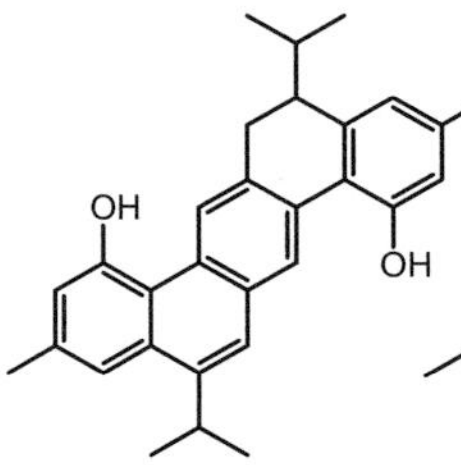

92 (parviflorene A)

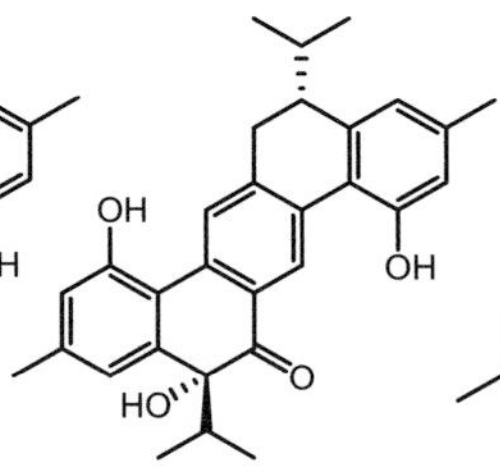

93 (parviflorene B)

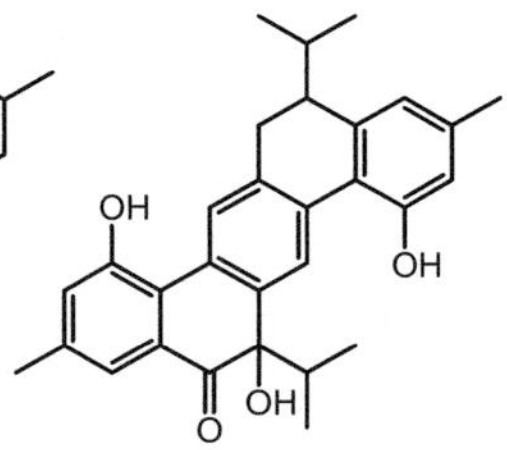

94 (parviflorene C)

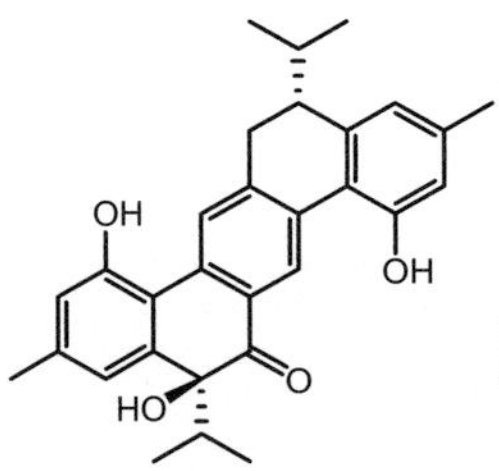

95 (parviflorene D)

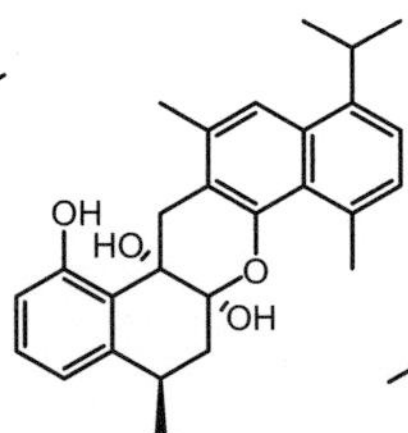

96 (parviflorene E)

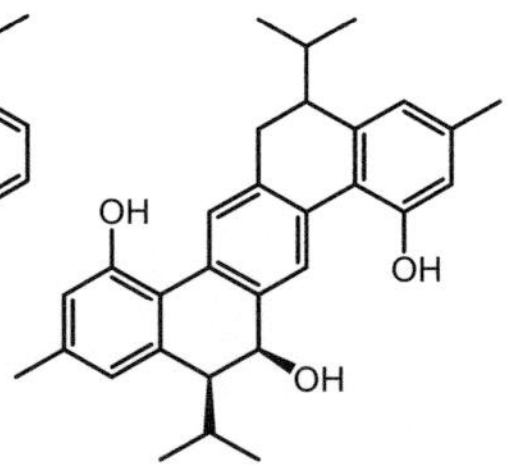

97 (parviflorene F)

Further investigation of this plant led to the isolation of nine DSs of related biological origin, parviflorenes B–J, from the underground parts of *C. parviflora* [111]. These compounds can be classified into six groups based on their carbon backbones: biscadinane, cadinane-isocadinane adduct, biscadinane with a different bond connection, biscadinane connected through a single bond, biscadinane minus an isopropyl unit, and biscadinane with a ring-contraction rearrangement [111]. All compounds, except parviflorene H ($IC_{50} > 100$ μg/cm^3) symmetrically formed by coupling of two identical cadinanes through a single bond, showed cytotoxicity against the KB and HeLa cell lines with IC_{50} values in the range of 1.5–14.8 μg/cm^3 [111–113, 243]. Parviflorene C (**94**), a cadinane-isocadinane adduct, showed a threefold reversal effect of VCR resistance against VCR-resistant KB cell lines. Parviflorenes C (**94**) and F (**97**) also exhibited cytotoxicity against the LNCaP (human prostate cancer) and TNF-related apoptosis inducing ligand (TRAIL)-resistant KOB (human adult T cell leukemia) cell lines [112]. Parviflorenes A (**92**) and F (**97**), the two most abundant DSs of the biscadinane type from this plant, were also evaluated in the human cancer 39-cell line panel assay of the Japanese Foundation for Cancer Research. Although parviflorenes A (**92**) and F (**97**) showed low differential cellular sensitivities, both compounds were cytotoxic against all these cell lines at considerably lesser concentrations with mean log GI_{50} (log concentration of compound for inhibition of cell growth at 50% compared to control) values of ca. 2.6 μ*M* [113]. COMPARE analysis showed no strong correlation (correlation index, $r < 0.5$) between parviflorene F (**97**) and standard anticancer drugs, which suggested a different mode of action for this DS [111]. Mechanistic studies showed that treatment with parviflorene F (**97**) changed mRNA expression of seven genes, with the cytotoxic effect of this compound possibly related to apoptosis induction through both the intrinsic and extrinsic apoptotic pathways [111], particularly by a caspase-dependent mechanism through TRAIL-R2 (tumor necrosis factor α-related apoptosis inducing ligand receptor 2) [244].

Twelve lindenane DSs, chloramultiols A–F (**98**–**103**), chloramultilides C and D, shizukaols C and D, spicachlorantin B and cycloshizukaol A, were isolated from the whole plant of *Chloranthus multistachys* [127]. The cytotoxic activities of these DSs were evaluated against five tumor cell lines (A-549, HL-60, PANC-1, SMMC-7721, and SK-BR-3), but none showed any discernible cytotoxic activity ($IC_{50} > 10$ μ*M*).

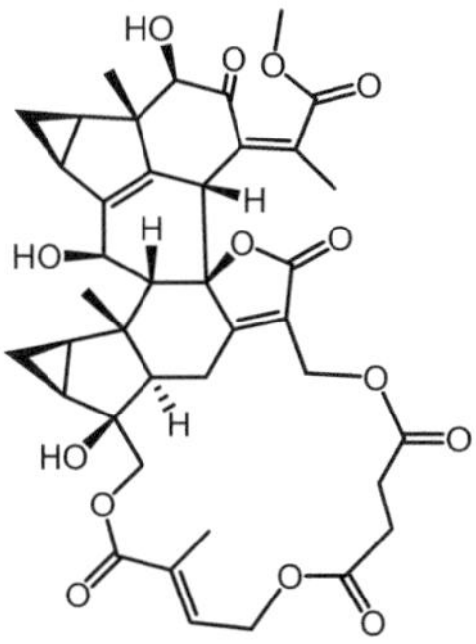

98 (chloramultiol A)

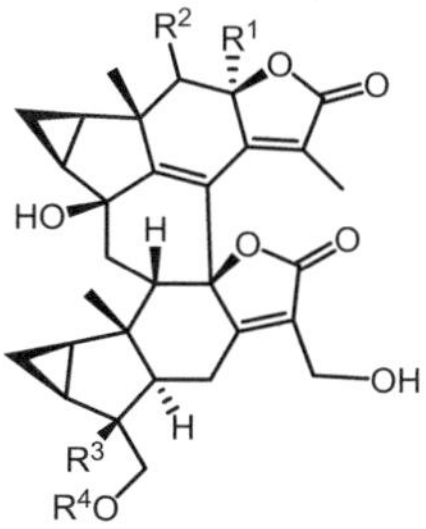

tigloyl group = 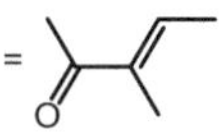

	R^1	R^2	R^3	R^4
99	OH	β-OH	OH	H (chloramultiol B)
100	OCH_3	β-OH	OH	Tig (chloramultiol C)
101	OH	=O	OH	Tig (chloramultiol D)
102	OH	β-OH	H	Ac (chloramultiol E)

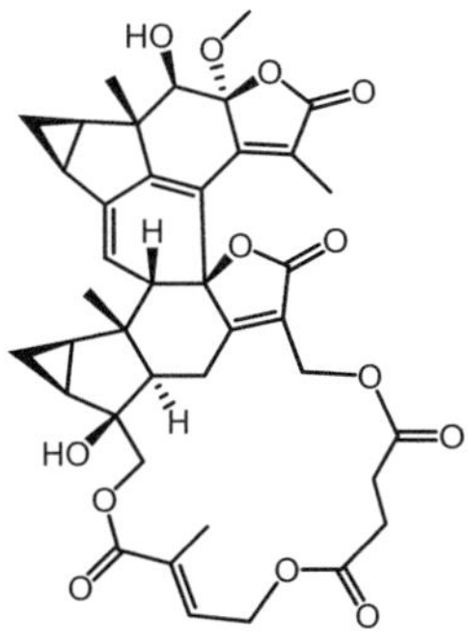

103 (chloramultiol F)

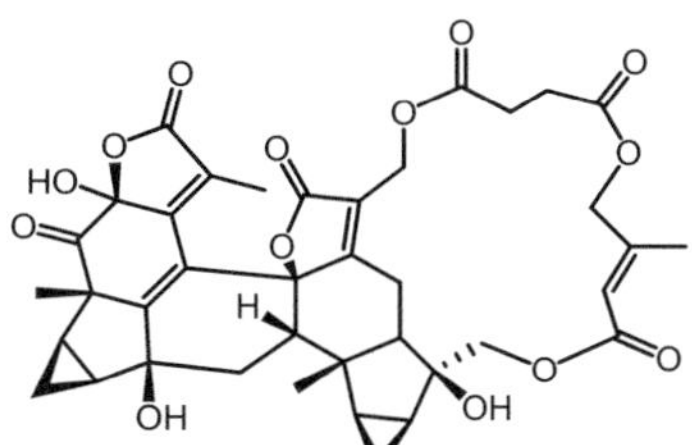

104 (chloraholide B)

Chloramultilides B–D and chlorahololide B (**104**), four lindenane DSs isolated from the whole plant of *Chloranthus spicatus*, were also evaluated for their cytotoxic activity against the P-388 and A-549 human cancer cell lines, but none showed any significant inhibitory activity [222].

Eleven lindenane DSs, sarcandrolides A–E (**105–109**), chlorahololide F, shizukaol B, shizukaol C, shizukaol E, shizukaol G, and cycloshizukaol A, were isolated from the whole plants of *Sarcandra glabra* [131]. The cytotoxic activities of **105–109** and two lindenane monomers were evaluated against the HL-60, A-549 and BEL-7402 cancer cell lines. The results revealed that only sarcandrolides A–C (**105–107**) showed inhibitory activities against the HL-60 cell line (IC_{50} values of 3.1, 8.4, and 8.5 μ*M*, respectively). In turn, sarcandrolides A (**105**) and C (**107**) inhibited the growth of A-549 cells (IC_{50} values of 7.2 and 4.7 μ*M*), but none of the DSs or monomers showed any inhibitory activity when evaluated against the BEL-7402 cell line.

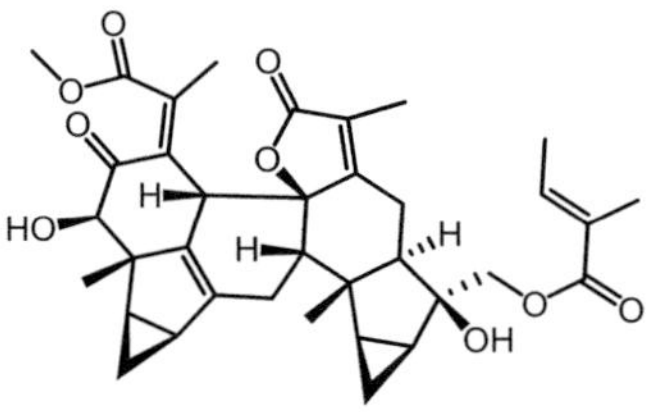

105 (sarcandrolide A)

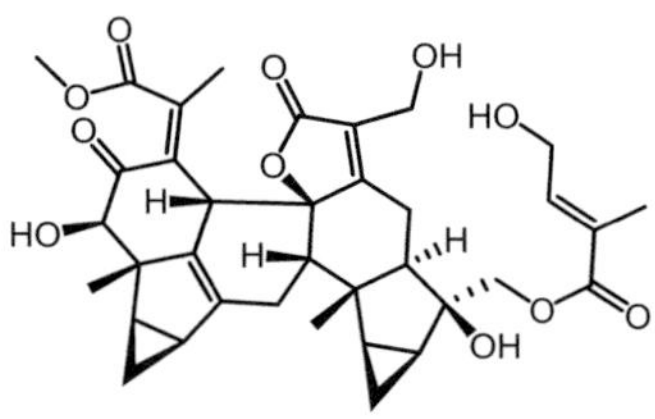

106 (sarcandrolide B)

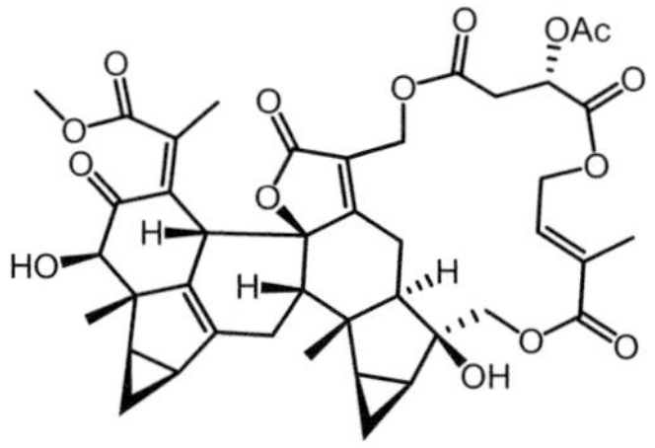

107 (sarcandrolide C)

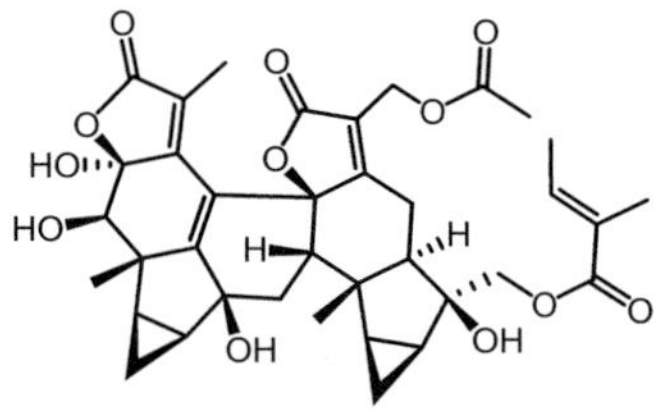

108 (sarcandrolide D)

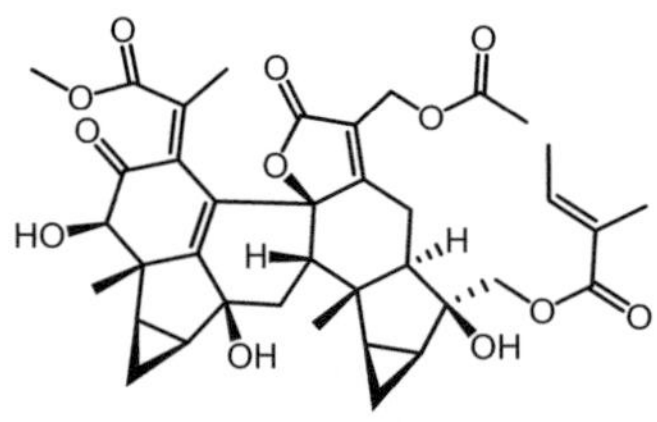

109 (sarcandrolide E)

Continuing investigation of the plant led to the isolation of five new DSs of the same type, sarcandrolides F–J [132]. These DSs were evaluated for cytotoxic activities against the HL-60 and A549 cancer cell lines. Sarcandrolides F (**110**) and H (**111**) exhibited cytotoxicity against the HL-60 cell line with IC_{50} values of 0.03 and 1.2 μ*M*, respectively, while the other three DSs and three monomers were inactive toward both the HL-60 and A-549 tumor cell lines ($IC_{50} > 10$ μ*M*).

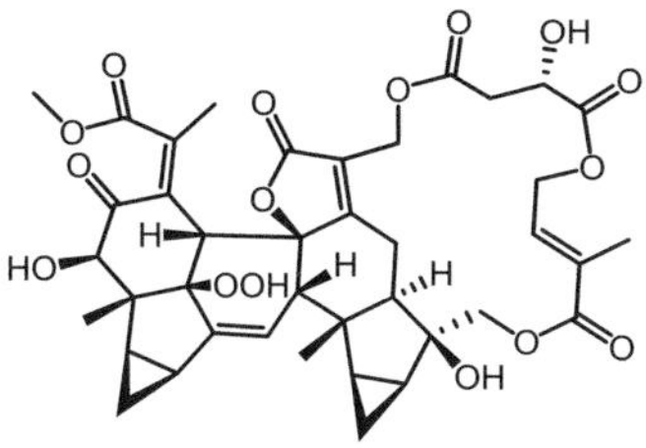

110 (sarcandrolide F)

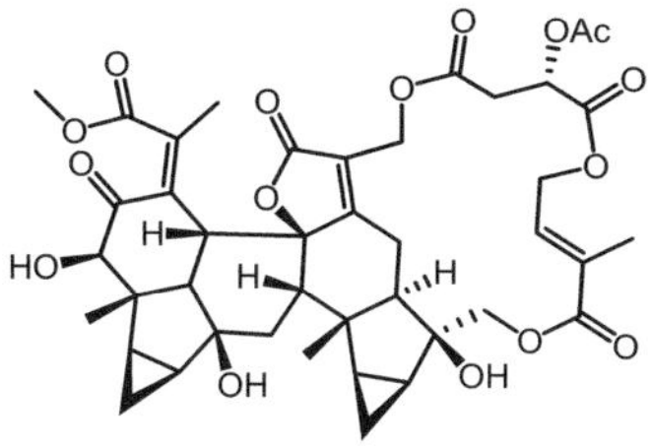

111 (sarcandrolide H)

Shizukaol B (**71**), cycloshizukaol A (**112**) and shizukaol F (**113**), three lindenane DSs isolated as the active principles of the MeOH extract of the roots of *Chloranthus japonicus*, were found to inhibit phorbol 12-myristate 13-acetate (PMA)-induced homotypic aggregation of HL-60 cells without cytotoxicity with *MIC* values of 34.1 n*M* (**71**), 0.9 μ*M* (**112**) and 27.3 n*M* (**113**), respectively [245]. Mechanistic study showed that they inhibited PMA-induced cell aggregation through down-regulation of ICAM-1 expression in HL-60 cells. The higher inhibitory activities of compounds **71** and **113** suggested that the effects of these compounds on ICAM-1 expression might be increased by the presence of a macrocyclic lactone ring.

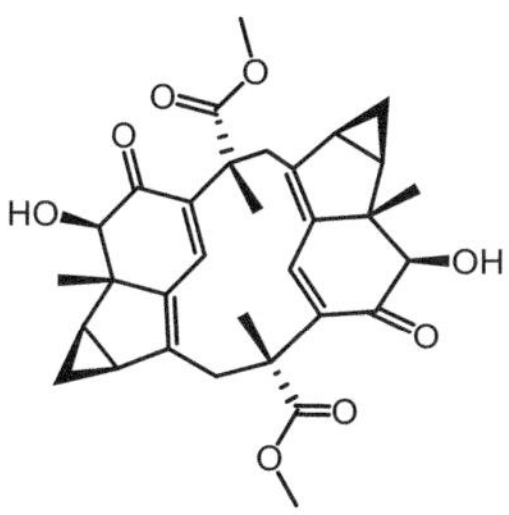

112 (cycloshizukaol A)

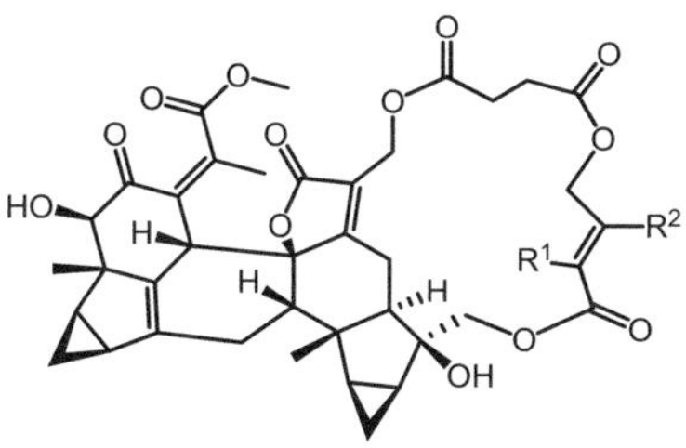

113 R^1 = H, R^2 = CH_3 (shizukaol F)

Aurisins A (**114**) and K (**115**) are two aristolane DSs isolated from two isolates of the luminescent mushroom *Neonothopanus nambi*, PW1 and PW2 [148]. Compounds **114** and **115** showed more potent cytotoxicity against the NCI-H187 cell line (IC_{50} values of 1.6 and 1.5 μ*M*, respectively) than the co-occurring aristolane monomer nambinone C ($IC_{50} > 16.4$ μ*M*). The two DSs were also active against the BC1 cell line (**114**, IC_{50} 3.7 μ*M*) or the KB cell line (**115**, IC_{50} 6.87 μ*M*). In addition, **114** showed cytotoxicity against four cholangiocarcinoma cell lines

(KKU-100, KKU-139, KKU-156, and KKU-213) (IC_{50} values in the range 1.6–2.8 μ*M*), comparable to or better than the standard drug ellipticine.

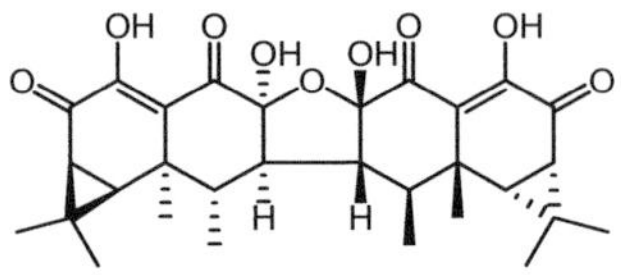

114 (aurisin A)

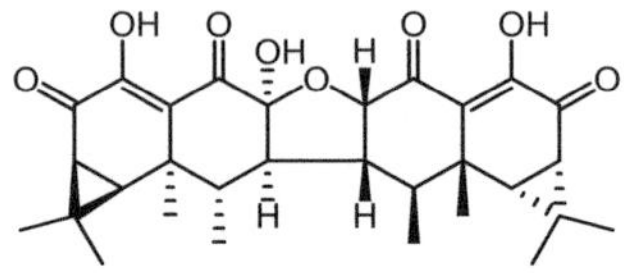

115 (aurisin B)

The elemane–eudesmane type compound DSs serratustones A (**75**) and B were evaluated against the HL-60 and A-549 tumor cell lines, but neither showed activity at the concentration levels used [171].

Foveolide B (**116**) is an eudesmane DS isolated from the stems of *Ficus foveolata* [121]. Cytotoxicity assays against five human cancer cell lines demonstrated that **116** was specifically cytotoxic toward the SW620 cell line, while its structurally related monomers, except one possessing an α,β-unsaturated moiety, were inactive against all these cell lines used.

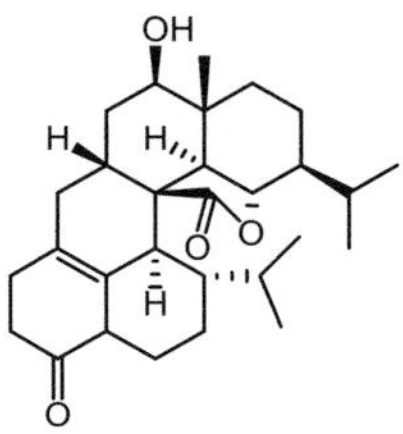

116 (foveolide B)

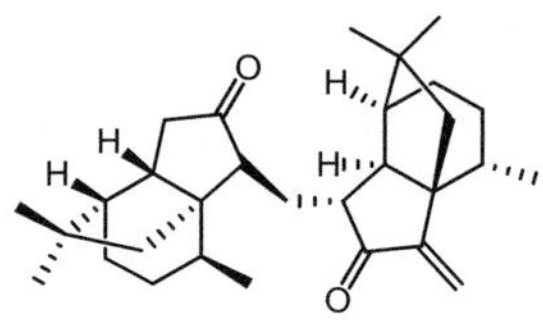

117 (alertenone)

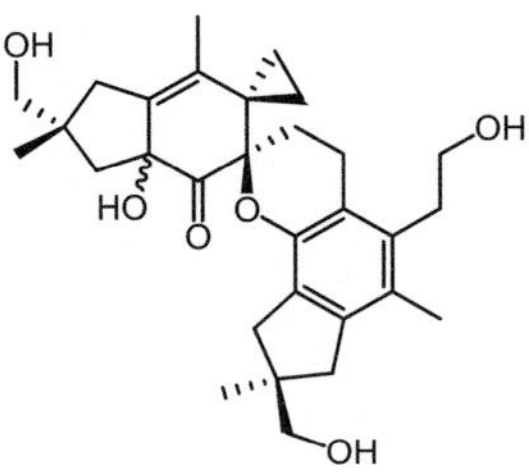

118 (bovistol)

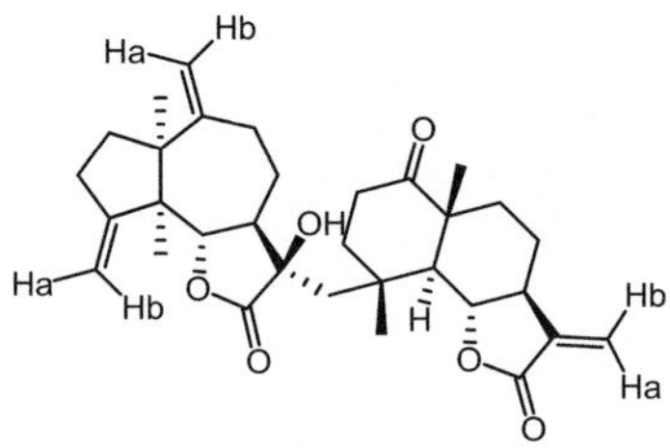

119 (lappadilactone)

Alertenone (**117**) is a quadrane DS isolated from *Alertigorgia* sp., a gorgonian [147]. In comparative testing against ten human tumor cell lines (A-549, HOP-92, SF-295, SF-539, SNB-19, LOX, M14, MALME-3 M, OVCAR-3, and MCF7), **117** was surprisingly inactive ($IC_{50} > 35$ μg/cm^3), as compared to its monomeric precursor suberosenone, which exhibited strong cytotoxicity (IC_{50} values of 0.002–1.6 μg/cm^3). However, the reason for this differential activity is still not properly understood.

Bovistol (**118**) is an illudane-type DS isolated from the basidiomycete *Bovista* sp. 96042 [152]. Compound **118** showed weak cytotoxic activities for HeLa S3 cells.

Japonicones A–C (eudesmane–guaiane-type compound DSs) and D (1,10-secoeudesmane–guaiane-type compound DS) were isolated from the aerial parts of *Inula japonica* [163]. The cytotoxic activities evaluated against four tumor cell lines (A549, LOVO, CEM and MDA-MB-435) indicated japonicone A (**63**) to be the most active DS (IC_{50} values of 1.6, 0.26, 0.001, and 0.20 μg/cm^3, respectively), and this compound exhibited more potent cytotoxicity than the positive control doxorubicin. Structure-activity relationship (SAR) analysis suggested that the $\Delta^{5,6}$ moiety (as in japonicone A) rather than a $\Delta^{4,5}$ unit (as in japonicone B) is preferable for this type of activity.

Another eudesmane–guaiane type DS, lappadilactone (**119**), was isolated from the dried roots of *Saussurea lappa* [166]. Cytotoxicity evaluation of all the purified compounds showed that **119** was more potent than its monomeric analogues with IC_{50} values of 2.4, 1.8, and 2.5 μg/cm^3, respectively, against the HepG2, HeLa, and OVCAR-3 cancer cell lines. The enhanced potency was ascribed in part to the presence of an α-methylene-γ-lactone group.

Virgaurin B (**120**) is a furanoeremophilane DS isolated from the roots of *Ligularia virgaurea* [75]. Cytotoxicity evaluation on seven cell lines (HSC-2, HCS-2, HeLa, RERF-LC-KJ, A549, A172, and K562) demonstrated that virgaurin B showed dose-dependent effects for A172 and RERF-LC-JK cells in the concentration range of 1–100 μ*M* and the IC_{50} value was determined as about 10 μ*M* for A172 cells. The growth of all cell lines except HCS-2 was approximately inhibited at 100 μ*M*.

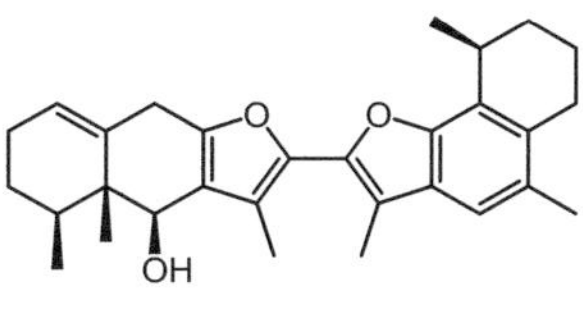

120 (virgaurin B)

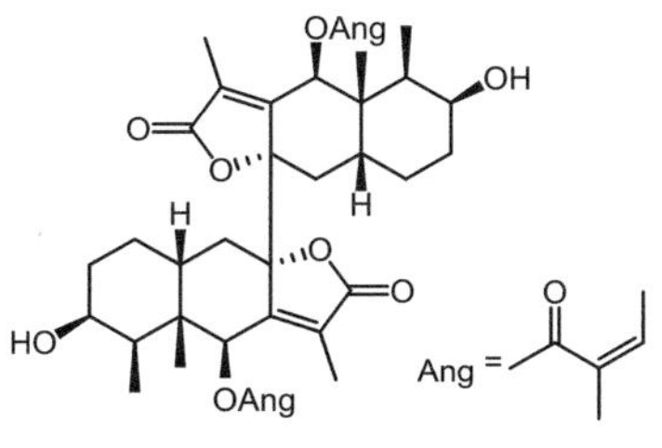

121 (bieremoligularolide)

Bieremoligularolide (**121**) is an eremophilane DS isolated from the roots of *Ligularia muliensis* [78]. Cytotoxic activity determination showed **121** to possess cytotoxicity against HL-60, SMMC-7721, and HeLa cells with IC_{50} values of 5.5, 16.1, and 8.9 μ*M*. However, the co-occurring monomer eremoligularin showed no discernible cytotoxicity against these three cell lines ($IC_{50} > 100$ μ*M*).

Artemilinin A (**122**), a eudesmane–guaiane-type pseudo-disesquiterpenoid DS isolated from *Artemisia argyi* [181], was also inactive in cytotoxicity assays against the HL-60, BGC-823, Bel-7402, and KB tumor cell lines at a final concentration of 10 μg/cm^3, in spite of the presence of an α-methylene-γ-lactone group present in this molecule.

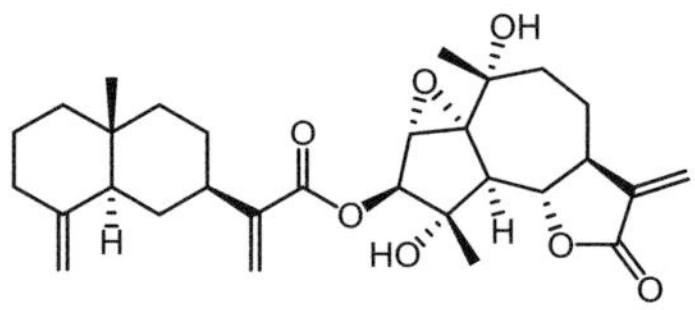

122 (artemilinin A)

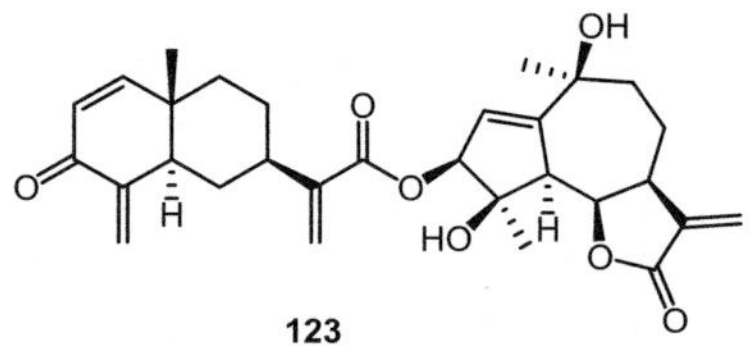

123

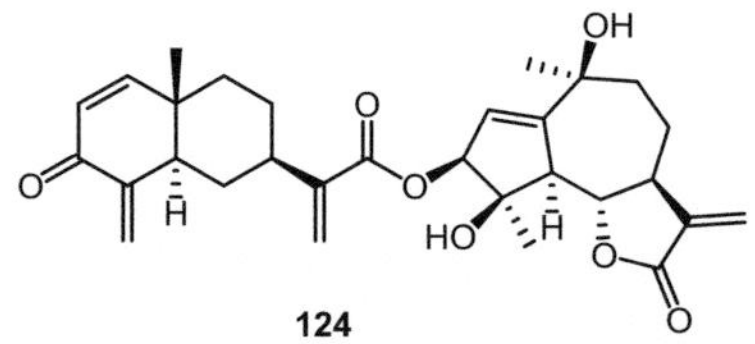

124

Two eudesmane–ester-guaiane type pseudodisesquiterpenoid DSs, **123** and **124**, were isolated from the leaves of *Warionia saharae* [246]. Both compounds showed similar cytotoxicities against HeLa (KB) cells, peripheral blood mononuclear cells (PBMCs), and human Jurkat T leukemia cells, with IC_{50} values in the range of 1.0–2.2 μ*M*, more potent than their co-occurring guaiane monomers. Along with its cytotoxic effect, **123** showed a potent down-regulation of the mRNA levels of the β-actin and GAP-DH house-keeping genes of PBMCs after 20 h using a concentration of 10 μ*M*.

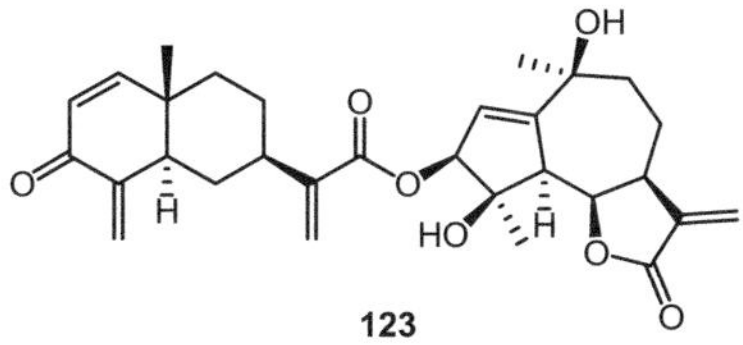

123

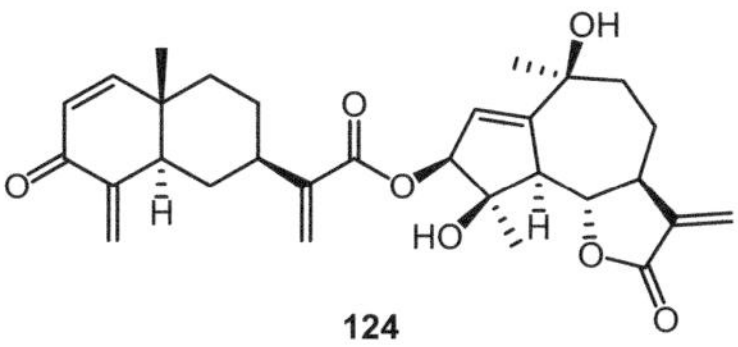

124

Vernodalidimers A (**125**) and B (**126**), two elemane-isobutanoic acid ester-elemane-type pseudo-disesquiterpenoids, isolated from the seeds of *Vernonia anthelmintica* [211], exhibited potent cell growth inhibitory activity against HL-60 cells with IC_{50} values of 0.72 and 0.47 μM, respectively.

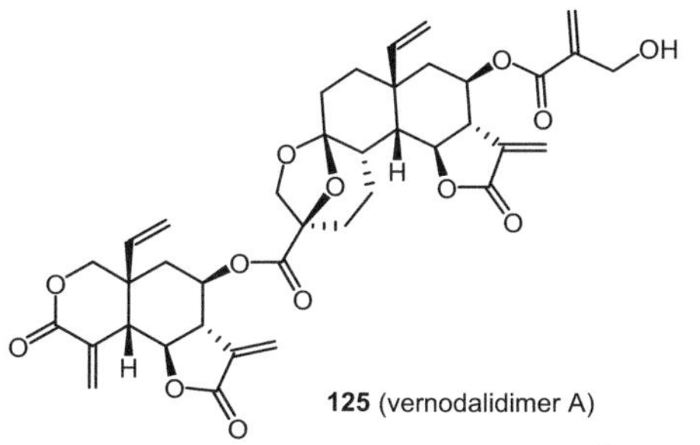

125 (vernodalidimer A)

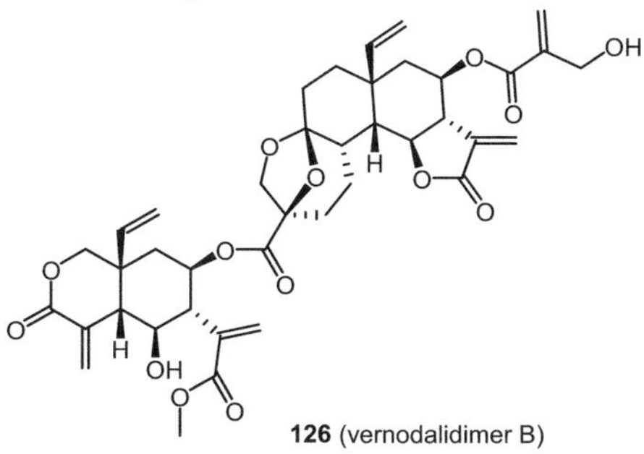

126 (vernodalidimer B)

Cryptoporic acids C–J were isolated from the fungus *Cryptoporus volvatus* [192–194, 247, 248]. Cryptoporic acid E (**57**, CPA-E) is a drimane-*O*-isocitric acid ester-drimane-type DS obtained from this organism [194]. Biological investigation demonstrated that **57** inhibited colon cancer development induced with *N*-methyl-*N*-nitrosourea in rats and 1,2-dimethylhydrazine in both rats and mice. Oral administration of this compound yielded a 50% or more reduction in the incidence and number of colon tumors [247]. Cyptoporic acid E was suggested as a potential candidate chemopreventive agent for colon cancer. This compound inhibited okadaic acid-induced tumor promotion in two-stage carcinogenesis experiments on mouse skin at concentrations of 1.2–5.9 μ*M* [248].

Ligulamulienins A (**127**) and B (**128**), two furanoeremophilane-*O*-12-noreremophilane type DSs isolated from the rhizomes of *Ligularia muliensis* [91], showed cytotoxicity against the MGC-803, HEP-G2, and S-180 tumor cell lines with IC_{50} values in the range of 11.9–65.4 μ*M*.

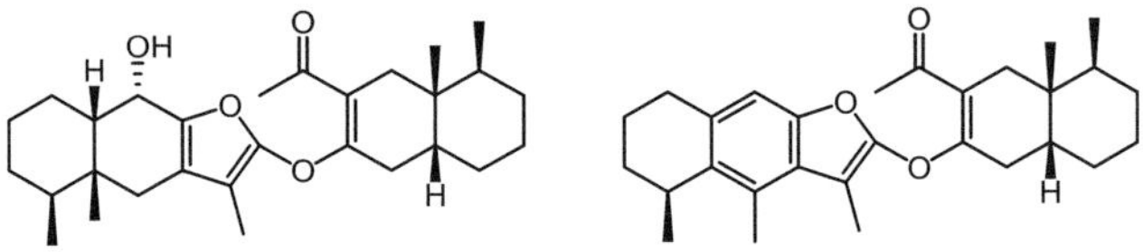

127 (ligulamulienin A) **128** (ligulamulienin B)

Nine dimeric aza-sesquiterpenoid-type pseudo-disesquiterpenoid DSs including 6-hydroxythiobinupharidine (**129**), 6,6′-dihydroxythiobinupharidine (**130**), and 6-hydroxythionuphlutine B (**131**) were isolated from the rhizomes of *Nuphar pumilum* [249]. Cytotoxicity evaluation against the U937, B16F10, and HT1080 tumor cell lines demonstrated that DSs with a 6-hydroxy group showed substantial cytotoxic activity at a concentration of 10 μ*M*, but DSs lacking a 6-hydroxy group and monomeric sesquiterpene alkaloids showed only weak activity.

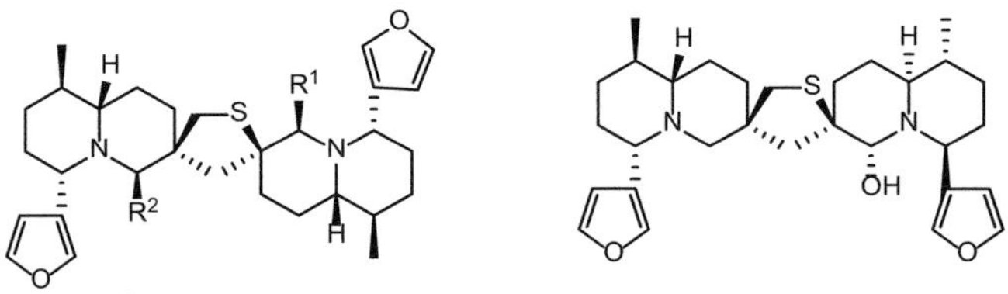

129 R^1 = OH, R^2 = H (6-hydroxythiobinupharidine)
130 R^1 = R^2 = OH (6,6′-dihydroxythiobinupharidine)
131 (6-hydroxythionuphlutine B)

The apoptosis-inducing activity of 6-hydroxythiobinupharidine on U937 cells examined by morphological observation and a DNA fragmentation assay (TUNEL method) showed that apoptosis of U937 cells was almost immediately observed within 1 h after treatment by 6-hydroxythiobinupharidine at 2.5–10 μ*M*.

Encouraged by the fact that the methanolic extract and its alkaloid fraction from the rhizomes of *Nuphar pumilum* inhibited invasion of B16 melanoma cells across collagen-coated filters in vitro, the isolated compounds were further evaluated for their antimetastatic activity [250]. The three apoptosis-inducing DSs that carry a 6-hydroxy group (6-hydroxythiobinupharidine, 6,6′-dihydroxythiobinupharidine, and 6-hydroxythionuphlutine B) also proved to be the most active antimetastatic agents, with IC_{50} values of 0.029, 0.087, and 0.36 μ*M*, respectively. The alkaloid fraction (20 mg/kg/day, po) and the principal DS, 6-hydroxythiobinupharidine (5 mg/kg/day, po), significantly inhibited lung tumor formation by more than 90% ten days after injection of B16 melanoma cells in mice. These observations suggested that the active DSs of this type could be a new candidate class for the development of oncology drugs that possess both apoptosis-inducing and antimetastatic activities.

Halichonadins G–I (**132–134**) were isolated from a marine sponge, *Halichondria* sp. [201]. Halichonadins G (**132**) and H (**133**) are eudesmane homo-dimers linked through a methyl 2-(1-(2-amino-2-oxoethyl)ureido) acetate fragment and a 2-hydroxymalonamide fragment, respectively, while halichonadin I (**134**) is a eudesmane hetero-dimer linked through a urea fragment. Halichonadins G (**132**) and I (**134**) showed cytotoxicity against L1210 cells (IC_{50} 5.9 and 6.9 μg/cm^3) and KB cells (IC_{50} 6.7 and 3.4 μg/cm^3), while halichonadin H and a non-DS isolate showed no discernible cytotoxicity against both cell lines ($IC_{50} > 10$ μg/cm^3).

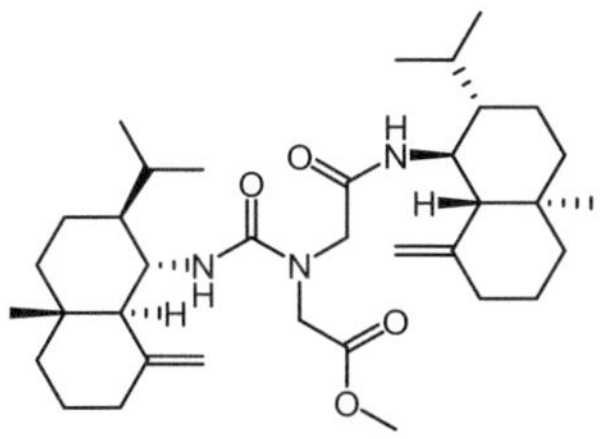

132 (halichonadin G)

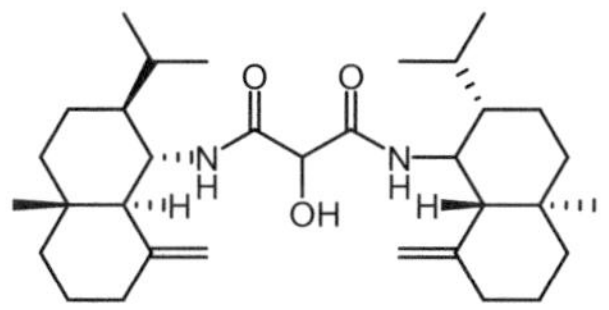

133 (halichonadin H)

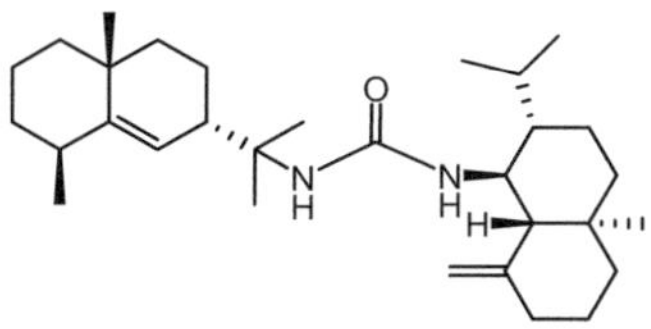

134 (halichonadin I)

Halichonadin K (**55**), a pseudo-disesquiterpenoid-type eudesmane homodimer linked with a piperidine ring through amide bonds isolated from an Okinawan marine sponge *Halichondria* sp. [202], showed cytotoxicity against KB cells with an IC_{50} value of 10.6 μg/cm^3.

The indolo-sesquiterpene type DSs, dixiamycins A–B (**59–60**), showed only weak growth inhibition ($IC_{50} \geq 57.6$ μM) of the MCF7, NCI-H460, SF268, and HepG2 tumor cell lines [198].

4.2 Anti-inflammatory Activity

Nitric oxide (NO), which is produced from oxidation of L-arginine by NO synthase (NOS), has been implicated in various physiological and pathological processes (e.g. vasodilation, nonspecific host defense, ischemia-reperfusion injury, and chronic or acute inflammation) [50]. As a response to proinflammatory agents such as interleukin-1β, tumor necrosis factor-α and LPS, NO may be overproduced in various cell types (e.g. macrophages, endothelial cells, and smooth muscle cells) by inducible NOS (iNOS). Therefore, inhibition of NO release might be an effective therapeutic strategy for the treatment of inflammatory diseases.

Ainsliadimer A (**135**) is a guaiane DS isolated from *Ainsliaea macrocephala* [29]. Biological evaluation showed that ainsliadimer A (**135**) exerted inhibitory activity against LPS-induced NO production in RAW264.7 cells with an IC_{50} value of 2.4 μg/cm^3.

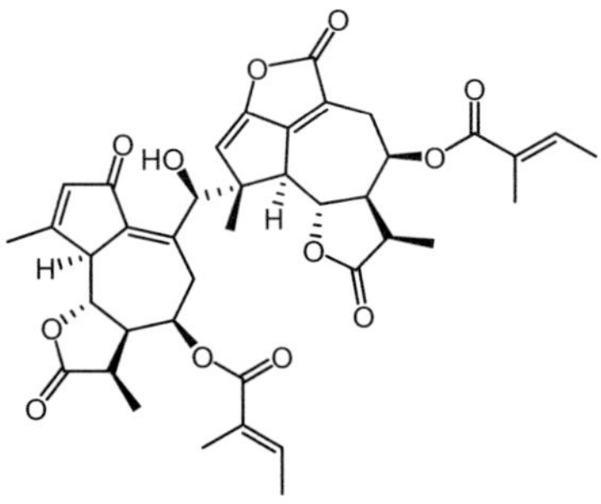

5 (diguaiaperfolin)

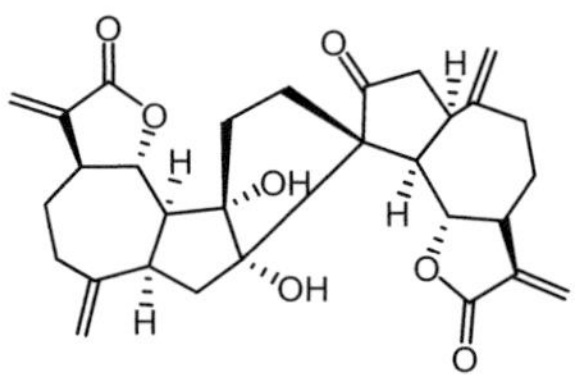

135 (ainsliadimer A)

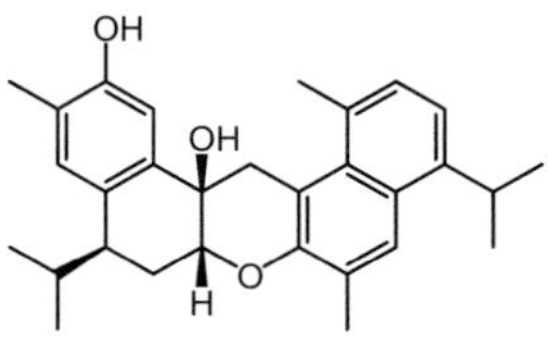

136 (dicadalenol)

The anti-inflammatory activities of the guaiane DSs artanomalide D (**81**), 8-*O*-acetylarteminolide, and artanomaloide were evaluated in a COX-2 inhibition assay [32], but only artanomalide D (**81**) and two monomeric sesquiterpenoids, artanomalic acid and artanomalide B, were reported to show activities at 50 μ*M*, with inhibitory rates of 99.5, 98.2, and 98.4%.

Another guaiane DS, diguaiaperfolin (**5**), was isolated from *Eupatorium perfoliatum* [51], a herb used traditionally for the treatment of fever, malaria and inflammation-associated diseases. Biological evaluation showed that this DS possesses anti-inflammatory activity against LPS-stimulated macrophages by inhibition of NO release (IC_{50} 16.5 μ*M*) and iNOS formation, and was more active than several monomers. However, its selectivity index was rather low (SI ca. 2). Mechanistic study showed that the anti-inflammatory activity resulted from significant down-regulation of cytokines CSF-3, IL-1α, IL-1β, TNF, and chemokines CCL2, CCL22, and CXCL10.

Dicadalenol (**136**) is a cadinane DS isolated together with various cadinane and isocadinane monomers from the aerial parts of *Heterotheca inuloides* [106]. The effects of these natural products on TPA-induced mouse ear edema when administered topically showed that although the isolates tested all displayed anti-inflammatory activity, **136** was the most active compound. Preliminary structure-activity analysis indicated that the phenolic or carboxylic groups had no effect on the antiinflammatory activities, but the dimeric feature of dicadalenol (**136**) might contribute to the overall increased anti-inflammatory effect.

Eight lindenane DSs, **16**, **17**, spicachlorantin A, spicachlorantin C, chloramultilide A, henriol A, and shizukaols B and D, were isolated from the whole plant of *Chloranthus serratus* [125]. These isolates were evaluated for their inhibitory effects on lipopolysaccharide-induced nitric oxide production in RAW264.7 cells. Compound **17** and shizukaols B and D showed significant anti-inflammatory activities, with IC_{50} values of 0.22, 0.15, and 7.2 μ*M*, respectively. Hexadecadrol was used as a positive control, with an IC_{50} value of 0.47 μ*M*). The other DSs tested were inactive in this assay ($IC_{50} > 10$ μ*M*).

Nine known lindenane DSs and two lindenane monomers isolated from the whole plant of *Chloranthus fortunei* were evaluated for their inhibitory effects on LPS-induced NO production in RAW264.7 cells with hexadecadrol as the positive control (IC_{50} 0.47 μ*M*) [129]. Henriol D and shizukaols E, G, M, and O showed significant anti-inflammatory activities with IC_{50} values of 1.9, 3.7, 2.0, 7.0, and 2.0 μ*M*. Shizukaols B and D from *Chloranthus serratus* also showed IC_{50} values of 0.15 and 7.2 μ*M*, respectively [129]. However, cycloshizukaol A, shizukaol C and two monomers were inactive ($IC_{50} > 10$ μ*M*) in this biological assay.

Two pseudoguaiane-type DSs, dichrocepholides D and E (**137**, **138**), were isolated from the aerial part of *Dichrocephala integrifolia* [50]. The effects of **137** and **138** and their co-occurring monomers on LPS-induced NO production in mouse peritoneal macrophages showed that the two DSs were not as active as the two monomers.

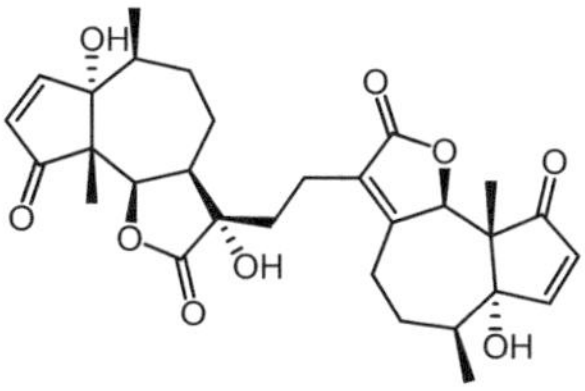

137 (dichrocepholide D)

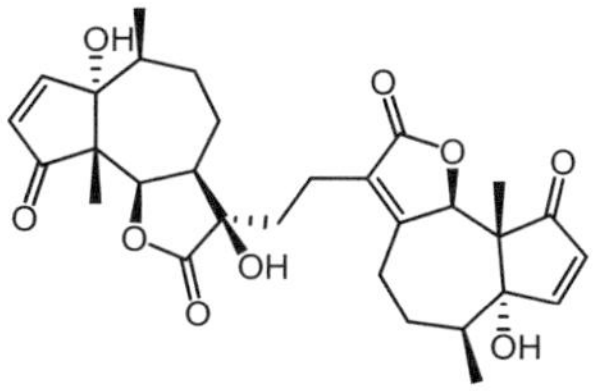

138 (dichrocepholide E)

Compound DSs, heterodimers composed of two different structural types of sesquiterpenoid, are perhaps the most important class to have shown anti-inflammatory activity. Neojaponicone A [165], and various japonicones [163–165], lineariifolianoids [167], and inulanolides [161] that are either eudesmane–guaiane dimers, 1,10-secoeudesmane–guaiane dimers, or germacrane-guaiane dimers, have been reported to exhibit anti-inflammatory activity.

Nuclear factor-κB (NF-κB) is a key regulator of the cellular inflammatory and immune response. NF-κB comprises mainly two proteins, p50 and p65. In non-stimulated cells, the heterodimer is held in the cytosol through interaction with the NF-κB inhibitory proteins, IκBα and IκBβ [161, 251]. Constitutive activation of NF-κB, which activates the expression of iNOS, COX-2, inflammatory cytokines, TNF-α, and cell adhesion molecules, may lead to inflammatory diseases. Inhibition of NF-κB activation is, therefore, a promising strategy for the development of anti-inflammatory agents [161].

Compound DSs seem to be good candidates for NF-κB inhibitory screening. In fact, since the discovery of NF-κB inhibitory compound DSs by Jin and coworkers in 2006 [161], subsequent efforts have shown that many DSs are promising NF-κB inhibitors.

Inulanolides A–D (**37–38**, **139–140**), four 1,10-secoeudesmane–guaiane (A, C–D) or germacrane-guaiane (B) DSs isolated from the aerial parts of *Inula britannica* var. *chinensis*, were reported to show potent inhibitory activities on LPS-induced NF-κB activation (IC_{50} 0.48–6.7 μ*M*) in a gene assay system as well as NO production (IC_{50} 1.5–7.6 μ*M*) and TNF-α production (IC_{50} 3.2–60.3 μ*M*) in RAW264.7 cells, with inulanolides B (**38**) and C (**139**) being the most active substances (IC_{50} 0.48–0.49 μ*M* for NF-κB activation; 1.59–1.52 μ*M* for NO production; 3.23–3.21 μ*M* for TNF-α production), and were comparable to or more active than the positive control, parthenolide (with IC_{50} values of 3.0, 2.5, and 3.0 μ*M*, respectively) [161]. Another promising aspect is that these DSs did not show significant cytotoxicity to the RAW264.7 cells at their effective concentrations. It is also very interesting that among the seven tested compounds, only one sesquiterpene monomer showed similar activity.

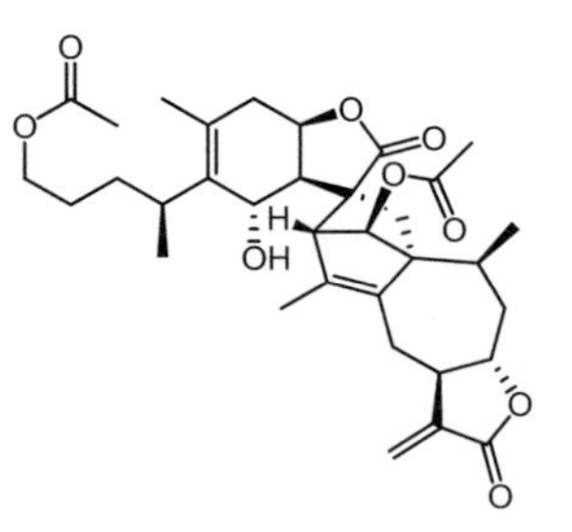

37 (inulanolide A)

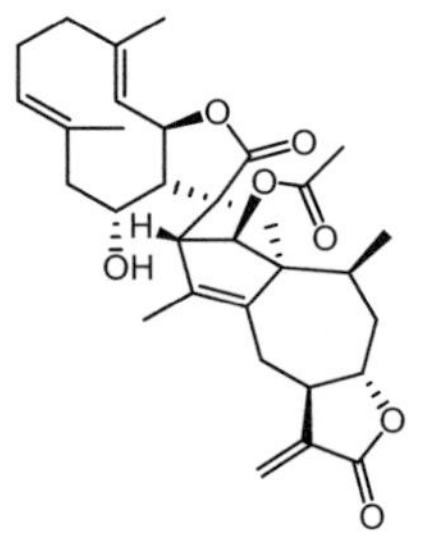

38 (inulanolide B)

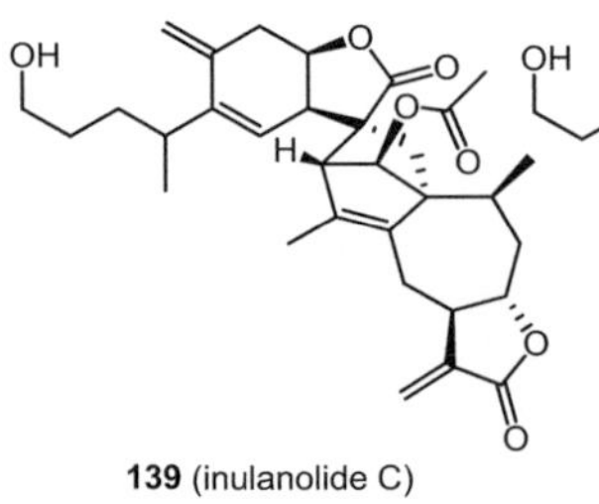

139 (inulanolide C)

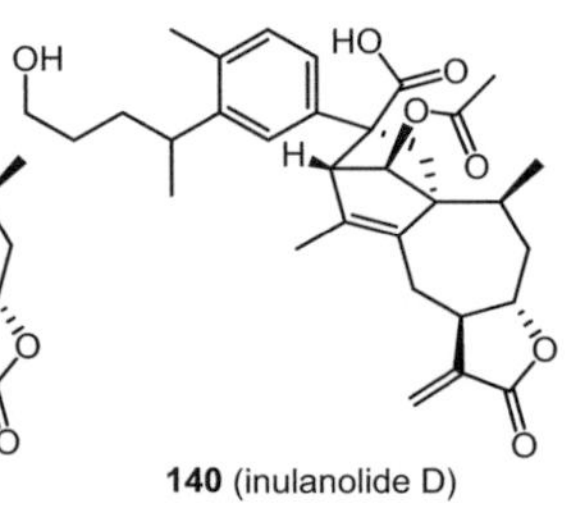

140 (inulanolide D)

Inula japonica is an herb used in traditional Chinese medicine that has long been used to cure inflammation. Chemical investigations of the aerial parts of *I. japonica* have afforded a series of compound DSs named japonicones E–T [162, 164, 165], inulanolides A and C [162], and neojaponicone A [165], which contain either an eudesmane–guaiane or a 1,10-secoeudesmane–guaiane sesquiterpenoid dimeric unit.

Japonicones E–L were tested for their inhibitory effects against LPS-induced NO production in RAW264.7 macrophages [164]. However, only japonicones E (**141**) and F (**142**), of the eudesmane–guaiane type, and japonicone J (**143**), of the 1,10-secoeudesmane–guaiane type, showed activity, with IC_{50} values of lower than 50 μg/cm^3 (20.8, 4.1, and 9.6 μg/cm^3, respectively, as compared to 48.7 μM for the positive control, aminoguanidine).

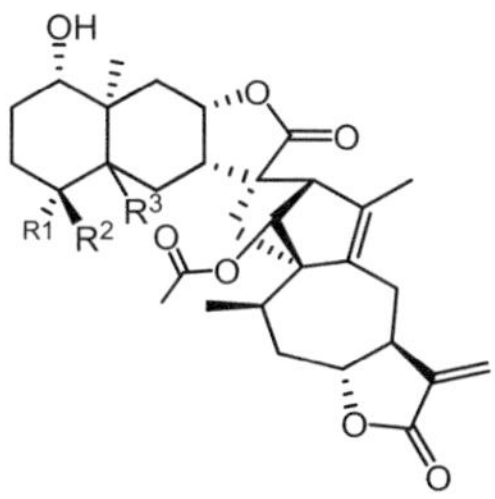

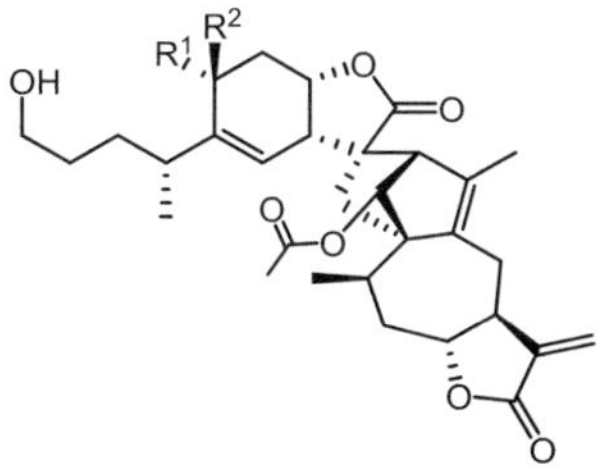

R¹ R² R³
141 – CH_2 β-OOH (japonicone E)
142 Me H α-OOH (japonicone F)

143 R¹ = O-COPr^i
R² = Me (japonicone J)

Neojaponicone A (**144**) and japonicones M–P are 1,10-secoeudesmane–guaiane type DSs [165]. Biological evaluation showed that neojaponicone A (**144**) and japonicone M (**145**) exhibited inhibitory activities against LPS-induced NO production in RAW264.7 macrophages, with IC_{50} values of 4.5 and 12.0 μg/cm^3.

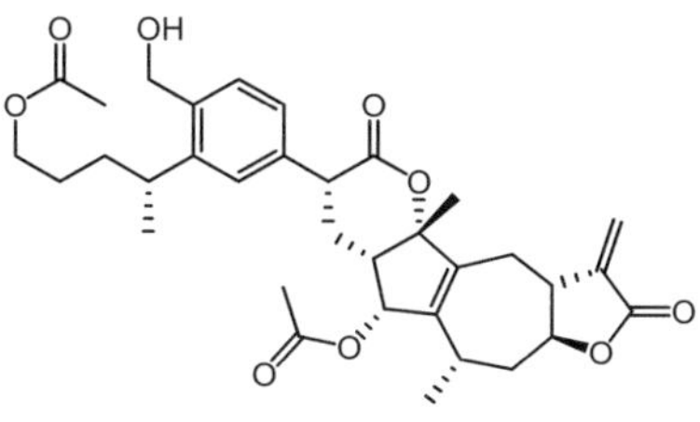

144 (neojaponicone A)

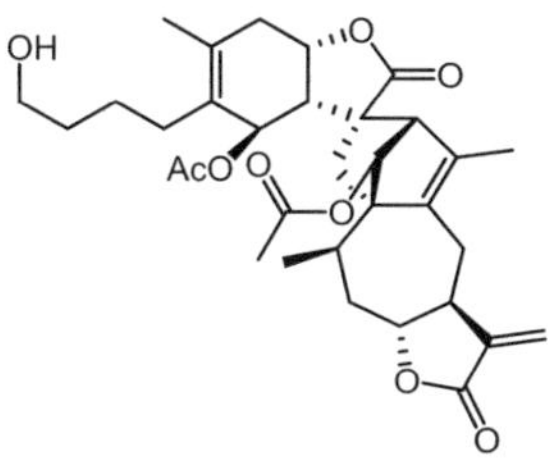

145 (japonicone M)

Six additional 1,10-secoeudesmane–guaiane type DSs, japonicones Q–T (**146**–**149**) and inulanolides A (**37**) and C (**150**), also showed significant inhibition against LPS-induced NO production in RAW264.7 cells, with IC_{50} values of 8.5, 8.9, 4.3, 4.3, 4.2, and 9.2 μM, respectively (as compared to 7.9 μM of the positive control, aminoguanidine) [162]. However, similar levels of cytotoxicity (the IC_{50} values were 17.1, 27.6, 13.4, 12.3, 14.6 and 13.7 μM, respectively) were also observed.

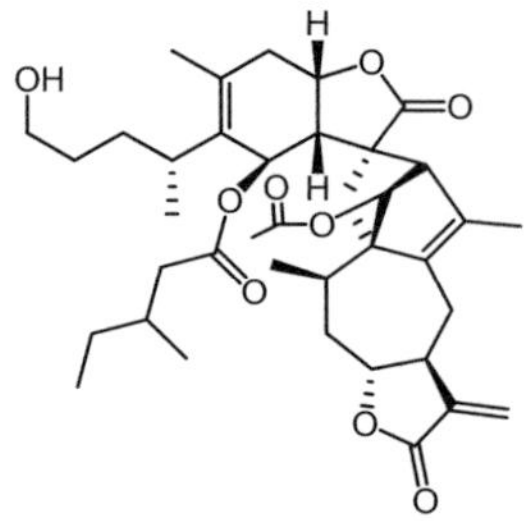

146 (japonicones Q)

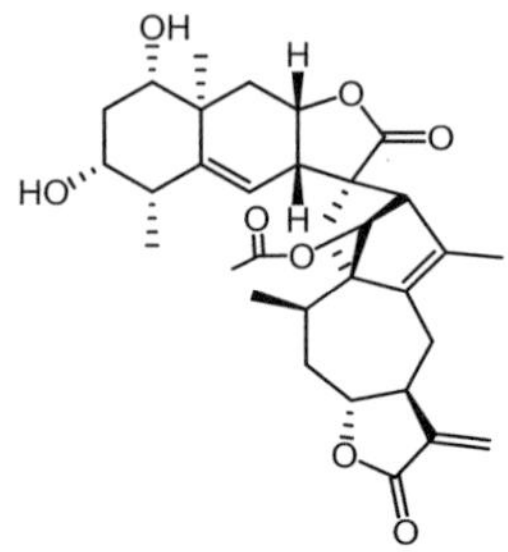

147 (japonicones R)

Although anti-TNF agents are very effective for the treatment of inflammatory diseases, increased risk of infections may result. Selectively inhibiting TNF receptor-1 (TNFR1)-mediated signaling but not TNFR2 signaling has been postulated to reduce inflammation without affecting the host immune response. Japonicone A (**63**), isolated from *Inula japonica*, was reported to be a TNF-α antagonist by reducing the TNF-α mediated cytotoxicity on L929 cells and inhibiting the binding of ^{125}I-labeled TNF-α to the L929 cell surface [252]. Mechanistic investigation showed that **63** can target directly TNF-α, and selectively disrupt its interaction with TNFR1, and antagonize its pro-inflammatory activities without compromising host defense against the virus. These observations suggested that **63** is a promising lead for the development of new anti-inflammatory drugs.

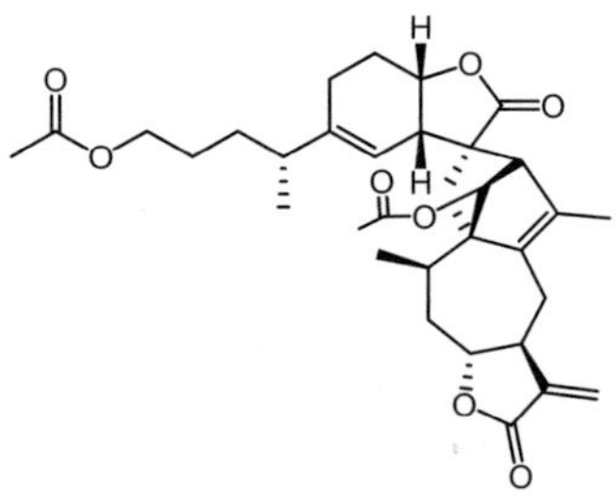

148 (japonicone S)

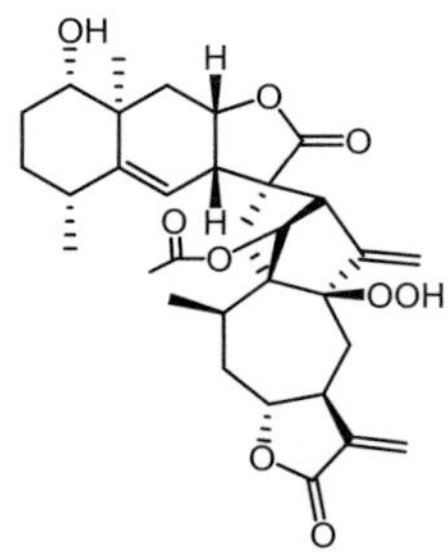

149 (japonicone T)

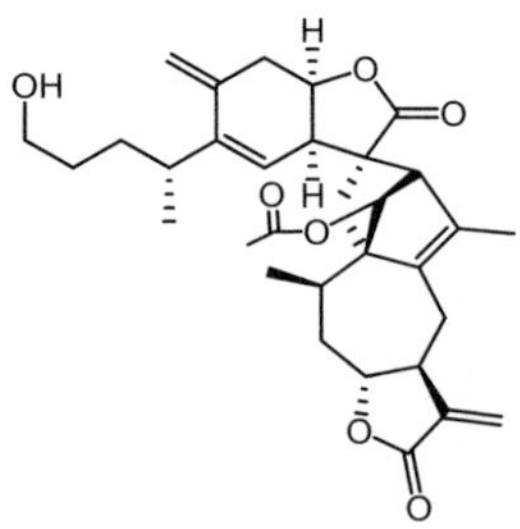

150 (inulanolide C)

For japonicones G, H, I, K, and L, which are also either eudesmane–guaiane-type (japonicone G) or 1,10-secoeudesmane–guaiane-type DSs that did not show strong inhibitory activity against LPS-induced NO production in RAW264.7 macrophages [164], the suggestion that the guaianolide moiety (but not the α-methylene-γ-lactone) might contribute to the enhanced activity [164] seems not to be correct. The structure-activity relationship of these types of DSs for NO production inhibitory activity is still not clear.

Lineariifolianoids A–D (**40**, **151**–**153**) are four xanthane–guaiane type compound DSs isolated from *Inula lineariifolia* [167]. Evaluation of the inhibitory effects on the TNF-α-sensitive L929 cells showed that the *exo*-adduct lineariifolianoid D (**153**) inhibited TNF-α-mediated cytotoxicity in a dose-dependent manner from 2.5 to 10 μ*M*, but the three *endo*-adducts lineariifolianoids A–C (**40**, **151**–**152**) did not exhibit such activity.

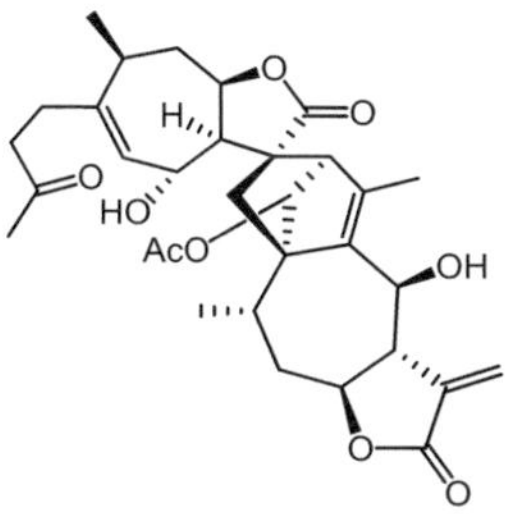

151 (lineariifolianoid B)

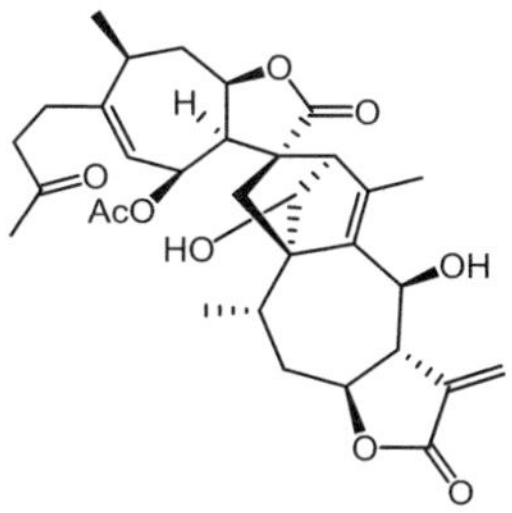

152 (lineariifolianoid C)

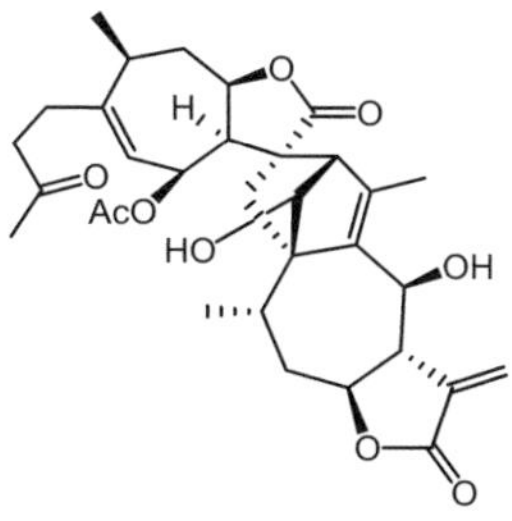

153 (lineariifolianoid D)

Two eudesmane–ester-guaiane type pseudodisesquiterpenoid DSs, **154** and **155**, were evaluated for their anti-inflammatory effects [246]. Compound **154** strongly inhibited NF-κB binding (IC_{50} 2.5 μ*M*) and luciferase activity (IC_{50} 1.0 μ*M*) in HeLa cells in a NF-κB electromobility shift assay (EMSA) and IL-6 reporter gene assay, respectively, indicating its anti-inflammatory potential.

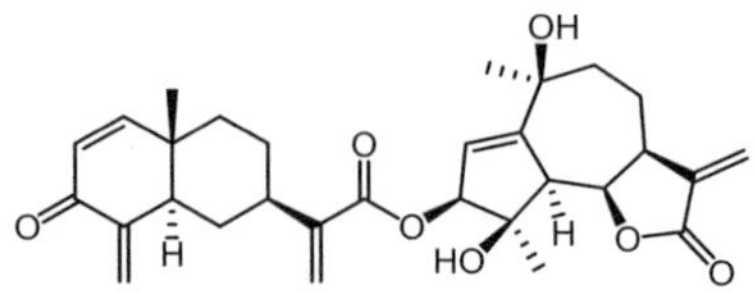

154

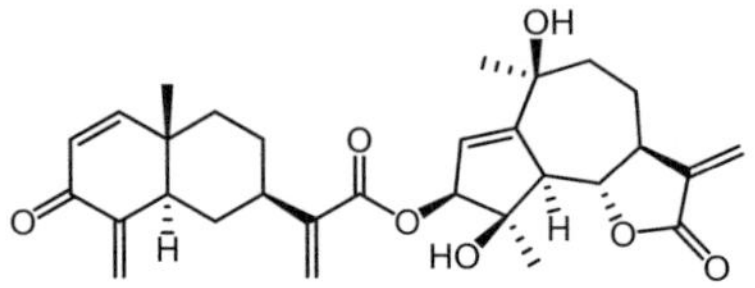

155

4.3 Immunosuppressive Activity

In addition to their cytotoxic activity (see Sect. 4.1), aza-sesquiterpenoid-type pseudo-disesquiterpenoid DSs have been reported to be potent immunosuppressants [253, 254]. Dimeric sesquiterpenes that possess an OH group in the quinolizidine ring [e.g. 6-hydroxythiobinupharidine (**130**), 6,6′-dihydroxythiobinupharidine (**131**), 6-hydroxythionuphlutine B (**132**), 6′-hydroxythionuphlutine B (**156**), and 6,6-′-dihydroxythionuphlutine B (**157**)] inhibited plaque-forming cell (PFC) formation of mouse spleen cells at 1 μ*M*. However, those DSs tested lacking a hydroxy group in the quinolizidine ring (e.g. thiobinupharidine, thionuphlutine B, and neothiobinupharidine) and several monomeric sesquiterpene alkaloids showed no significant suppression. It should be noted that at 1 μ*M*, **130**, **132**, and **156** were nontoxic to mouse splenocytes, and **131** showed only minor or minimal cytotoxicity. These results suggest that an OH group at the 6-position in the quinolizidine ring and dimerization of the aza-sesquiterpenoids is essential for the immunosuppressive effect of this type of DSs. In addition, increase of the number of OH groups seems likely to improve the activity.

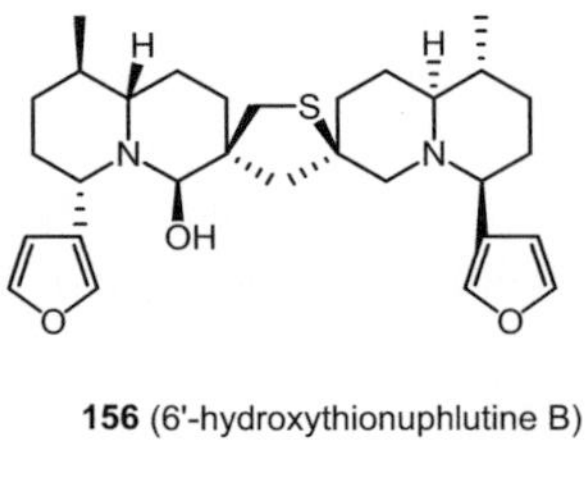

156 (6'-hydroxythionuphlutine B)

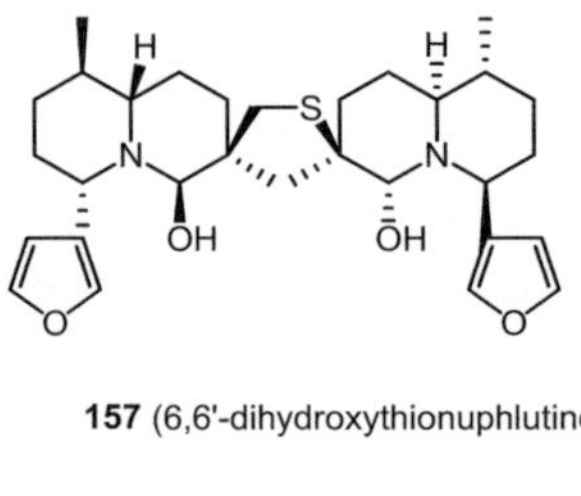

157 (6,6'-dihydroxythionuphlutine B)

The immunosuppressive activity of the lindenane DS shizukaol B (**71**) was evaluated on four types of immune cells [B cells, T cells, macrophages, and dendritic cells (DCs)] to determine cell-type selectivity [135]. The results showed that **71** strongly inhibited LPS-induced B cell proliferation with an IC_{50} value of 137 ng/cm^3, but only slightly affected ConA-induced T cell proliferation at 1000 ng/cm^3 and did not inhibit LPS-induced macrophage NO production and LPS-induced MHC-II expression in DCs. Since B-cell depletion with anti-CD20 antibodies has become an accepted second line modality of therapy for patients with autoimmune diseases (e.g. rheumatoid arthritis, lupus, and diabetes), compound **71** might therefore be used in anti-B-cell therapy.

4.4 Potassium Channel Blocking and Cardiovascular Activity

Potassium channels are dynamic pore-forming transmembrane proteins known to play critical roles in cellular signaling processes regulating neurotransmitter release, heart rate, insulin secretion, neuronal excitability, epithelial electrolyte transport, smooth muscle contraction, and cell volume regulation [255]. Voltage-gated K^+ channels (K_V) are involved in diverse physiological processes ranging from repolarization of neuronal and cardiac action potentials. Targeting K_V therefore offers tremendous opportunities for the development of new drugs to treat cancer, autoimmune diseases and metabolic, neurological and cardiovascular disorders [256]. Blockage of the delayed rectifier potassium current of K_V, which is important in regulating cellular excitability, will inhibit potassium efflux, prolong the duration of action potentials, and may provide potential for the treatment of various diseases (e.g. hypertension, peripheral vascular disease, and neurodegenerative disorders).

Six lindenane DSs, chloraholides A–F (**158**, **104**, **159–162**), were isolated from *Chloranthus holostegius* [221, 228]. Their effects on K_V were examined using whole cell voltage-clamp recordings in rat dissociated hippocampal neurons with tetraethylammonium chloride as the positive control. Chloraholides A–F exerted potent inhibition on the delayed rectifier K^+ current (I_K), with negligible effects on the fast transient K^+ current (I_A). The IC_{50} values of chloraholides A–F on the delayed rectifier K^+ current (I_K) were 10.9 ± 12.3, 18.6 ± 2.5, 3.6 ± 10.1, 2.7 ± 0.3, 27.5 ± 5.1 and 57.5 ± 6.1 μ*M*, and these compounds were 18–388-fold more potent than the positive control tetraethylammonium chloride (IC_{50} 1.05 ± 0.21 m*M*), a classical blocker of the delayed rectifier K^+ current. These observations showed that chloraholides A–F are potent and selective blockers of potassium channels and it was suggested by the investigators that lindenane DSs might have potential in the development of new drugs to treat cancer, autoimmune diseases, and metabolic, neurological and cardiovascular disorders.

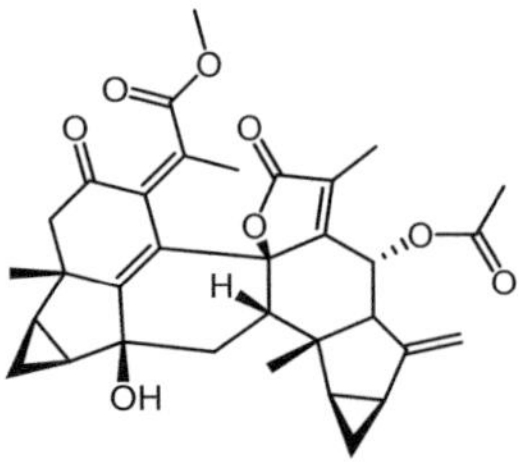

158 (chlorahololide A)

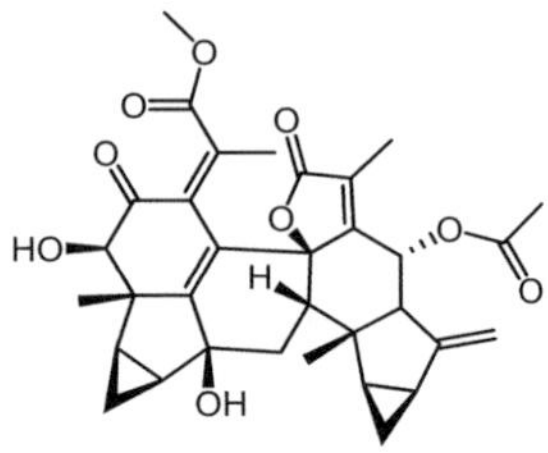

159 (chlorahololide C)

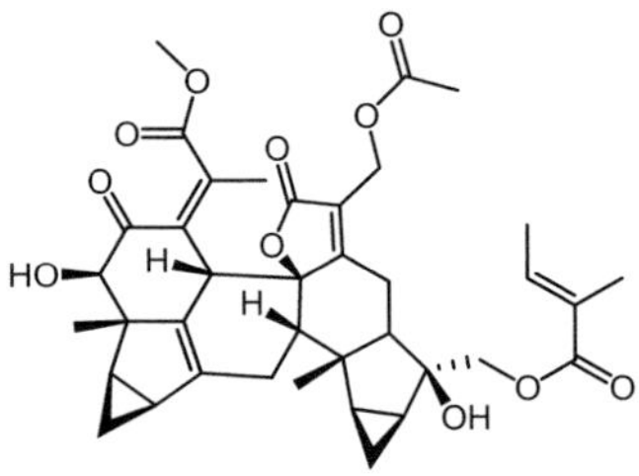

160 (chlorahololide D)

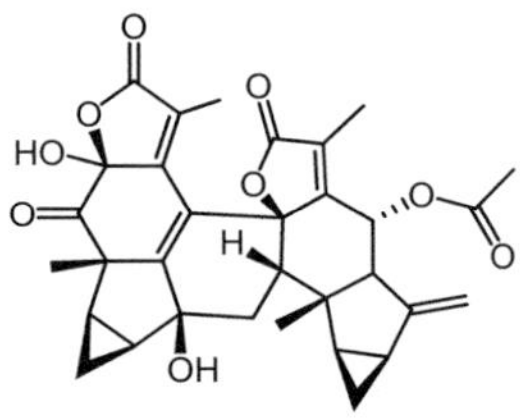

161 (chlorahololide E)

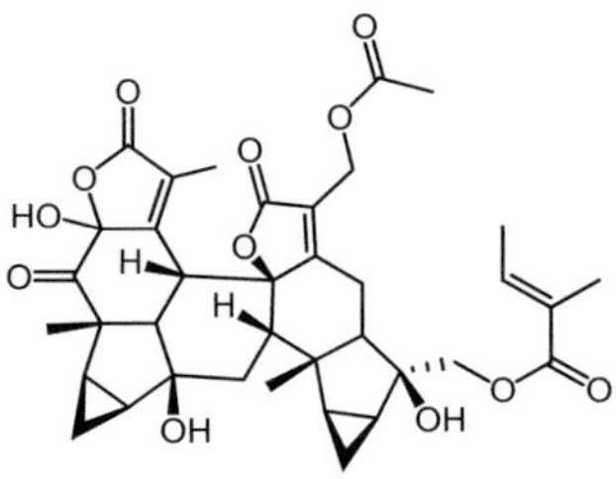

162 (chlorahololide F)

Biatractylolide (**163**) is an eudesmane DS isolated from *Atractylodes macrocephala* [115] and *Trattinickia rhoifolia* [116]. Its effects on the isolated guinea pig atrium showed that **163** could prominently inhibit the contractile force and decrease the heart rate of the isolated right atrium of the guinea pig [257]. These actions could be blocked by atropine completely. Compound **163** was able to inhibit the positive staircase phenomenon of the isolated left atrium of guinea atrium but had no effect on the post-rest potentiation (PRP) of the isolated left atrium. The results indicated that **163** possesses significant negative inotropic and chronotropic effects, suggesting it as a potential blood pressure-lowering agent.

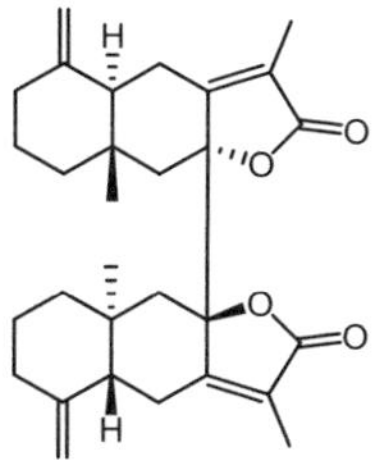

163 (biatractylolide)

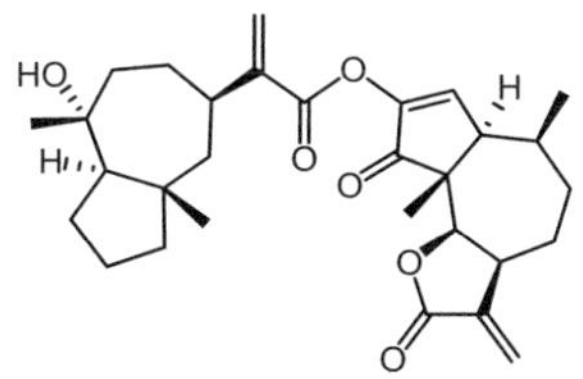

164 (arrivacin B)

Arrivacins A (**62**) and B (**164**) are two pseudoguaianolide DSs isolated from the CH_2Cl_2 extract of *Ambrosia psilostachya* [180]. Bioassay work showed that **62** clearly inhibited angiotensin II binding to receptors from the bovine adrenal cortex with an IC_{50} 3 μ*M*, while **164** was found to be a weaker inhibitor (IC_{50} 13 μ*M*).

Shizukaol B (**71**), cycloshizukaol A (**112**) and shizukaol F (**113**) were also shown to prevent ICAM-1/LFA-1 mediated cell aggregation and monocyte adhesion to HUVEC through the inhibition of ICAM-1, VCAM-1 and E-selectin expression, which suggested that they might be useful for the design of antiatherosclerotic agents relevant to endothelial activation [245].

4.5 Antimalarial, Antiprotozoal, Antibacterial, Antifungal, and Antiviral Activity

Artemisinin and its derivatives are a potent class of antimalarial drugs. Various studies have shown that artemisinin is a safe sesquiterpene lactone effective against a range of protozoal-caused diseases [236]. Further investigations showed that artemisinin dimers possess more drug-like properties as compared to the monomers. Detailed information of artemisinin-derived dimers as potential antimalarial and antiprotozoal agents can be found in two recent review articles [240, 241].

The two aristolane DSs, aurisins A (**114**) and K (**115**), were also shown to exhibit antimalarial activity against *Plasmodium falciparum* (IC_{50} 0.80 and 0.61 μ*M*) [148].

(5*S*,6*R*,7*R*,8*R*,11*R*)-2-Oxo-8-tigloyloxyguaia-1(10),3-dien-6,12-olide-14-carboxylic acid (**165**) is a guaiane DS isolated from *Eupatorium perfoliatum* [67]. Although this DS was found to be potent against all of the protozoa in which it was evaluated, it also exhibited high toxicity. The highest activity was found against *Plasmodium falciparum*, with an IC_{50} of 2.0 μ*M* and a selectivity index of about 8. In contrast, two co-occurring monomeric sesquiterpene lactones with this DS exhibited weak to no discernible activity against the test protozoa used. A preliminary SAR analysis indicated that although this DS does not contain an α-methylene-γ-lactone partial structure, the α,β,γ,δ-unsaturated lactone structure present in the molecule represents a Michael acceptor system and hence might be all or in part responsible for the activity.

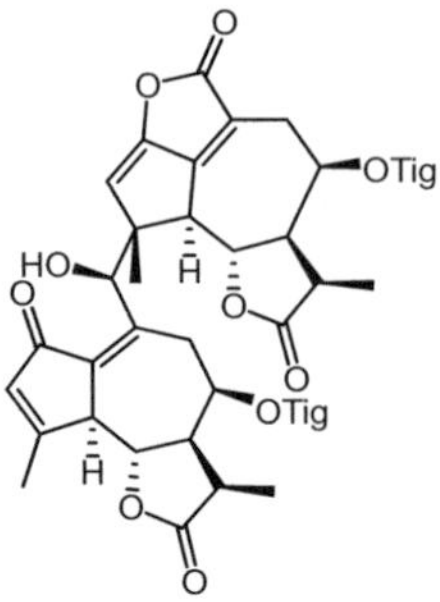

165 (5*S*,6*R*,7*R*,8*R*,11*R*)-2-oxo-8-tigloyloxyguaia-1(10),3-diene-6,12-olide-14-carboxylic acid)

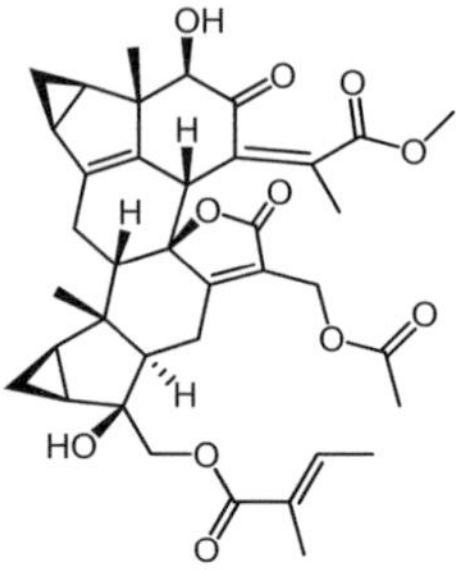

166 (CHE-23C)

The indolo-sesquiterpene type DSs, dixiamycins A–B (**59–60**), were reported to show better antibacterial activities than the co-occurring monomers against four indicator strains (*Escherichia coli* ATCC 25922, *Staphylococcus aureus* ATCC 29213, *Bacillus subtilis* SCSIO BS01 and *Bacillus thuringiensis* SCSIO BT01) with *MIC* values in the range of 4–16 μg/cm^3), and **59** was more active than **60** against *S. aureus* and *B. thuringiensis* [198].

The illudane DS bovistol (**27**) showed very weak antibacterial (*MIC*: 100 μ*M* against *Micrococcus luteus*) [152]. The aristolane DSs, aurisins A (**114**) and K (**115**), were also shown to exhibit antimycobacterial activity against *Mycobacterium tuberculosis* with *MIC*s of 92.6 and 23.8 μ*M* [148].

The lindenane DS CHE-23C (**166**) was isolated from methanol extracts of the stems and roots of *Chloranthus henryi* [258]. The compound showed potent antifungal activities in vitro against various phytopathogenic fungi such as *Alternaria kikuchiana*, *Botrytis cinerea*, *Colletotrichum lagenarium*, *Magnaporthe grisea*, *Pythium ultimum*, and *Phytophthora infestans* with *MIC*s of 8, 8, 8, 4, 1, 32 μg/cm^3, respectively. In particular, it exhibited 91 and 100% disease-control activity in vivo against tomato late blight (*P. infestans*) and wheat leaf rust (*Puccinia recondita*) at concentrations of 33 and 100 μg/cm^3, respectively. The activities found were more potent than that of the commercially available fungicide chlorothalonil, but weaker than that of dimethomorph. Based on these observations, this DS is a potential lead compound for the development of effective antifungal agents.

Four lindenane DSs, chloramultilides B–D and chlorahololide B, were also evaluated for their antifungal activity [222]. However, only chloramultilide B showed inhibitory activities against *Candida albicans* and *C. parapsilosis* with *MIC* values of 0.068 μ*M* in each case. The *MIC* values of other DS were greater than 0.16 μ*M*.

Bovistol (**27**) also showed very weak antifungal (*MIC*: 100 μ*M* against *Mucor miehei*) activities [152].

The 8,9-*seco*-lindenane DS chloramultiol G (**18**) was evaluated for its antifungal activity against four microorganisms but showed no activity in this test [128].

Chrysanolide C (**167**) is a guaiane DS isolated from the flowers of *Chrysanthemum indicum* [47]. Anti-HBV activity evaluation showed that **167** exhibited potent inhibitory activities against the secretion of HBsAg ($IC_{50} = 33.91$ μ*M*) and HBeAg ($IC_{50} = 30.09$ μ*M*), with selectivity indices of 0.95 and 1.07. Although this DS was more potent than its monomer and less potent than the trimer, its cytotoxic potency was similar to its anti-HBV activity.

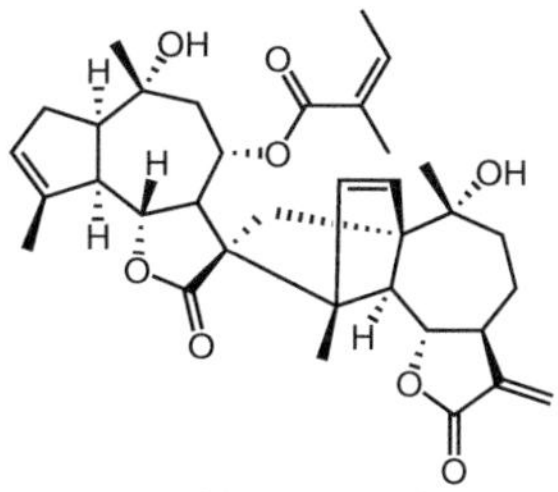

167 (chrysanolide C)

The lindenane DS, shizukaol F (**113**), was shown to inhibit HIV-1 replication (IC_{50} of 6.12 μ*M*) and LTR/Gag production of HIV-1 reverse transcription (IC_{50} of 9.11 μ*M*) [259]. Mechanistic studies showed that shizukaol F inhibited HIV-1 RT-RNase H with an IC_{50} value of 26.4 μ*M*, but had no effect on HIV-1 RT RNA-dependent DNA polymerase activity.

Cryptoporic acid C (**168**) and cryptoporic acid F (**169**) showed 99.6% (IC_{50} 61.0 μg/cm^3) and 99.7% (IC_{50} 42.2 μg/cm^3) inhibition of HIV-1 at a concentration of 200 μg/cm^3, while the other cryptoporic acids were inactive in the assay procedure used [247, 248].

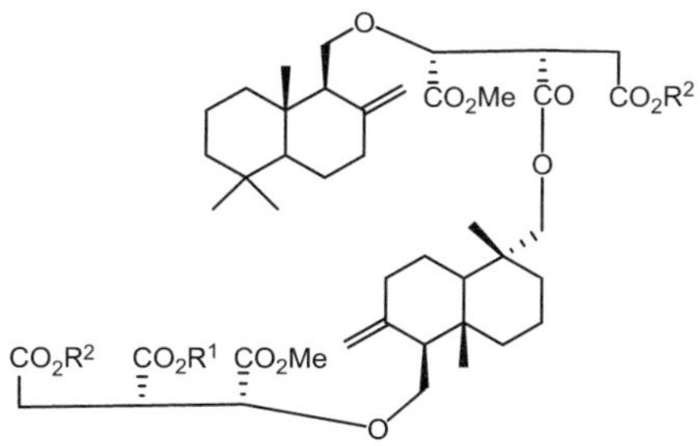

168 R^1 = Me, R^2 = H (CPA-C)
169 $R^1 = R^2$ = H (CPA-F)

4.6 Neurotrophic Activity

Mastigophorenes A–D (**22**, **170–172**) are herbertane (isocuparane) DSs isolated from the liverwort *Mastigophora diclados* [143]. Mastigophorenes A, B, and D have been found to accelerate neuritic sprouting at 0.1–10 μ*M* in a neuritic cell culture of a fetal rat cerebral hemisphere. However, mastigophorene C and the three monomeric isocuparenes only suppressed neuritic differentiation. The neurotrophic activity of mastigophorenes A, B, and D may be attributable to their dimerized form and specific coupling properties.

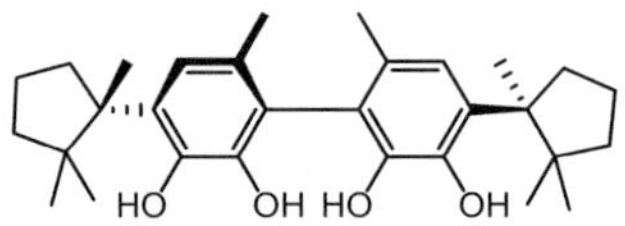

170 (mastigophorene B)

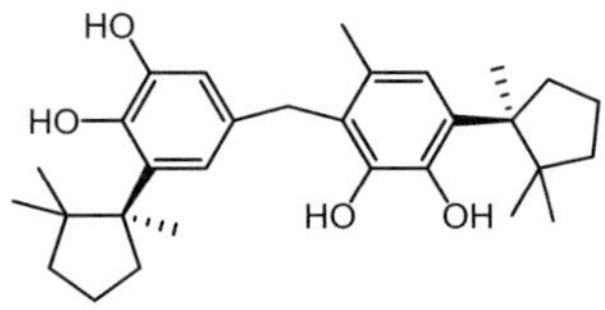

171 (mastigophorene C)

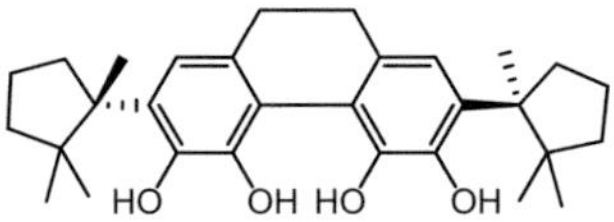

172 (mastigophorene D)

Encouraged by the above work, a simplified mastigophorene analog and its dimethyl ether were screened for potential neurotrophic effects on primary dopaminergic cell cultures in vitro [260]. Although differing by only two *O*-methyl groups, the two simplified synthetic mastigophorene analogs **173** and **174** exhibited different effects on the growth of dopaminergic neurons in primary cell culture. Compound **174** not only enhanced the survival of already differentiated dopaminergic neurons in primary cell cultures, but also exerted a growth-promoting effect on cell area and length of cell processes. In turn, compound **173** showed significant stimulatory effects only on cell area and cell number and the number of cell processes was even reduced at concentrations above 0.5 μ*M*. The differential activities of the two analogs were in agreement with those of mastigophorenes A–D. The increased lipophilic property of **174** by the two *O*-methyl groups could account for its enhanced cell permeability, which may eventually lead to improved activity.

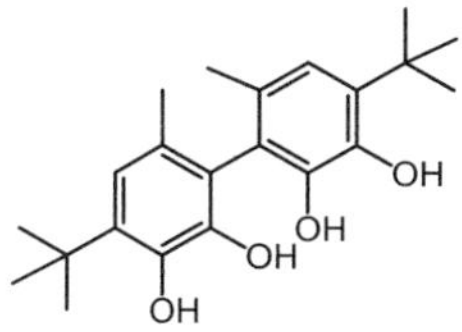

173

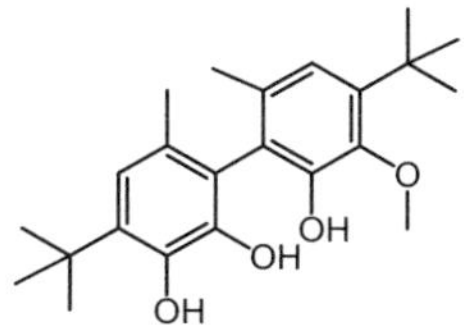

174

The effects of biatractylolide (**163**) on the memory of AD rats induced by Aβ_{1-40} [261] and $AlCl_3$ [262] were also investigated, and the results showed that **163** improved the memory function of the AD rats and reduced the AChE activity in the brain of both types of rats.

4.7 *Miscellaneous Activities*

Tianmushanol (**175**) and 8-*O*-methyltianmushanol (**176**), two lindenane DSs isolated from the leaves of *Chloranthus tianmushanensis*, were reported to possess tyrosinase inhibition with IC_{50} values of 358 ± 3 and 312 ± 3 μ*M*, respectively, close to that of the standard tyrosinase inhibitor, kojic acid (IC_{50} 243 ± 4 μ*M*).

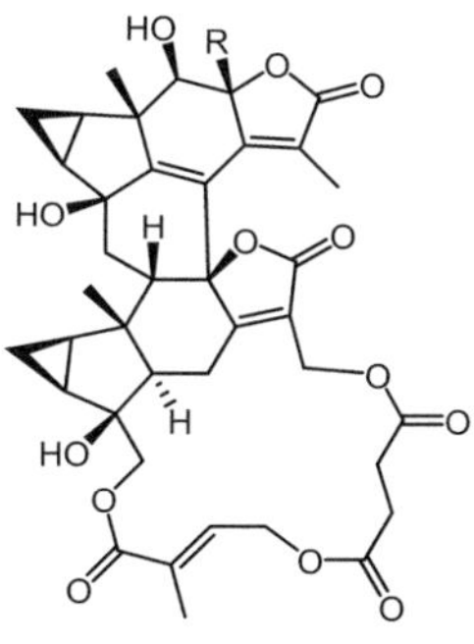

175 R = OH (tianmushanol)
176 R = OCH_3 (8-*O*-methyltianmushanol)

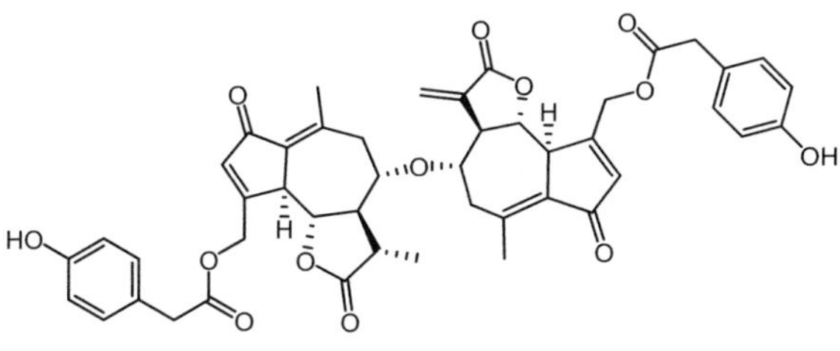

177 (latucain C)

Lactucains A, B, and C (**177**) are guianane DSs isolated from an extract of *Lactuca indica* as active principles with potential antidiabetic activity [207]. The four DSs, together with two known monomeric guaiane-type sesquiterpene lactones, were evaluated for antihyperglycemic activity in vivo using STZ-diabetic rats. Lactucain C (**177**) (Δ $-22.7 \pm 12\%$) showed moderate lowering of plasma glucose at a dose of 1 mM/kg.

The superoxide anion radical ($O_2^{-\bullet}$), which is produced by inflammation and injury due to radiation and other reasons, has been recognized as an important factor in hepatic, cardiac, renal, and pancreatic insufficiency and injury. Inhibition of superoxide anion radical release and radical scavenging plays an important role in the prevention against ischemia and inflammation. Cryptoporic acids D (**56**) and G (**178**) strongly inhibited the release of superoxide anion radical from guinea pig peritoneal macrophages induced by the $O_2^{-\bullet}$ stimulant FMLP (formyl methionyl leucyl phenylalanine) at concentrations from 0.05 to 25 μg/cm^3 [247, 248]. In addition, the inhibitory activity of cryptoporic acids C–G with dimeric structures was about 100–500 times as potent as the monomers cryptoporic acids A and B. Cryptoporic acid C (**168**) also inhibited the release of $O_2^{-\bullet}$ from rabbit polymorphonuclear leukocytes induced by FMLP at a concentration of 2 μg/cm^3.

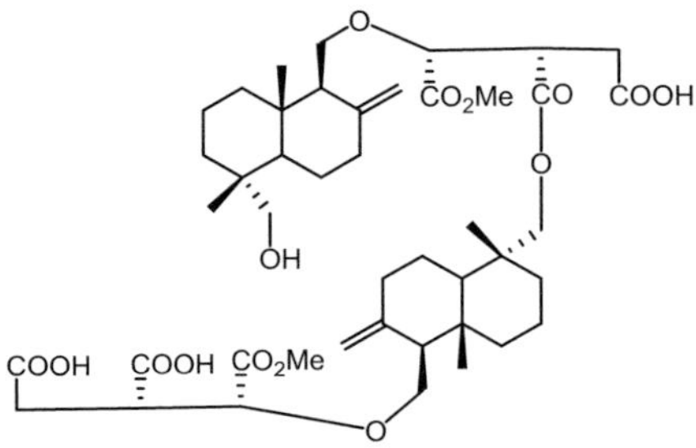

178 (cryptoporic acid G)

5 Synthesis

The promising biological activities, complex structures, and diversified coupling patterns of DSs, have attracted in the past several decades considerable scientific interest as evidenced by efforts made toward understanding their biogenesis and developing procedures for their chemical synthesis. As a result, biogenetic routes for several DSs have been postulated and synthetic efforts toward construction of various sesquiterpenoids ready for convergence and dimerization strategies for synthesis of DSs from two sesquiterpenoid monomers have been reported. The proposed biosynthetic pathways for dimerization of various types of compounds have been reviewed [7, 263, 264]. In this section, the biosynthesis methods employed by Nature will be summarized and each type of proposed dimerization reactions will be illustrated. Efforts toward chemical dimerization strategies for synthesis of DSs will also be briefly reviewed.

5.1 Biogenesis

Since dimeric sesquiterpenoids (DSs) are generated biogenetically from the coupling of two sesquiterpenoid molecules, biosynthetic pathways containing a key dimerization reaction of their co-occurring precursors or derivatives of their precursors have been postulated for the syntheses of various natural DSs. Enzymatic catalysis, should regio- or diastereoisomeric DSs co-occur, newly generated chiral centers (as compared to their precursors) be present, and non-enzymatical reactions be feasible, are all often involved in the dimerization process [263]. A straightforward analysis of the coupling patterns of two constitutional sesquiterpenoid units may indicate that DSs can be formed either through direct cyclization, oxidative coupling, esterification, etherification, aldol reaction, Michael-type reaction, or via various types of linkers.

5.1.1 [4 + 2] Diels–Alder Reactions

Most DSs are biosynthesized from two monomeric sesquiterpenoids via regular Diels–Alder (D–A) cycloaddition reactions to generate a six-membered cyclohexene. The cyclopentadiene present in one monomer and the electron-deficient carbon–carbon double bond (usually the exomethylene of an α-methylene butenolide) present in another, generally react as diene and dienophile, respectively.

Guaiane sesquiterpenoids as possible monomeric precursors usually occur as guaianolides, which possess both cyclopentadiene and α-methylene butenolide moieties in forms ready for a Diels–Alder reaction. Achillinin C (**179**), a 1,10-secoguaiaolide-guaiaolide DS isolated from the flowers of *Achillea millefolium* [28], was suggested to be a Diels–Alder reaction product of the electron-deficient 11′,13′-double bond of iso-*seco*-tanapartholide (**179a**) with the 1,3-diene moiety of the guaianolide **179b** (Scheme 3). The stereoselectivity was ascribed as a result of the approach of the double bond of **179a** from the less hindered convex side of the diene **179b** in an *endo*-addition.

The biomimetic synthesis of meiogynin A (**76**) may involve enzymatic catalysis of an *exo*-Diels–Alder reaction between a bisabolatriene acid **76a** and a zingiberene-type monomer **76b** (Scheme 4) [14]. A similar reaction can be proposed for the biosynthesis of foveolide B (**116**) [121].

Weariifolianoids A–D (**40, 151–153**) are formed presumably by [4 + 2] cycloaddition of xanthane and guaiane catalyzed by Diels–Alder-ase (Scheme 5) [167]. Lineariifolianoids A–C (**40, 151–152**) were constructed by a common

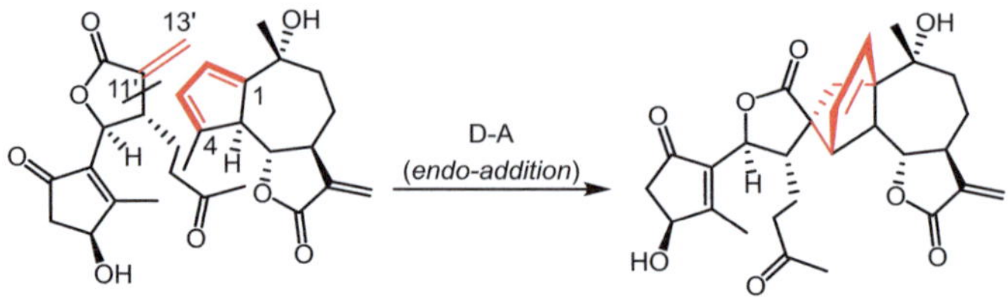

179a 179b 179 (achillinin C)

Scheme 3 Biosynthesis pathway of achillinin C (**178**)

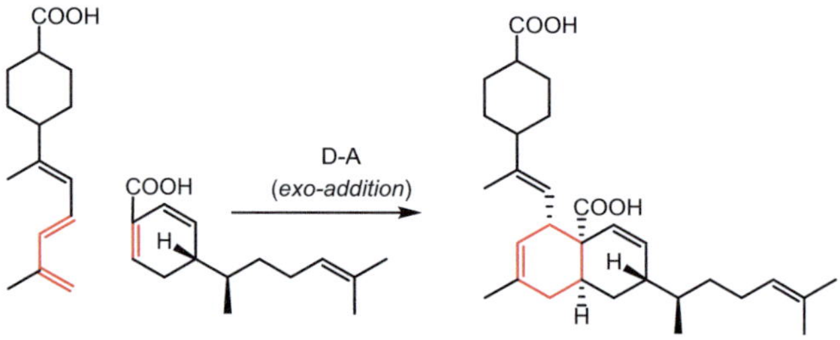

76a 76b 76 (meiogynin A)

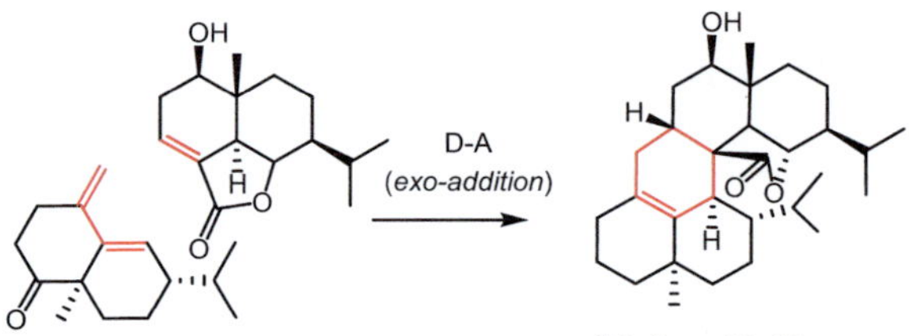

116 (foveolide B)

Scheme 4 Biosynthesis pathway of meiogynin A (**76**) and foveolide B (**116**)

endo-selective Diels–Alder reaction, while lineariifolianoid D (**153**) was biosynthesized in an unusual *exo* fashion.

Sterostrein A (**180**) is an illudalane-dinorilludalane DS isolated from cultures of the fungus *Stereum ostrea* BCC 22955 [160]. Its biosynthesis may involve a reaction sequence starting from the [4 + 2] Diels–Alder cycloaddition of an illudalane and a norilludalane precursor (Scheme 6). In turn, *endo*-Diels–Alder cycloaddition of two lindenane precursors followed by oxidation and rearrangement would generate chloraholide C (**159**) (Scheme 7).

Three diastereoisomeric bisabolene DSs, *cis*-dimer A (**181**), *cis*-dimer B (**182**), and *trans*-dimer C (**183**), and their potential precursor, dehydrotheonelline (**181a**), have been isolated from the South China Sea sponges *Axinyssa variabilis* and *Lipastrotethya ana* [10]. A plausible biosynthesis pathway to the three isomers involving an intermolecular [4 + 2] Diels–Alder cycloaddition of two molecules of **181a** was postulated (Scheme 8). The *trans*- and *cis*-dimers were formed according to *exo*- and *endo*-Diels–Alder coupling, respectively.

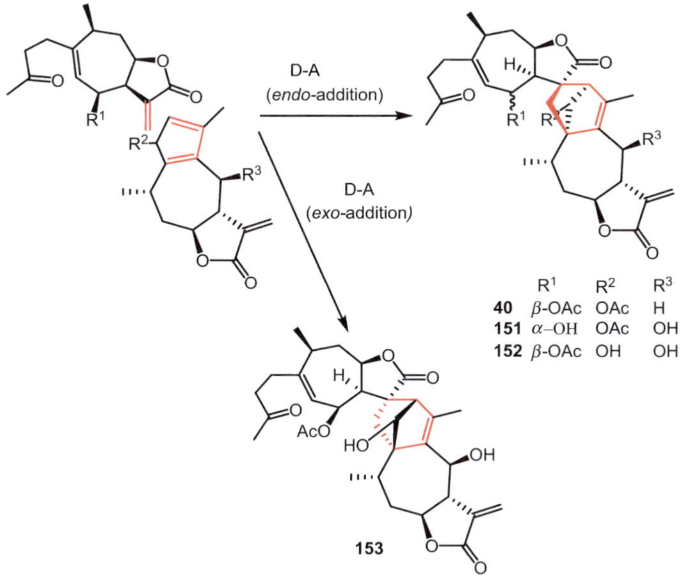

Scheme 5 Plausible biogenesis pathway of lineariifolianoids A–D (**40**, **151**–**153**)

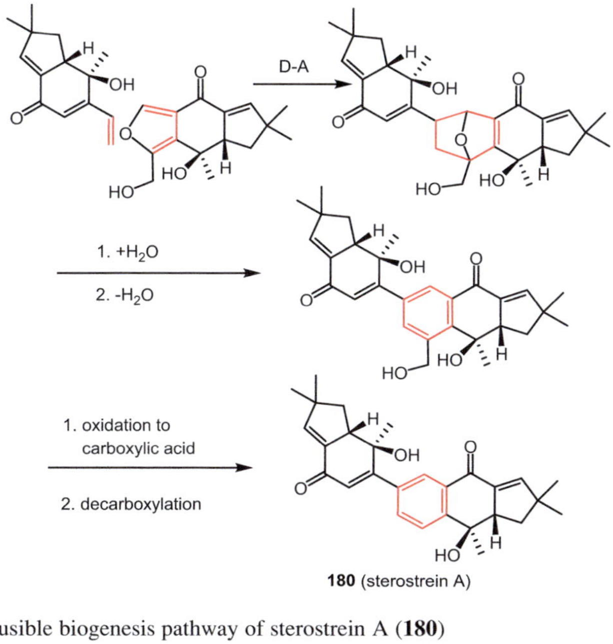

Scheme 6 Plausible biogenesis pathway of sterostrein A (**180**)

Two novel lindenane-type sesquiterpenoid dimers, sarcanolides A (**184**) and B (**185**), were isolated from the whole plants of *Sarcandra hainanensis* [133]. It was proposed that enzymatic Diels–Alder cycloaddition of two lindenane-type

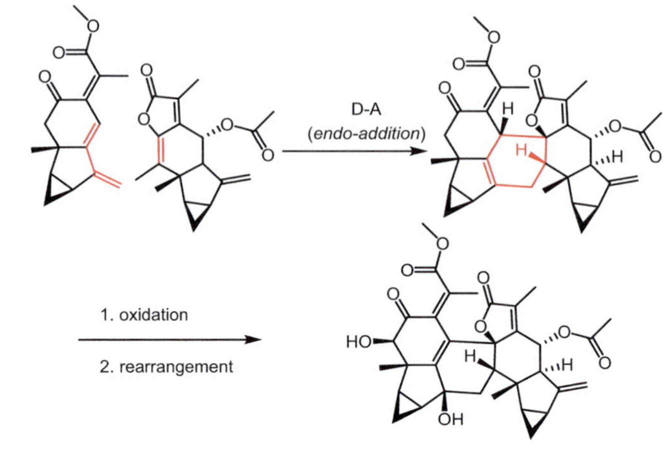

Scheme 7 Plausible biogenesis pathway of chlorahololide C (**159**)

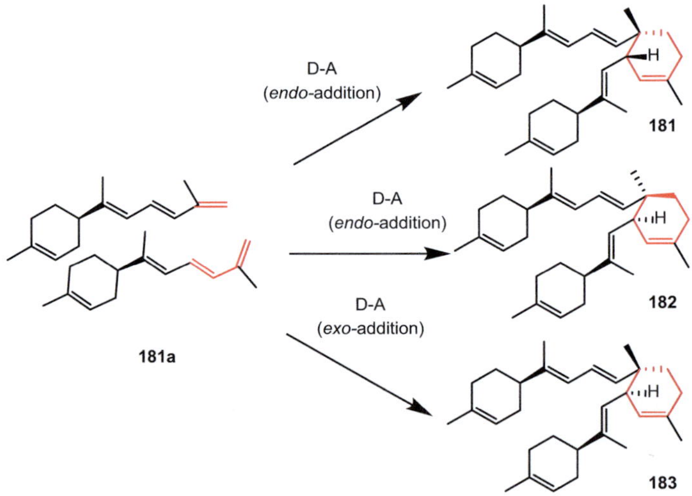

Scheme 8 Plausible biogenesis pathway of *cis*-dimer A (**181**), *cis*-dimer B (**182**) and *trans*-dimer C (**183**)

sesquiterpenoids **184a** and **184b** to form intermediate **184c** is the key step in the biosynthesis of this class of DSs (Scheme 9). After acid-catalyzed rearrangement, protonation and discharge, oxidation, acylation, and loss of water, sarcanolides A (**184**) and B (**185**) could be generated.

During the biosynthesis of DSs, when one of the atoms other than carbon is found among the reactants of the Diels–Alder reaction, the biosynthetic reaction is therefore referred to as a hetero-Diels–Alder (HDA) cycloaddition. Cyclopentadienes and α-methylene butenolide units are frequently found in this type of intermolecular

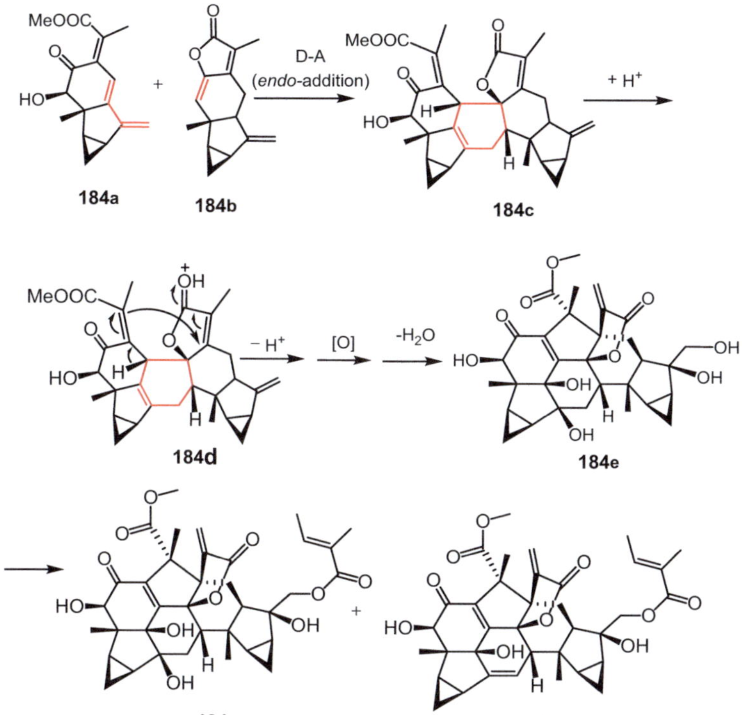

Scheme 9 Plausible biogenesis pathway of sarcanolides A (**184**) and B (**185**)

[4 + 2] addition for the formation of DSs, and the 3,4-dihydro-2*H*-pyran ring is generally the key structural unit formed in the intermediates.

Bialantolactone is an eudesmanolide DS isolated from the roots of *Inula helenium* [114]. With the isolated double bond of one molecule as the electron donor and the α,β-unsaturated carbonyl of another as the acceptor, HDA of the co-occurring precursor alantolactone, followed by oxidation of the newly generated double bond to an oxirane and its subsequent rearrangement, could generate bialantolactone (**13**) (Scheme 10).

Ainsliadimer A (**135**) is a guaiane DS that contains a five-membered cyclopentane ring and not a commonly encountered six-membered ring at the convergence of the two sesquiterpenoid monomers, and a HDA cycloaddition reaction followed by ring opening through acid hydrolysis and ring reformation via an aldol reaction were proposed for the dimerization of two monomeric molecules (Scheme 11) [29].

Macrophyllidimer A (**42**), an elemane–eudesmane DS that possesses a single C–C bond as linkage of the two sesquiterpenoid subunits, was proposed to be produced through an HDA reaction between the α,β-unsaturated ketone group of one monomer and the olefin group of the other, and subsequent hydrolysis and rearrangement at the heterocycle (Scheme 12) [169].

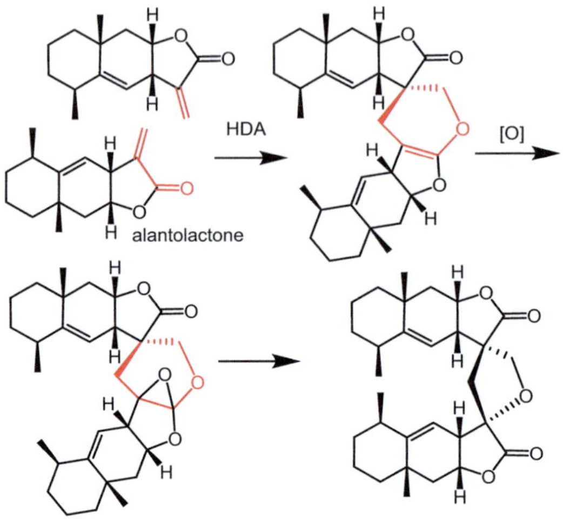

Scheme 10 Plausible biogenesis pathway of bialantolactone (**13**)

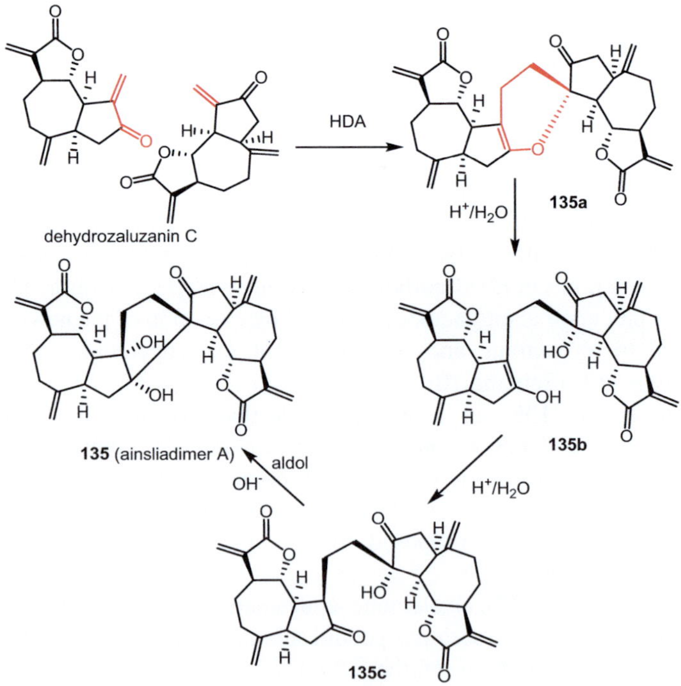

Scheme 11 Proposed biogenesis pathway for ainsliadimer A (**135**)

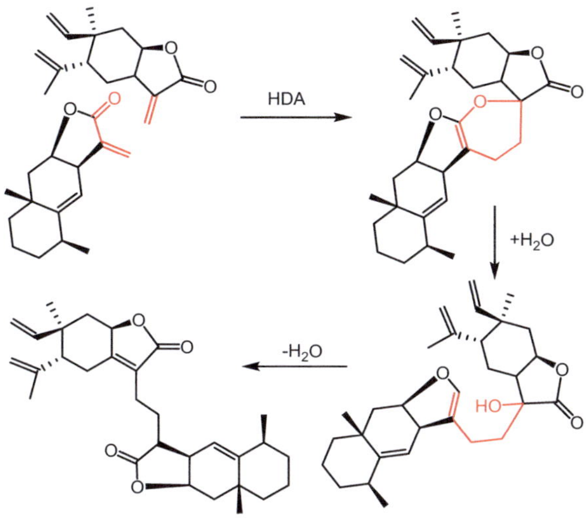

42 (macrophyllidimer A)

Scheme 12 Proposed biogenesis pathway for macrophyllidimer A (**42**)

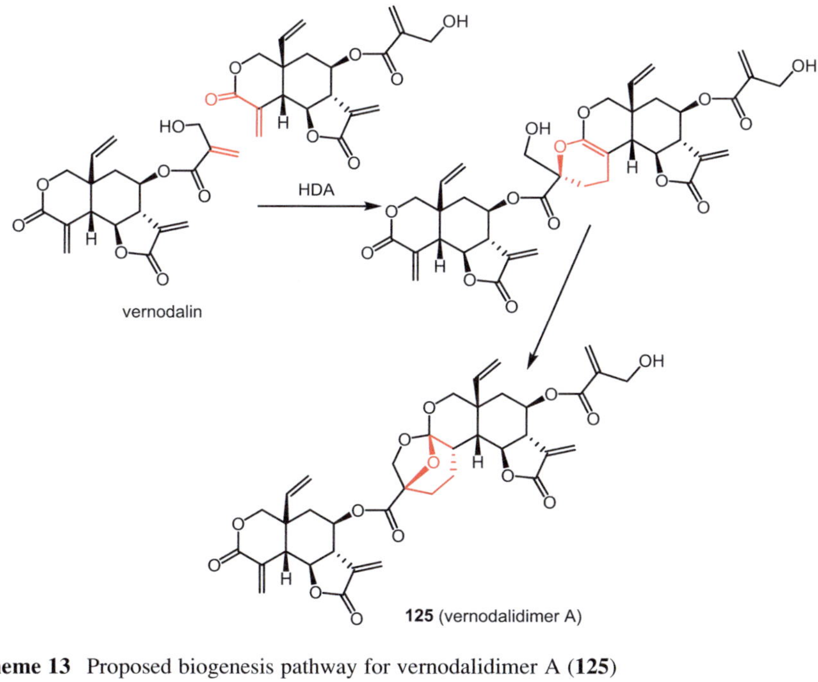

125 (vernodalidimer A)

Scheme 13 Proposed biogenesis pathway for vernodalidimer A (**125**)

A plausible biogenetic pathway of an orthoester elemanolide DS, vernodalidimer A (**125**) from vernodalin, was proposed as shown in Scheme 13 [211]. Unlike other

DSs, the formation of **1** was not from a direct connection of the sesquiterpenoid backbones but was considered to be derived through regio- and stereospecific Diels–Alder cycloaddition between the enone in the sesquiterpenoid backbone of one monomer (vernodalin) and the methylene of the acyl group of another.

5.1.2 [2 + 2] Cycloaddition and [6 + 6] Cycloaddition

Although not very frequent, [2 + 2] cycloaddition is occasionally found in the biosynthesis of DSs that contain a cyclobutane unit linking the two constitutional sesquiterpenoid units. Vielanin B (**186**), from the leaves of *Xylopia vielana* [71], and chloranthalactone A photodimer (**68**), from the leaves of *Chloranthus glaber*, are two representative DSs formed via [2 + 2] cycloaddition [231]. In turn, the lindenane DS cycloshizukaol A (**112**), which possesses a cyclododecatetraene ring, is a C_2-symmetrical [6 + 6] cycloaddition product (Fig. 34) [265].

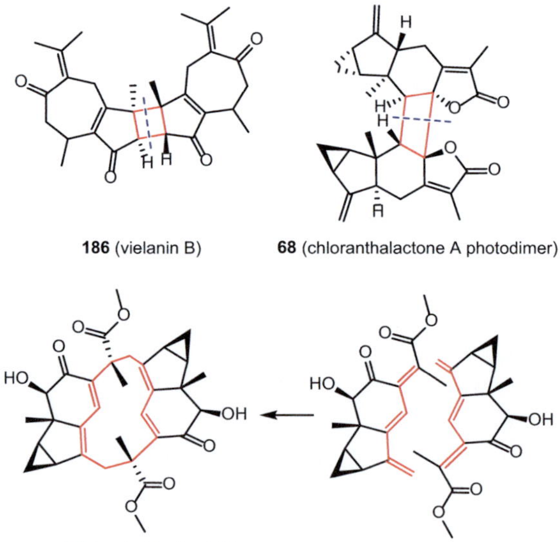

186 (vielanin B)

68 (chloranthalactone A photodimer)

112 (cycloshizukaol A)

Fig. 34 [2 + 2] Cycloaddition and [6 + 6] cycloaddition products

5.1.3 Radical Reactions

As radicals are important factors in many biological processes, dimerization of two sesquiterpenoid monomers through radical pathways to form DSs can be found in many naturally occurring examples of these compounds.

The furoeremophilanolide DS, 8β-[eremophil-3′,7′(11′)-dien-12′,8′α;15′,6′α-diolide]-eremophil-3,7(11)-dien-12,8α;15,6α-diolide (**187**), isolated from *Ligularia atroviolacea*, was formed with two symmetric eremophilanolide units by direct C–C bond connection. A reasonable free radical mechanism was proposed for its biosynthesis (Scheme 14) [86]. For aromatic radicals that may be transferred from one atom to another (and even to form the unstable benzyl radical), condensation may occur through homocoupling or heterocoupling. The two sesquiterpenoid units can therefore be joined to one another via C–C bonds or through bridging groups as in the biosynthesis of mastigophorenes A–D (**22**, **170–172**), isolated from the liverwort *Mastigophora didados* (Scheme 15) [143].

Scheme 14 Proposed biogenesis pathway for **187**

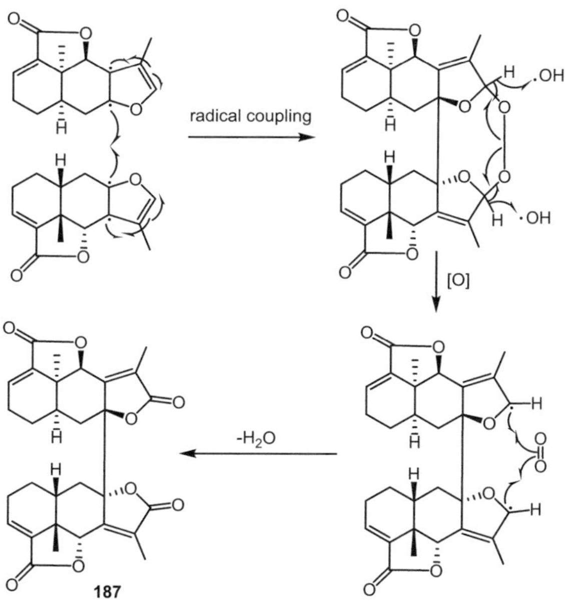

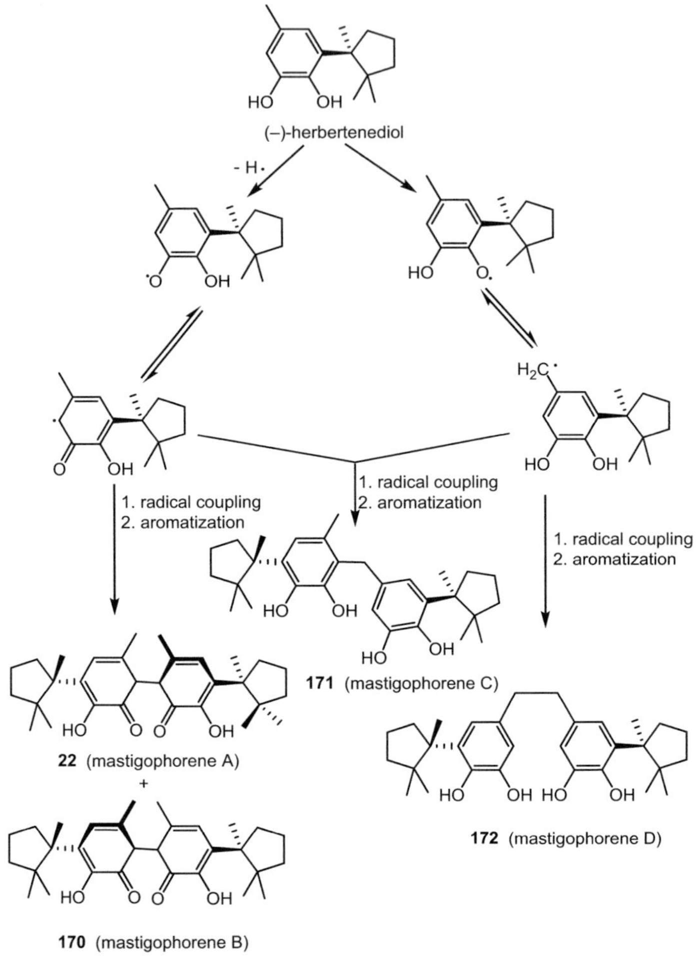

Scheme 15 Plausible biosynthesis route to mastigophorenes A–D (**22**, **170–172**) based on radical coupling from (−)-herbertenediol

5.1.4 Aldol Reactions

Serratustones A (**75**) and B (**188**), which were isolated from *Chloranthus serratus* [171], were proposed as dimers of an elemane sesquiterpenoid (**75a**) and an eudesmane unit (**75b**). Aldol condensation was involved as the key step in the dimerization (Scheme 16). Formation of a semiketal **75d** (from **75c**) and **75f** (from **75e**) followed by dehydration and oxidation would produce **75** and **188**.

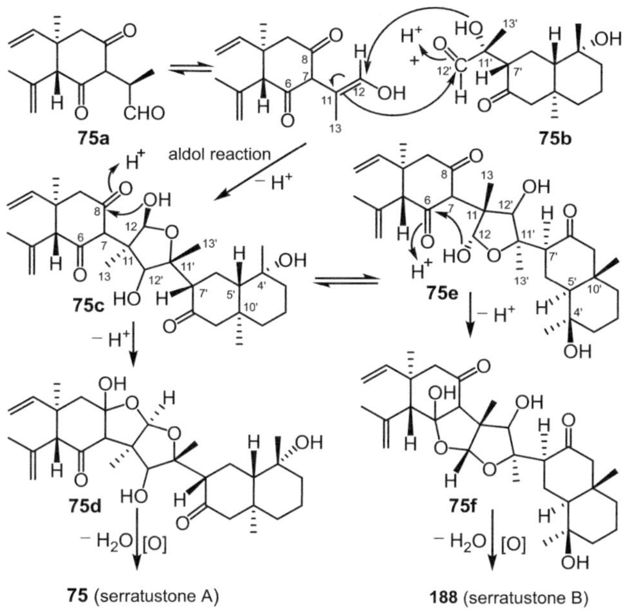

Scheme 16 Plausible biosynthesis pathway to serratustones A (**75**) and B (**188**)

5.1.5 Esterification, Etherification, and Acetal-Formation Reactions

As can be seen in the non-carbon–carbon-connected pseudo-disesquiterpenoids (see Sect. 2.2.2), quite a number of DSs can be formed by dimerization of two sesquiterpenoid monomers through esterification and etherification. Occasionally, the two units may be linked by two separate isocitric acids with one esterifying at one end and another etherifying at the other (Fig. 18). Acetal-formation reactions can be found in cinnafragrins A and B, and capsicodendrin [188].

5.1.6 Dimerization Through a Linker

DSs of the non-carbon–carbon-connected pseudo-disesquiterpenoids can also be formed by connecting the two sesquiterpenoid units via a linker like a disulfide, a urea, or a diamide (see Sect. 2.2.2). Scheme 17 illustrates how halichonadin K (**55**) could be biosynthesized by amidization of two identical eudesmane sesquiterpenoid monomers linked with a piperidine ring [202].

Scheme 17 Plausible biosynthesis pathway of halichonadin K (**55**)

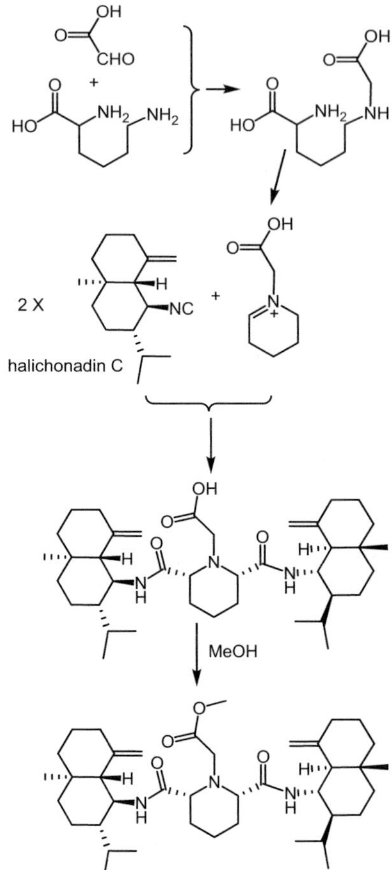

5.1.7 Michael-Type Reactions

The guaiane DS diguaiaperfolin (**5**) was suggested to be a dimer of **5a** and a putative dienol-intermediate (**5b**) formed through a Michael-type reaction and a subsequent semiacetal formation reaction (Scheme 18) [51]. After a Michael-type attack of the C-10–C-14-enol double bond of the dienediol intermediate on the electron-deficient C-4′ in **5a**, and subsequent loss of a water molecule, the lactone ring between C-2–O and C-14′ could be formed to generate diguaiaperfolin (**5**).

Scheme 18 Plausible biosynthesis pathway of diguaiaperfolin (**5**)

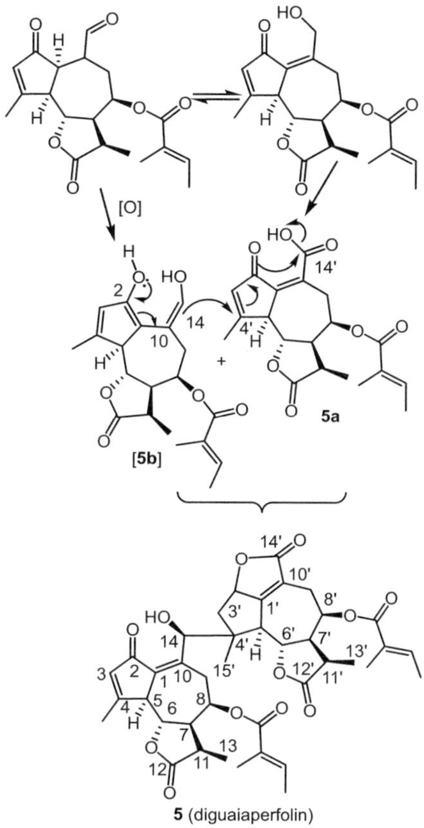

5.2 *Chemical Synthesis*

Based on the pathways proposed for the biosynthesis of DSs, various dimerization strategies have been developed for the synthesis of DSs. Among these strategies, Diels–Alder cycloaddition and oxidative coupling have been employed widely.

5.2.1 Diels–Alder Cycloaddition

Bisacutifolones A–B (**25** and **189**) are two pinguisane-type DSs originally isolated from *Porella acutifolia* subsp. *tosana* [149]. In their biogenesis, they can be assembled by Diels–Alder reactions between two monomeric sesquiterpenoids. After completion of the synthesis of the monomeric precursor acutifolone A (**25a**) using the Mukaiyama aldol reaction as the key step, an intermolecular Diels–Alder reaction at 120°C in a sealed tube in the presence of 2,6-di-*tert*-butyl-4-methylphenol ("butylated hydroxytoluene", BHT) successfully led to stereoselective dimerization of **25a**, which eventually afforded bisacutifolones A

Scheme 19 Synthesis of bisacutifolones A–B (**25** and **189**) by Diels–Alder reaction

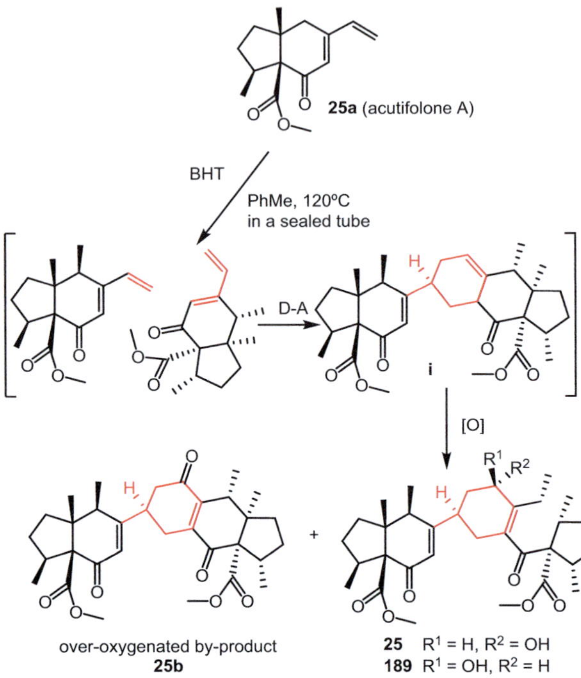

(**25**) and B (**189**) and the over oxygenated by-product (**25b**) in 30, 35, and 11% yields, respectively (Scheme 19) [266]. Theoretical calculations revealed that the dimerization reaction proceeded through the most stable transition state, in which the regioselectivity of the coupling reaction was rationalized. It should be noted that the plausible dimeric intermediate **i** was not obtained owing to rapid autoxidation of the newly generated olefinic moiety.

Encouraged by the proposed biogenesis, an effort has been made toward construction of the heptacyclic core of chlorahololides through a flexible synthetic strategy [267]. After a sequence of chemical transformations including an S_N2-type intramolecular nucleophilic substitution, the desired diene **190b** and dienophile **190a** were obtained. Extensive experiments revealed that treatment of diene **190b** (2.5 equiv.) and dienophile **190a** (1.0 equiv.) with butylated hydroxytoluene (BHT) in reflux toluene (160°C, sealed tube) would furnish the desired *endo*-Diels–Alder cyclization product **190** in 90% yield and have good diastereoselectivity ($dr = 10{:}1$ by ^{1}H NMR) in terms of the direction of the angular methyl and cyclopropyl groups (Scheme 20).

The unique architectural features and potent inhibitory effect of (+)-ainsliadimer A (**135**) against LPS-induced NO production in RAW264.7 cells have attracted the interest of Lei and coworkers in its total synthesis [268]. Although spontaneous hetero-Diels–Alder cycloadditions of α-alkylidene ketones have been reported in the literature, the desired dimer **135b** was not observed when monomer **135a** was allowed to stand at 20°C without solvent for two weeks. Inspired by the observation that hydrogen bonding catalysis reactions that can mimic the action of enzymes or antibodies have achieved great success in organic synthesis, the investigators

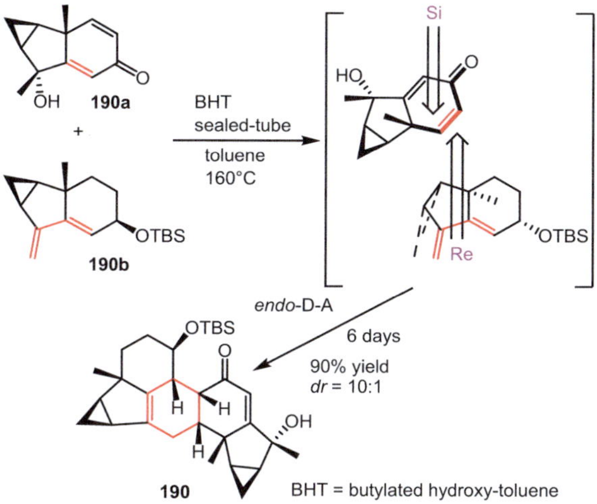

Scheme 20 Synthesis of the heptacyclic core **190** of chloraholides via *endo*-Diels–Alder cycloaddition

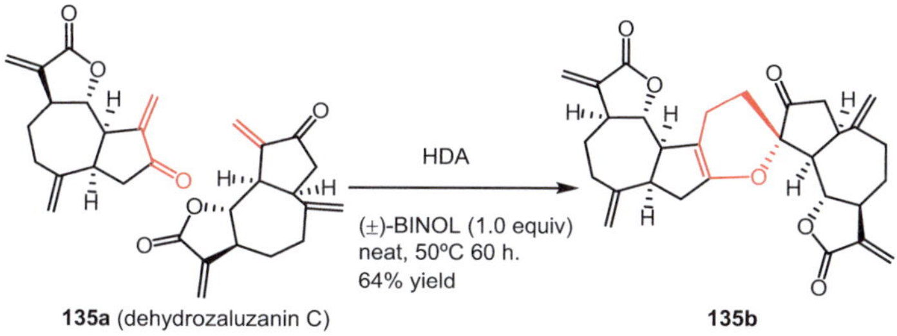

Scheme 21 Hydrogen bonding-promoted HDA dimerization for the synthesis of **135b**

carried out extensive investigations on hydrogen bond-promoted hetero-Diels–Alder reactions. It was observed the cycloadditions were accelerated to a much greater extent if β-naphthol or (±)-BINOL were used as hydrogen bond donor catalysts. The reactivity was ascribed to the formation of a hydrogen bond between the OH group and the ketone moiety, while the high facial selectivity could be explained as a result of the preferential approach of the diene to the dienophile from the less hindered α-face (Scheme 21). The higher yield using BINOL rather than β-naphthol indicated that the bidentate nature of BINOL might help orient two monomers to facilitate the [4 + 2]-hetero-Diels–Alder cycloaddition.

Completion of the total synthesis of **135** is illustrated in Scheme 22 [268]. Hydrolysis of compound **135b** under mild acidic conditions afforded compound **135c**, for which an intramolecular aldol reaction by treatment with a large excess of DBU in dilute CH_2Cl_2 (0.0008 *M*) at 20°C efficiently generated **135** in 89% yield.

(−)-Gochnatiolides A–C (**191**, **74**, **192**) and (−)-ainsliadimer B are guaiane DSs that possess a complex heptacyclic ring system with an intriguing spiro[4,5]decane moiety. Guided by the proposed biosynthesis pathway (Scheme 23), biomimetic

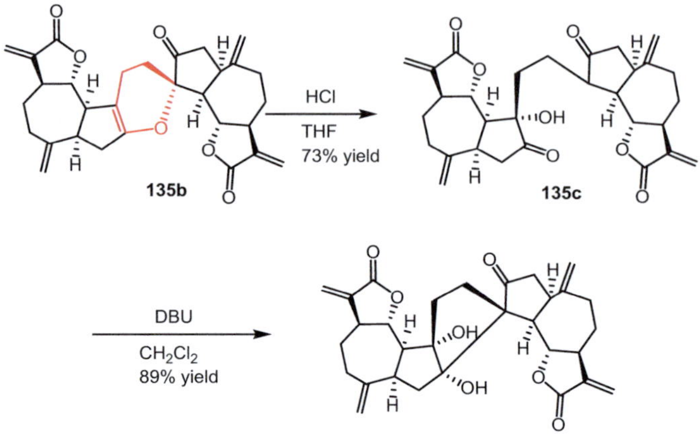

135 (ainsliadimer A)

Scheme 22 Total synthesis of (+)-ainsliadimer A (**135**)

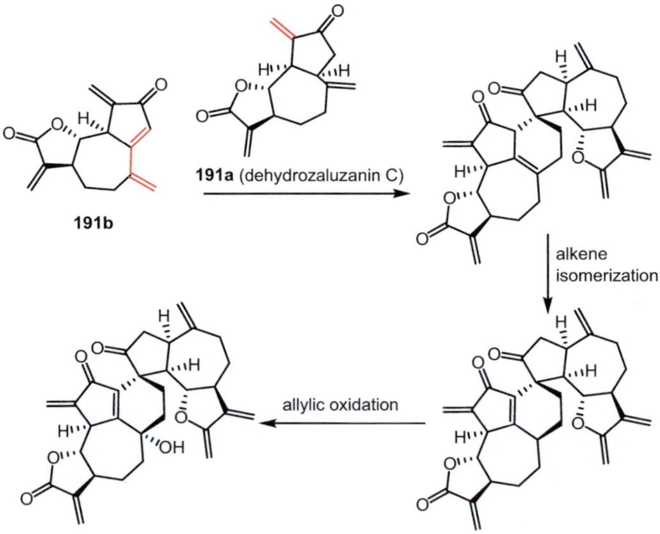

191 (gochnatiolide A) **192** (gochnatiolide C)

Scheme 23 Proposed biogenetic pathways for gochnatiolides A and C (**191**–**192**)

syntheses of these DSs have been carried out [232]. Investigations showed that the use of dimethyl sulfoxide (DMSO) solution of silyl enol ether **191c** that was transformed from **191a** could prevent homodimerization of the diene **191a** (Scheme 24), which was prone to homo-Diels–Alder dimerization to generate the undesired homodimer **191d**. Thus, treatment of a dimethyl sulfoxide solution of silyl enol ether **191c** and **191a** with $Pd(OAc)_2$ under air gave gochnatiolides A–C (**191**, **74**, **192**) with isolated yields of 16, 2, and 6%, along with homodimer **191d** in 20% yield (Scheme 24). It was also discovered that increasing the amount of

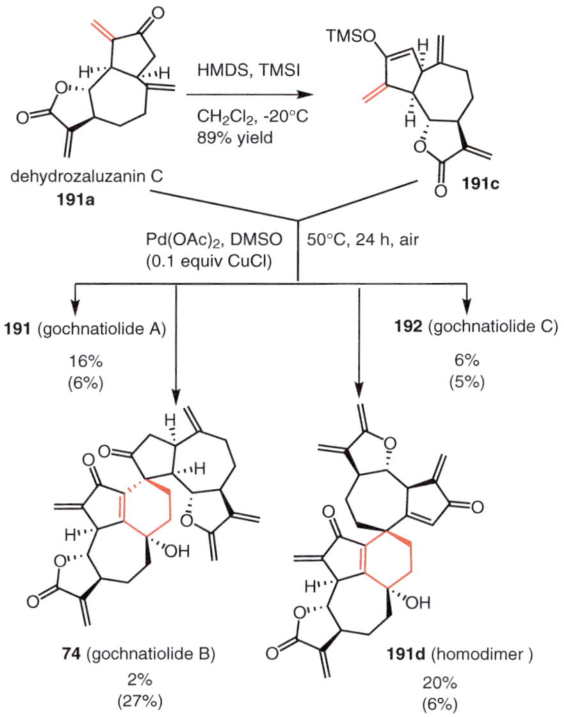

Scheme 24 Synthesis of gochnatiolides A–C (**191**, **74**, **192**) without (with) copper catalyst

dienophile **191a** to disfavor the homodimerization of diene **191c** improved the total yields of gochnatiolides A–C. However, when the reaction was performed in an anerobic glovebox, only gochnatiolide C (**192**) was isolated in 14% yield, suggesting that oxygen is a prerequisite for the allylic oxidation. Meanwhile, improved selectivity could be achieved in the presence of a catalytic amount of copper additives (e.g. CuCl). Addition of 0.1 equiv. of CuCl improved the yield of gochnatiolide B (**74**) to 27% and reversed the ratio of **191** to **74** from 6.6:1 to 1:4.5. This "copper effect" on the stereochemical outcome of radical mediated allylic oxidation was rationalized as a result of the chelating effect between Cu and the ketone carbonyl as well as the alkene moiety in the transition state.

In efforts toward the total synthesis of the bisabolane DS, bacchopetiolone (**193**), a tandem aromatic oxidation/Diels–Alder reaction of aryl propionic acids has been developed [269]. Treatment of **193a** with BTIB (bis(trifluoroacetoxy)-iodobenzene) resulted in smooth aromatic oxidation and subsequent Diels–Alder cycloaddition was conducted to provide dimer **193b**, which possesses the same relative stereochemistry of **193**, as a single diastereomer in 60% yield (Scheme 25). However, bis-decarbonylation via Hofmann rearrangement on both amide functional groups transformed from its lactones failed.

In the total synthesis of (+)-aquaticol (**19**), a biomimetic phenol dearomatization approach has been employed [270]. Dearomatizing *ortho*-selective hydroxylation

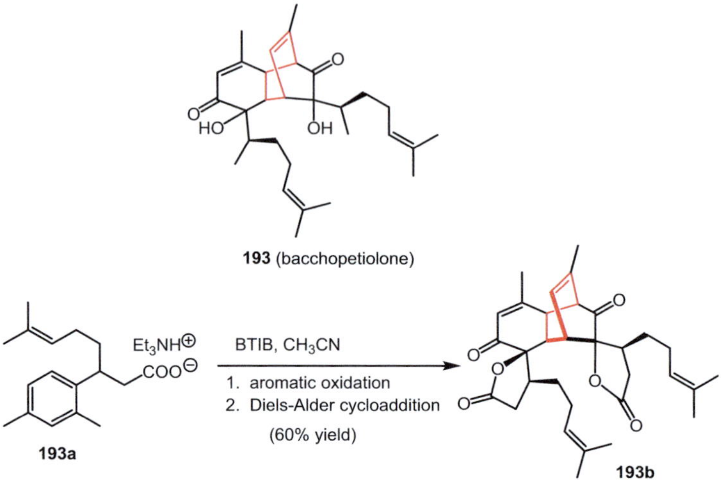

Scheme 25 Tandem phenolic oxidation/Diels–Alder reaction of aryl propionic acids toward total synthesis of the bisabolane DS, bacchopetiolone (**193**)

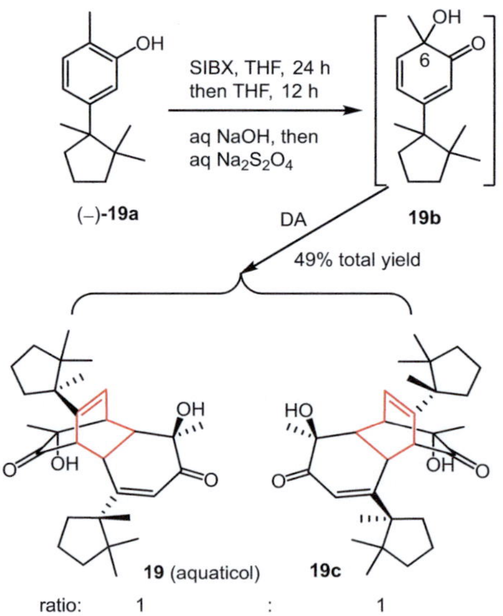

Scheme 26 SIBX-mediated oxidation and subsequent dimerization for the synthesis of (+)-aquaticol (**19**)

of (−)-hydroxycuparene with the λ^5-iodane SIBX (stabilized IBX) gave the orthoquinol (6*R*,7*S*) (**19a**). *endo*-Diels–Alder dimerization of this compound yielded the two homodimers, (+)-aquaticol (**19**) and **19c**, in a ratio of 1:1 and a total yield of 49% (Scheme 26). The outcome was explained as a result of double diastereofacial differentiation in the Diels–Alder dimerization of orthoquinols with a C_2-symmetric transition, in which only orthoquinols with the same configuration at their stereogenic C-6 center combined with each other to furnish the expected *endo* cyclo dimers. Computational analysis showed that a double "Cieplak–Fallis" hyperconjugation appears to be the determining factor in this stereoselectivity, which was also observed in all cases reported to date of the kinetically controlled [4 + 2] dimerization of chiral orthoquinols.

5.2.2 Oxidative Coupling

Like Diels–Alder reactions, oxidative coupling has been widely used in the chemical synthesis of DSs. The most notable examples are those of mastigophorenes, for which their biosynthesis features a radical coupling-type of dimerization (see Scheme 15). Several efforts have been devoted to their total syntheses [271–276]. The first synthesis of mastigophorenes A (**22**) and B (**170**) was achieved in 1999 by Bringmann and Connolly via the phenolic coupling of natural herbertenediol (Scheme 27) [271]. After transformation of herbertenediol to a chemically appropriate monophenolic coupling precursor (**22a**), the oxidative dehydrodimerization was brought about using (*tert*-BuO)$_2$. Subsequent deprotection gave **22** and **170** in their "natural" atropisomeric ratio (ca. $dr = 40{:}60$), as isolated from the liverwort, which suggested that mastigophorenes are formed biosynthetically either by a non-enzymatic reaction or with the enzyme not leading to any additional stereoselectivity besides the internal asymmetric induction exerted by the chiral cyclopentyl residue.

An oxazoline-mediated asymmetric Ullmann coupling was utilized to establish chirality about the biaryl axis of mastigophorenes A (**22**) and B (**170**) (Scheme 28) [272]. After a cascade of reactions, the acid **22c**, which was obtained by bromination of the arene **22b** followed by oxidation, was transformed into the (*S*)-*tert*-leucinol-derived oxazoline, **22d**, in 82% yield. Asymmetric Ullmann coupling of **22d** afforded **22e** and **22f** in a ratio of 3:1 as a result of thermodynamic distribution of products. A sequence of six chemical reactions of the **22e** (or **22f**) led to completion of the total synthesis.

Scheme 27 Oxidative dehydrodimerization for the synthesis of mastigophorenes A (**22**) and B (**170**)

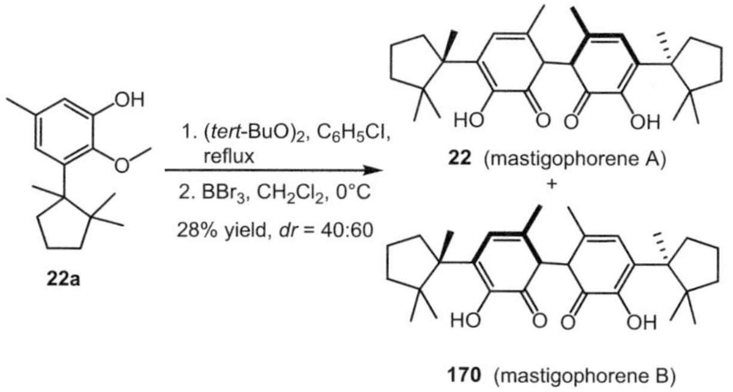

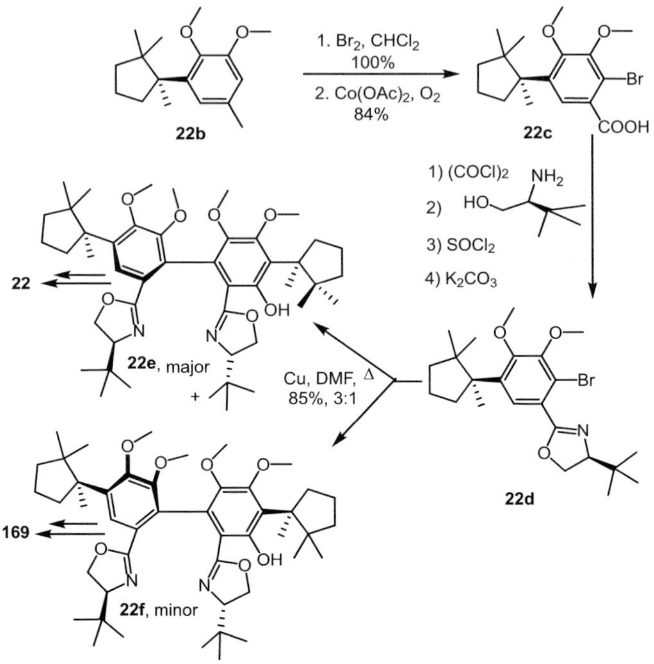

Scheme 28 Asymmetric Ullmann coupling for the synthesis of mastigophorenes A (**22**) and B (**170**)

It was found that when (−)-chloranthalactone A was irradiated with a Hg lamp for 12 h, (−)-chloranthalactone F was isolated in 69% yield without any other diastereomers detected (Scheme 29). The diastereoselectivity was possibly due to the fact that the beneficial electrical and orbital interaction between γ-alkylidenebutenolide segments in the superposition conformation of the transient state leading to chloranthalactone F contributed greatly to the photodimerization [277].

In the biomimetic synthesis of biatractylolide (**163**) and biepiasterolide (**194**), a cobalt-mediated radical dimerization of chloro lactones have been developed. Thus, treatment of chloroatractylolide **163a** with freshly prepared $Co(PPh_3)_3Cl$ under Yamada's conditions (PhH, rt, 2 h) afforded the desired dimers **163** and **194** in good yield (Scheme 30) [278, 279]. In turn, the use of DTBP (di-*tert*-butyl peroxide) as a radical generator did not succeed as it did in the synthesis of their simplified analogues [280].

As mentioned in the preceding Sect. 3.5, dimerization of debromolaurinterol to laurebiphenyl (**72**) was achieved via its oxidative coupling with manganese dioxide [140].

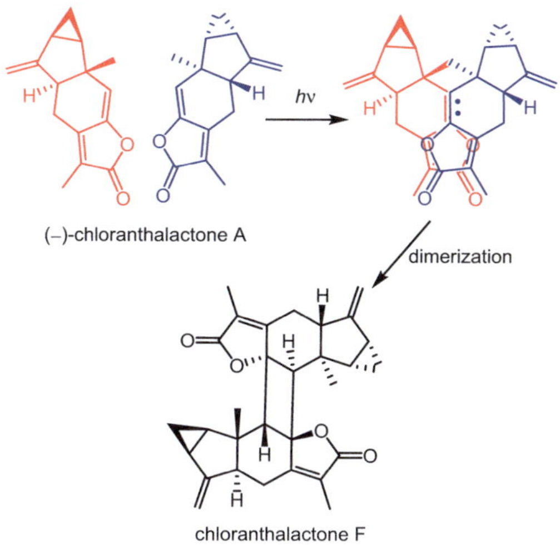

Scheme 29 Oxidative coupling for the synthesis of chloranthalactone F (chloranthalactone A photodimer, **68**)

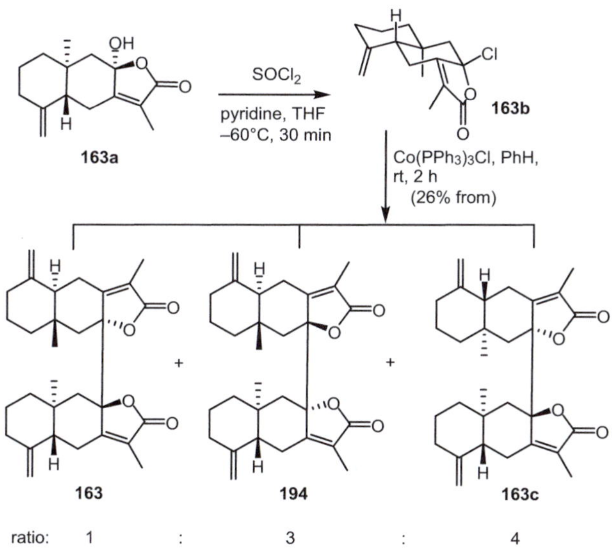

Scheme 30 Cobalt-mediated radical dimerization of chloro lactones for the biomimetic synthesis of (±)-biatractylolide (**163**) and (±)-biepiasterolide (**194**)

5.2.3 Dimerization with Linkers

As can be seen in many pseudosesquiterpenoid DSs, various linkers are present in the DS molecule to connect the two sesquiterpenoid units. Their syntheses must therefore involve a dimerization of two monomeric sesquiterpenoid precursors via a corresponding linker. Synthesis procedures for artemisinin dimers can be found in

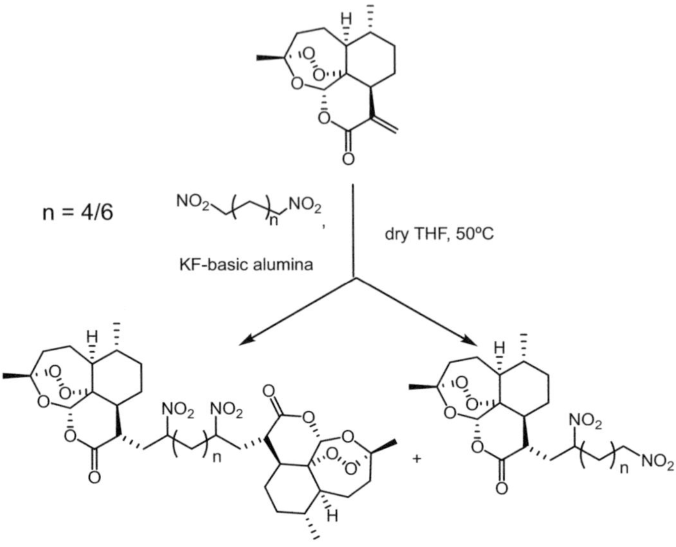

Scheme 31 Use of nitroaliphatics as linkers for the synthesis of novel artemisinin carba-dimer

the references cited in the reviews [240, 241] focusing on their biological activities. Michael addition has been widely used. The synthesis of artemisinin carbadimers at C-16 has been developed (Scheme 31) [281], where dinitroaliphatics was used as linkers and KF on alumina employed to enhance the basic property of the catalyst to promote addition between nitroparaffins and reactive Michael acceptors.

5.2.4 Miscellaneous Dimerization Methods

In the synthesis of mastigophorenes A (*P*, **22**) and B (*M*, **170**), high diastereoselectivity was obtained in their stereoselective total synthesis [273, 274]. Following the "lactone concept", the configuration at the biaryl axis was atropo-divergently induced to be (*P*) or, optionally, (*M*), by stereocontrolled reductive ring cleavage (diastereomeric ratio up to 97:3) of the configurationally unstable joint biaryl lactone precursor **22j** using the oxazaborolidine–borane system (Scheme 32).

For synthesis of the asymmetric core-side chain linked mastigophorenes C (**171**) and D (**172**), McMurry coupling of the desired aldehyde followed by catalytic hydrogenation of the resulting stilbene was used to construct the sidechain-sidechain coupled product (Scheme 33) [276]. Subsequent O-demethylation of the product gave **172**. Transformation of a desired aryl bromide to the corresponding aryl lithium reagent, reaction of which with **172a** gave the (racemic) diarylcarbinol readily transformed to **171**.

It should be noted that mastigophorenes A (*P*, **22**) and B (*M*, **170**) could also be obtained from the biotransformation of herbertenediol by *Penicillium sclerotiorum* [282].

In the total synthesis of vannusal B (**30**) and its analogues [218, 219] efficient dimerization and cyclization strategies have been developed (Scheme 34).

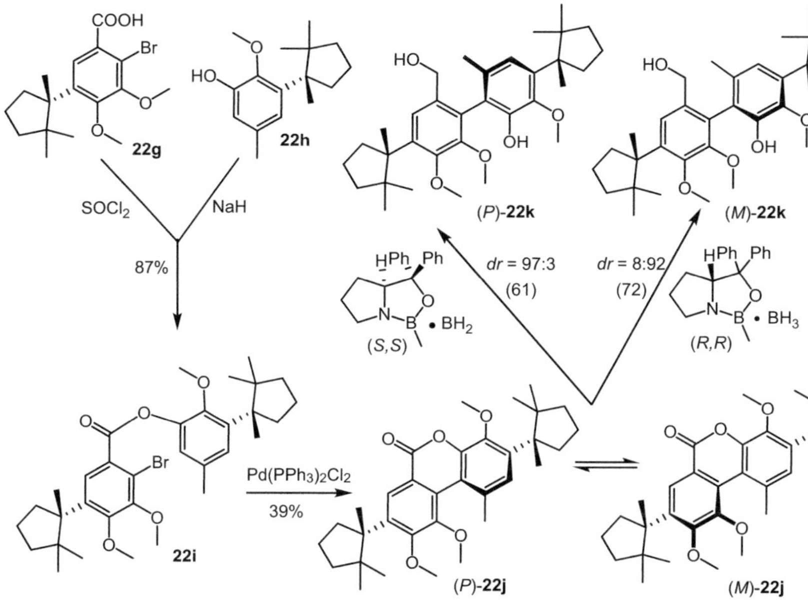

Scheme 32 Stereocontrolled reductive ring cleavage by a lactone toward total synthesis of mastigophorenes A (*P*, **22**) and B (*M*, **170**)

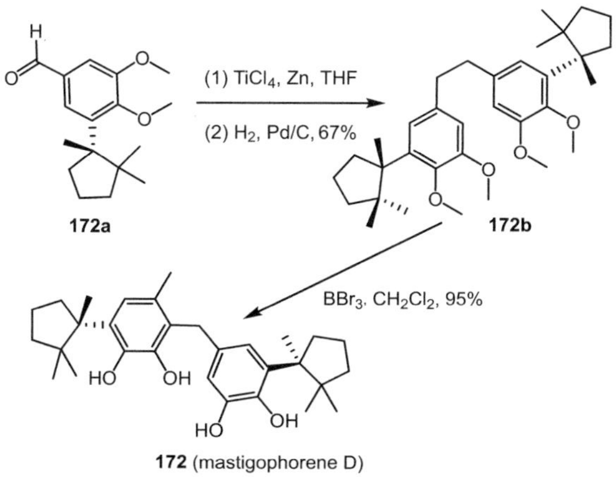

Scheme 33 McMurry coupling for the synthesis of mastigophorene D (**172**)

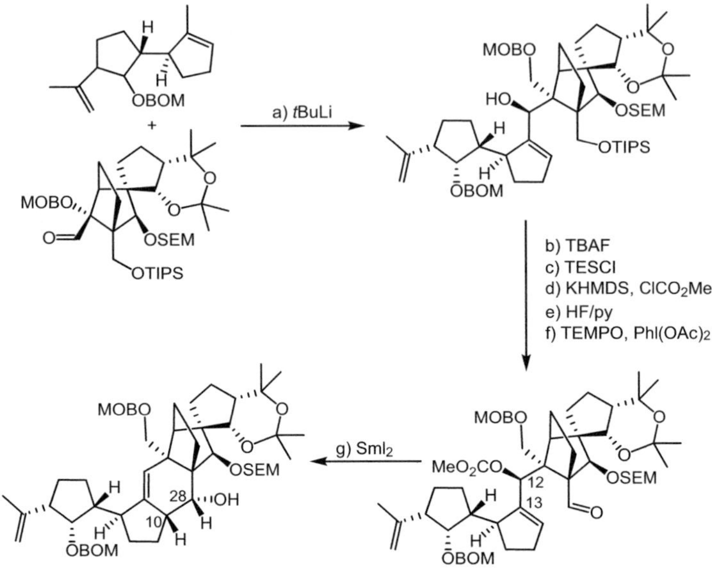

Scheme 34 Dimerization and cyclization methods developed for the total synthesis of vannusal B (**30**) and its analogues

6 Conclusions

As shown in the preceding sections, the dimeric and sesquiterpenoid structural features of DSs render to this class of compounds in most cases more "drug-like" properties as compared to their monomeric precursors. The interesting biological activity of DSs might result from simultaneous interaction with both moieties of a target dimeric protein. Several DSs have been demonstrated to possess potent biological activities and are potential candidates for further drug development. It can be expected that the strategies developed already for the structure elucidation and chemical synthesis of DSs will pave the way for their further development.

Acknowledgments The authors would like to express their gratitude to Dr. Zha-Jun Zhan for his kind help in literature discussions and to Dr. Neng-Lin Zhang for her help in reference editing.

References

1. Hadden MK, Blagg BSJ (2008) Dimeric approaches to anti-cancer chemotherapeutics. Anticancer Agents Med Chem 8:807
2. Mei G, Di Venere A, Rosato N, Finazzi-Agrò A (2005) The importance of being dimeric. FEBS J 272:16

3. Alexander LD, Sellers RP, Davis MR, Ardi VC, Johnson VA, Vasko RC, McAlpine SR (2009) Evaluation of di-sansalvamide A derivatives: synthesis, structure–activity relationship, and mechanism of action. J Med Chem 52:7927
4. Ni F, Kota S, Takahashi V, Strosberg AD, Snyder JK (2011) Potent inhibitors of hepatitis C core dimerization as new leads for anti-hepatitis C agents. Bioorg Med Chem Lett 21:2198
5. Jervis PJ, Moulis M, Jukes J-P, Ghadbane H, Cox LR, Cerundolo V, Besra GS (2012) Towards multivalent CD1d ligands: synthesis and biological activity of homodimeric α-galactosyl ceramide analogues. Carbohydr Res 356:152
6. Zhan Z-J, Ying Y-M, Ma L-F, Shan W-G (2011) Natural disesquiterpenoids. Nat Prod Rep 28:594
7. Lian G, Yu B (2010) Naturally occurring dimers from chemical perspective. Chem Biodivers 7:2660
8. Zdero C, Bohlmann F, Niemeyer HM (1991) An unusual dimeric sesquiterpene and other constituents from Chilean *Baccharis* species. Phytochemistry 30:1597
9. Jolad SD, Timmermann BN, Hoffmann JJ, Bates RB, Camou FA, Cole JR (1988) Sesquiterpenoids from *Coreocarpus arizonicus*. Phytochemistry 27:3545
10. Mao S-C, Manzo E, Guo Y-W, Gavagnin M, Mollo E, Ciavatta ML, van Soest R, Cimino G (2007) New diastereomeric bis-sesquiterpenes from Hainan marine sponges *Axinyssa variabilis* and *Lipastrotethya ana*. Tetrahedron 63:11108
11. Cichewicz RH, Clifford LJ, Lassen PR, Cao X, Freedman TB, Nafie LA, Deschamps JD, Kenyon VA, Flanary JR, Holman TR, Crews P (2005) Stereochemical determination and bioactivity assessment of (*S*)-(+)-curcuphenol dimers isolated from the marine sponge *Didiscus aceratus* and synthesized through laccase biocatalysis. Bioorg Med Chem 13:5600
12. Joseph-Nathan P, Hernández JD, Román LU, García EG, Mendoza V, Mendoza S (1982) Coumarin and terpenoids from *Perezia alamani* var. *oolepis*. Phytochemistry 21:1129
13. Sun L-L, Shao C-L, Chen J-F, Guo Z-Y, Fu X-M, Chen M, Chen Y-Y, Li R, de Voogd NJ, She Z-G, Lin Y-C, Wang C-Y (2012) New bisabolane sesquiterpenoids from a marine-derived fungus *Aspergillus* sp. isolated from the sponge *Xestospongia testudinaria*. Bioorg Med Chem Lett 22:1326
14. Litaudon M, Bousserouel H, Awang K, Nosjean O, Martin M-T, Dau METH, Hadi HA, Boutin JA, Sevenét T, Guéritte F (2009) A dimeric sesquiterpenoid from a Malaysian *Meiogyne* as a new inhibitor of Bcl-xL/BakBH3 domain peptide interaction. J Nat Prod 72:480
15. Mao S-C, Guo Y-W, van Soest R, Cimino G (2011) Trans-dimer D, a novel dimeric sesquiterpene with a bis-bisabolene skeleton from a Hainan sponge *Axinyssa variabilis*. J Asian Nat Prod Res 13:770
16. Marco JA, Sanz JF, Yuste A, Carda M, Jakupovic J (1991) Sesquiterpene lactones from *Artemisia barrelieri*. Phytochemistry 30:3661
17. Cai Y (1997) Stereochemistry of difurocumenone. J Beijing Med Univ 29:229
18. Triana J, López M, Rico M, González-Platas J, Quintana J, Estévez F, León F, Bermejo J (2003) Sesquiterpenoid derivatives from *Gonospermum elegans* and their cytotoxic activity for HL-60 human promyelocytic cells. J Nat Prod 66:943
19. Macias FA, Lopez A, Varela RM, Molinillo JMG, Alves PLCA, Torres A (2004) Helivypolide G. A novel dimeric bioactive sesquiterpene lactone. Tetrahedron Lett 45:6567
20. Bohlmann F, Adler A, Jakupovic J, King RM, Robinson H (1982) Naturally occurring terpene derivatives. Part 412. A dimeric germacranolide and other sesquiterpene lactones from *Mikania* species. Phytochemistry 21:1349
21. Makhmudov MK, Tashkhodzhaev B, Abduazimov BK (1993) Conformation of the dimeric sesquiterpene germacrane lactone mycoguaianolide. Khim Prir Soedin: 213
22. Xue HZ, Zhang J, He LX, He CH, Zheng QT, Feng R (1989) The structure of versicolactone D. Yaoxue Xuebao 24:917
23. Vokáč K, Samek Z, Herout V, Šorm F (1968) The structure of artabsin and absinthin. Tetrahedron Lett 9:3855

24. Beauhaire J, Fourrey JL, Vuilhorgne M, Lallemand JY (1980) Dimeric sesquiterpene lactones: structure of absinthin. Tetrahedron Lett 21:3191
25. Ma C-M, Nakamura N, Hattori M, Zhu S, Komatsu K (2000) Guaiane dimers and germacranolide from *Artemisia caruifolia*. J Nat Prod 63:1626
26. Beauhaire J, Fourrey JL, Guittet E (1984) Structure of absintholide, a new guaianolide dimer of *Artemisia absinthium* L. Tetrahedron Lett 25:2751
27. Stefani R, Schorr K, Tureta JM, Vichnewski W, Merfort I, Da Costa FB (2006) Sesquiterpene lactones from *Dimerostemma* species (Asteraceae) and in vitro potential anti-inflammatory activities. Z Naturforsch C 61:647
28. Li Y, Zhu M-C, Zhang M-L, Wang Y-F, Dong M, Shi Q-W, Huo C-H, Sauriol F, Kiyota H, Gu Y-C, Cong B (2012) Achillinin B and C, new sesquiterpene dimers isolated from *Achillea millefolium*. Tetrahedron Lett 53:2601
29. Wu Z-J, Xu X-K, Shen Y-H, Su J, Tian J-M, Liang S, Li H-L, Liu R-H, Zhang W-D (2008) Ainsliadimer A, a new sesquiterpene lactone dimer with an unusual carbon skeleton from *Ainsliaea macrocephala*. Org Lett 10:2397
30. Wang Y, Shen Y-H, Jin H-Z, Fu J-J, Hu X-J, Qin J-J, Liu J-H, Chen M, Yan S-K, Zhang W-D (2008) Ainsliatrimers A and B, the first two guaianolide trimers from *Ainsliaea fulvioides*. Org Lett 10:5517
31. Kasymov SZ, Abdullaev ND, Sidyakin GP, Yagudaev MR (1979) Anabsin—a new diguaianolide from *Artemisia absinthium*. Chem Nat Compd 15:430
32. Wen J, Shi H, Xu Z, Chang H, Jia C, Zan K, Jiang Y, Tu P (2010) Dimeric guaianolides and sesquiterpenoids from *Artemisia anomala*. J Nat Prod 73:67
33. Ullah N, Ahmed S, Ahmed Z, Mohammad P, Malik A (1999) Dimeric guaianolides from *Daphne oleoides*. Phytochemistry 51:559
34. Zan K, Chai X-Y, Chen X-Q, Wu Q, Fu Q, Zhou S-X, Tu P-F (2012) Artanomadimers A–F: six new dimeric guaianolides from *Artemisia anomala*. Tetrahedron 68:5060
35. Jakupovic J, Sun H, Geerts S, Bohlmann F (1987) New pseudoguaianolides from *Ambrosia maritima*. Planta Med 53:49
36. Mallabaev A, Tashkhodzhaev B, Saitbaeva IM, Yagudaev MR, Sidyakin GP (1986) The structure of artelein — a dimeric lactone of a new type from *Artemisia leucodes*. Chem Nat Compd 22:42
37. Ovezdurdyev A, Abdullaev ND, Yusupov MI, Kasymov SZ (1987) Artenolide, a new disesquiterpenoid from *Artemisia absinthium*. Khim Prir Soedin:667
38. Bohlmann F, Ang W, Trinks C, Jakupovic J, Huneck S (1985) Dimeric guaianolides from *Artemisia sieversiana*. Phytochemistry 24:1009
39. Hu JF, Feng XZ (1998) Artselenoide, a new dimeric guaianolide from *Artemisia selengensis*. Chin Chem Lett 9:829
40. Lee S-H, Kang H-M, Song H-C, Lee H, Lee UC, Son K-H, Kim S-H, Kwon B-M (2000) Sesquiterpene lactones, inhibitors of farnesyl protein transferase, isolated from the flower of *Artemisia sylvatica*. Tetrahedron 56:4711
41. Achenbach H, Benirschke G, Lange J (1996) Assufulvenal, a novel bis-sesquiterpene from *Joannesia princeps*. J Nat Prod 59:93
42. Gao F, Wang H, Mabry TJ (1990) Sesquiterpene lactone aglycones and glycosides and inositol derivatives from *Hymenoxys biennis*. Phytochemistry 29:3875
43. Tikhonova EV, Atazhanova GA, Raldugin VA, Bagryanskaya IY, Gatilov YV, Shakirov MM, Adekenov SM (2006) 2,12-Bis-hamazulenyl from *Ajania fruticulosa* essential oil. Chem Nat Compd 42:298
44. Ali MS, Ahmed W, Armstrong AF, Ibrahim SA, Ahmed S, Parvez M (2006) Guaianolides from *Salvia nubicola* (Lamiaceae). Chem Pharm Bull 54:1235
45. Ali MS, Ibrahim SA, Ahmed S, Lobkovsky E (2007) Guaiane sesquiterpene lactones from *Salvia nubicola* (Lamiaceae). Chem Biodivers 4:98
46. Ibrahim SA, Ali MS, Ahmad F, Moazzam M, Tareen RB (2007) Bistaraxacin: a dimeric-guaianolide from *Salvia nubicola* (Lamiaceae). J Chem Soc Pak 29:394

47. Gu Q, Chen Y, Cui H, Huang D, Zhou J, Wu T, Chen Y, Shi L, Xu J (2013) Chrysanolide A, an unprecedented sesquiterpenoid trimer from the flowers of *Chrysanthemum indicum* L. RSC Adv 3:10168
48. Castro V, Ciccio F, Alvarado S, Bohlmann F, Schmeda-Hirschmann G, Jakupovic J (1983) Decathieleanolide, a dimeric guaianolide from *Decachaeta thieleana*. Liebigs Ann Chem 1983:974
49. Bohlmann F, Zdero C, Schmeda-Hirschmann G, Jakupovic J, Dominguez XA, King RM, Robinson H (1986) Dimeric guaianolides and other constituents from *Gochnatia* species. Phytochemistry 25:1175
50. Morikawa T, Abdel-Halim OB, Matsuda H, Ando S, Muraoka O, Yoshikawa M (2006) Pseudoguaiane-type sesquiterpenes and inhibitors on nitric oxide production from *Dichrocephala integrifolia*. Tetrahedron 62:6435
51. Maas M, Deters AM, Hensel A (2011) Anti-inflammatory activity of *Eupatorium perfoliatum* L. extracts, eupafolin, and dimeric guaianolide via iNOS inhibitory activity and modulation of inflammation-related cytokines and chemokines. J Ethnopharmacol 137:371
52. Zdero C, Bohlmann F (1989) Sesquiterpene lactones and other terpenes from *Geigeria* species. Phytochemistry 28:3105
53. Todorova M, Trendafilova A, Mikhova B, Vitkova A, Duddeck H (2007) Terpenoids from *Achillea distans* Waldst. & Kit. ex Willd. Biochem Syst Ecol 35:852
54. Ali MS, Jahangir M, Uzair SS, Erian AW, Tareen RB (2002) Gnapholide: a new guaiac-dimer from *Pulicaria gnaphalodes* (Asteraceae). Nat Prod Lett 16:179
55. Bohlmann F, Ahmed M, Jakupovic J, King RM, Robinson H (1983) Dimeric sesquiterpene lactones and kolavane derivatives from *Gochnatia paniculata*. Phytochemistry 22:191
56. Tarasov VA, Kasymov SÉ, Sidyakin GP (1976) Sesquiterpene lactones of *Handelia trichophylla*. Chem Nat Compd 12:105
57. Jakupovic J, Zdero C, Grenz M, Tsichritzis F, Lehmann L, Hashemi-Nejad SM, Bohlmann F (1989) Twenty-one acylphloroglucinol derivatives and further constituents from South African *Helichrysum* species. Phytochemistry 28:1119
58. Martins D, Osshiro E, Roque NF, Marks V, Gottlieb HE (1998) A sesquiterpene dimer from *Xylopia aromatica*. Phytochemistry 48:677
59. Staneva J, Trendafilova-Savkova A, Todorova MN, Evstatieva L, Vitkova A (2004) Terpenoids from *Anthemis austriaca* Jacq. Z Naturforsch C 59:161
60. Beauhaire J, Fourrey JL, Lellemand JY, Vuilhorgne M (1981) Dimeric sesquiterpene lactone. Structure of isoabsinthin. Acid isomerization of absinthin derivatives. Tetrahedron Lett 22:2269
61. Qin J-J, Jin H-Z, Huang Y, Zhang S-D, Shan L, Voruganti S, Nag S, Wang W, Zhang W-D, Zhang R (2013) Selective cytotoxicity, inhibition of cell cycle progression, and induction of apoptosis in human breast cancer cells by sesquiterpenoids from *Inula lineariifolia* Turcz. Eur J Med Chem 68:473
62. Romo de Vivar A, Delgado G (1985) Structure and stereochemistry of mexicanin F, a novel dimeric nor-sesquiterpene lactone from *Helenium mexicanum*. Tetrahedron Lett 26:579
63. Lee K-H, Imakura Y, Sims D, McPhail AT, Onan KD (1976) Structure and stereochemistry of microlenin, a novel antitumor dimeric sesquiterpene lactone from *Helenium microcephalum*; X-ray crystal structure. J Chem Soc Chem Commun:341
64. Imakura Y, Lee KH, Sims D, Hall IH (1978) Antitumor agents. XXVIII: Structural elucidation of the novel antitumor sesquiterpene lactone, microlenin, from *Helenium microcephalum*. J Pharm Sci 67:1228
65. Imakura Y, Lee KH, Sims D, Wu RY, Hall IH, Furukawa H, Itoigawa M, Yonaha K (1980) Antitumor agents XXXVI: structural elucidation of sesquiterpene lactones microhelenins-A, B, and C, microlenin acetate, and plenolin from *Helenium microcephalum*. J Pharm Sci 69:1044
66. Li Y, Ni Z-Y, Zhu M-C, Zhang K, Wu Y-B, Dong M, Shi Q-W, Huo C-H, Sauriol F, Kiyota H, Gu Y-C, Cong B (2012) Millifolides A-C. New 1,10-seco-guaianolides from the flowers of *Achillea millefolium*. Z Naturforsch B 67:438

67. Maas M, Hensel A, Batista da Costa F, Brun R, Kaiser M, Schmidt TJ (2011) An unusual dimeric guaianolide with antiprotozoal activity and further sesquiterpene lactones from *Eupatorium perfoliatum*. Phytochemistry 72:635
68. Ahmed AA, Mahmoud AA, El-Gamal AA (1999) A xanthanolide diol and a dimeric xanthanolide from *Xanthium* species. Planta Med 65:470
69. Ahmed AA, Jakupovic J, Bohlmann F, Regaila HA, Ahmed AM (1990) Sesquiterpene lactones from *Xanthium pungens*. Phytochemistry 29:2211
70. Wang L, Wang J, Li F, Liu X, Chen B, Tang Y-X, Wang M-K (2013) Cytotoxic sesquiterpene lactones from aerial parts of *Xanthium sibiricum*. Planta Med 79:661
71. Kamperdick C, Phuong NM, Van Sung TV, Adam G (2001) Guaiane dimers from *Xylopia vielana*. Phytochemistry 56:335
72. Kamperdick C, Phuong NM, Adam G, Van Sung T (2003) Guaiane dimers from *Xylopia vielana*. Phytochemistry 64:811
73. Zhou J, Wang J-S, Zhang Y, Wang P-R, Guo C, Kong L-Y (2012) Disesquiterpenoid and sesquiterpenes from the flos of *Chrysanthemum indicum*. Chem Pharm Bull 60:1067
74. Kuroyanagi M, Naito H, Noro T, Ueno A, Fukushima S (1985) Furanoeremophilane-type sesquiterpenes from *Cacalia adenostyloides*. Chem Pharm Bull 33:4792
75. Saito Y, Takashima Y, Kamada A, Suzuki Y, Suenaga M, Okamoto Y, Matsunaga Y, Hanai R, Kawahara T, Gong X, Tori M, Kuroda C (2012) Chemical and genetic diversity of *Ligularia virgaurea* collected in northern Sichuan and adjacent areas of China: isolation of 13 new compounds. Tetrahedron 68:10011
76. Abdo S, de Bernardi M, Marinoni G, Mellerio G, Samaniego S, Vidarit G, Vita Finzit P (1992) Furanoeremophilanes and other constituents from *Senecio canescens*. Phytochemistry 31:3937
77. Li Y-S, Li S-S, Wang Z-T, Luo S-D, Zhu D-Y (2006) A novel bieremophilanolide from *Ligularia lapathifolia*. Nat Prod Res 20:1241
78. Wu Q-H, Wang C-M, Cheng S-G, Gao K (2004) Bieremoligularolide and eremoligularin, two novel sesquiterpenoids from *Ligularia muliensis*. Tetrahedron Lett 45:8855
79. Huang H-L, Xu Y-J, Liu H-L, Liu X-Q, Shang J-N, Han G-T, Yao M-J, Yuan C-S (2011) Eremophilane-type sesquiterpene lactones from *Ligularia hodgsonii* Hook. Phytochemistry 72:514
80. Liu J-Q, Zhang M, Zhang C-F, Qi H-Y, Bashall A, Bligh SWA, Wang Z-T (2008) Cytotoxic sesquiterpenes from *Ligularia platyglossa*. Phytochemistry 69:2231
81. Kurihara T, Suzuki S (1981) Studies on the constituents of *Farfugium japonicum* (L.) KITAM. IV. On the components of the rhizome and the leaves. Yakugaku Zasshi 101:35
82. Bohlmann F, Le-Van N (1978) Naturally occurring terpene derivatives. Part 142. New sesqui- and diterpenes from *Bedfordia salicina*. Phytochemistry 17:1173
83. Bohlmann F, Zdero C (1978) Naturally occurring terpene derivatives, 144. A dimeric furanoeremophilane and new cacalohastin derivatives from *Senecio crispus* Thunb. and *Senecio macrospermus* DC. Chem Ber 111:3140
84. Lewis DE, Massy-Westropp RA, Ingham CF, Wells RJ (1982) The structure determination of two related eremophilone dimers. Aust J Chem 35:809
85. Zhao Y, Jiang H, MacLeod M, Parsons S, Rankin DWH, Wang P, Cheng CHK, Shi H, Hao X, Guéritte F (2004) Isomeric eremophilane lactones from *Senecio tsoongianus*. Chem Biodivers 1:1546
86. Zhao J, Wu H, Huang KX, Shi SY, Peng H, Sun XF, Chen LR, Zheng QX, Zhang QJ, Hao XJ, Stöckigt J, Li XK, Zhao Y, Qu J (2008) One chloro-furoeremophilanoid and two new natural dimers from *Ligularia atroviolacea*. Chin Chem Lett 19:1319
87. Wang X, Sun L, Huang K, Shi S, Zhang L, Xu J, Peng H, Sun X, Wang L, Wu X, Zhao Y, Li X, Stöckigt J, Qu J (2009) Phytochemical investigation and cytotoxic evaluation of the components of the medicinal plant *Ligularia atroviolacea*. Chem Biodivers 6:1053
88. Xie W-D, Weng C-W, Li X, Row K-H (2010) Eremophilane sesquiterpenoids from *Ligularia fischeri*. Helv Chim Acta 93:1983
89. Xie W-D, Liu Y-H, Weng C-W, Zhao H, Row KH (2011) Fischelactone B: a new eremophilane dimer from *Ligularia fischeri*. J Chin Chem Soc 58:412

90. Sun X-B, Xu Y-J, Qiu D-F, Yuan C-S (2007) Sesquiterpenoids from the rhizome of *Ligularia virgaurea*. Helv Chim Acta 90:1705
91. Fei D-Q, Wu Q-H, Li S-G, Gao K (2010) Two new asymmetric sesquiterpene dimers from the rhizomes of *Ligularia muliensis*. Chem Pharm Bull 58:467
92. Liu X, Wu Q-X, Wei X-N, Shi Y-P (2007) Novel sesquiterpenes from *Ligularia virgaurea* spp. *oligocephala*. Helv Chim Acta 90:1802
93. Wu QX, Liu X, Shi YP (2005) A novel dimeric eremophilane from *Ligularia virgaurea* spp. *oligocephala*. Chin Chem Lett 16:1477
94. Lewis DE, Massy-Westropp RA, Snow MR (1979) *cis*, *trans*-Tetrahydromitchelladione. Acta Crystallogr B 35:2253
95. Zhang Z-X, Wang C-M, Fei D-Q, Jia Z-J (2008) Two novel asymmetric eremophilane dimers from the roots of *Ligularia virgaurea*. Chem Lett 37:346
96. Chen H-M, Wang B-G, Jia Z-J (1996) Novel sesquiterpenes from *Ligularia virgaurea*. Indian J Chem B 35B:1304
97. Wang BG, Jia ZJ (1997) A new benzofuranosesquiterpene dimer from *Ligularia virgaurea*. Chin Chem Lett 8:315
98. Wang BG, Jia ZJ, Yang XP (1997) Two minor benzofuranosesquiterpene dimers from *Ligularia virgaurea*. Planta Med 63:577
99. Paula VF, Rocha ME, Barbosa LCA, Howarth OW (2006) Aquatidial, a new bis-norsesquiterpenoid from *Pachira aquatica* Aubl. J Braz Chem Soc 17:1443
100. Nishizawa M, Inoue A, Sastrapradja S, Hayashi Y (1983) (+)-8-Hydroxycalamenene: a fish-poison principle of *Dysoxylum acutangulum* and *D. alliaceum*. Phytochemistry 22:2083
101. Nishizawa M, Yamada H, Sastrapradja S, Hayashi Y (1985) Structure and synthesis of bicalamenene. Tetrahedron Lett 26:1535
102. David JP, Yoshida M (1998) Bicalamenenes from *Ocotea corymbosa*. Rev Latinoam Quim 26:91
103. Cambie RC, Lal AR, Ahmad F (1990) Sesquiterpenes from *Heritiera ornithocephala*. Phytochemistry 29:2329
104. El-Seedi H, Ghia F, Torssell KBG (1994) Cadinane sesquiterpenes from *Siparuna marcrotepala*. Phytochemistry 35:1495
105. He L, Hou J, Gan M, Shi J, Chantrapromma S, Fun H-K, Williams ID, Sung HHY (2008) Cadinane sesquiterpenes from the leaves of *Eupatorium adenophorum*. J Nat Prod 71:1485
106. Delgado G, del Socorro OM, Chávez MI, Ramírez-Apan T, Linares E, Bye R, Espinosa-García FJ (2001) Antiinflammatory constituents from *Heterotheca inuloides*. J Nat Prod 64:861
107. Stipanovic RD, Bell AA, Mace ME, Howell CR (1975) Antimicrobial terpenoids of *Gossypium*: 6-methoxygossypol and 6,6′-dimethoxygossypol. Phytochemistry 14:1077
108. Triplett BA, Moss SC, Bland JM, Dowd MK, Phillips GC (2008) Induction of hairy root cultures from *Gossypium hirsutum* and *Gossypium barbadense* to produce gossypol and related compounds. In Vitro Cell Dev Biol Plant 44:508
109. Jagt DLV, Deck LM, Royer RE (2000) Gossypol prototype of inhibitors targeted to dinucleotide folds. Curr Med Chem 7:479
110. Takahashi M, Koyano T, Kowithayakorn T, Hayashi M, Komiyama K, Ishibashi M (2003) Parviflorene A, a novel cytotoxic unsymmetrical sesquiterpene-dimer constituent from *Curcuma parviflora*. Tetrahedron Lett 44:2327
111. Ishibashi M, Ohtsuki T (2008) Studies on search for bioactive natural products targeting TRAIL signaling leading to tumor cell apoptosis. Med Res Rev 28:688
112. Toume K, Takahashi M, Yamaguchi K, Koyano T, Kowithayakorn T, Hayashi M, Komiyama K, Ishibashi M (2004) Parviflorenes B-F, novel cytotoxic unsymmetrical sesquiterpene-dimers with three backbone skeletons from *Curcuma parviflora*. Tetrahedron 60:10817
113. Toume K, Sato M, Koyano T, Kowithayakorn T, Yamori T, Ishibashi M (2005) Cytotoxic dimeric sesquiterpenoids from *Curcuma parviflora*: isolation of three new parviflorenes and absolute stereochemistry of parviflorenes A, B, D, F, and G. Tetrahedron 61:6700

114. Jiang H-L, Chen J, Jin X-J, Yang J-L, Li Y, Yao X-J, Wu Q-X (2011) Sesquiterpenoids, alantolactone analogues, and seco-guaiene from the roots of *Inula helenium*. Tetrahedron 67:9193
115. Lin Y, Jin T, Wu X, Huang Z, Fan J, Chan WL (1997) A novel bisesquiterpenoid, biatractylolide, from the Chinese herbal plant *Atractylodes macrocephala*. J Nat Prod 60:27
116. Rosquete C, Del Olmo E, Sanz F, San Feliciano A (2002) The crystal structure of biatractylolide, an 8,8′ (C-C) linked dimeric 12,8-eudesmanolide from the resin of *Trattinickia rhoifolia* Willd. Chem Pharm Bull 50:964
117. Wang B-D, Yu Y-H, Teng N-N, Jiang S-H, Zhu D-Y (1999) Structural elucidation of biepiasterolide. Huaxue Xuebao 57:1022
118. Rosquete C, Del Olmo E, Del Corral JMM, Lopez JL, Gordaliza M, San Feliciano A (2010) Eudesmanolides and other terpenoids from *Trattinickia rhoifolia*. Adv Quim 5:123
119. Kouno I, Hirai A, Jiang Z, Tanaka T (1997) Bisesquiterpenoid from the root of *Lindera strychnifolia*. Phytochemistry 46:1283
120. Matusch R, Haeberlein H (1987) Novel bissesquiterpene lactones from *Helenium autumnale* L. Liebigs Ann Chem 1987:455
121. Somwong P, Suttisri R, Buakeaw A (2013) New sesquiterpenes and phenolic compound from *Ficus foveolata*. Fitoterapia 85:1
122. Jakupovic J, Schuster A, Bohlmann F, Dillon MO (1988) Lumiyomogin, ferreyrantholide, fruticolide and other sesquiterpene lactones from *Ferreyranthus fruticosus*. Phytochemistry 27:1113
123. Quijano L, Gómez-Garibay F, Trejo RBIB, Rios T (1991) Hydroxy-bis-dihydroencelin, a dimeric eudesmanolide and other eudesmanolides from *Montanoa speciosa*. Phytochemistry 30:3293
124. Kraut L, Mues R, Sim-Sim M (1994) Sesquiterpene lactones and 3-benzylphthalides from *Frullania muscicola*. Phytochemistry 37:1337
125. Zhang M, Iinuma M, Wang J-S, Oyama M, Ito T, Kong L-Y (2012) Terpenoids from *Chloranthus serratus* and their anti-inflammatory activities. J Nat Prod 75:694
126. Sun C-L, Yan H, Li X-H, Zheng X-F, Liu H-Y (2012) Terpenoids from *Chloranthus elatior*. Nat Prod Bioprospect 2:156
127. Ran X-H, Teng F, Chen C-X, Wei G, Hao X-J, Liu H-Y (2010) Chloramultiols A-F, lindenane-type sesquiterpenoid dimers from *Chloranthus multistachys*. J Nat Prod 73:972
128. Liu H-Y, Ran X-H, Gong N-B, Ni W, Qin X-J, Hou Y-Y, Lu Y, Chen C-X (2013) Sesquiterpenoids from *Chloranthus multistachys*. Phytochemistry 88:112
129. Zhang M, Wang J-S, Oyama M, Luo J, Guo C, Ito T, Iinuma M, Kong L-Y (2012) Anti-inflammatory sesquiterpenes and sesquiterpene dimers from *Chloranthus fortunei*. J Asian Nat Prod Res 14:708
130. Zhang S, Yang S-P, Yuan T, Lin B-D, Wu Y, Yue J-M (2010) Multistalides A and B, two novel sesquiterpenoid dimers from *Chloranthus multistachys*. Tetrahedron Lett 51:764
131. He X-F, Yin S, Ji Y-C, Su Z-S, Geng M-Y, Yue J-M (2010) Sesquiterpenes and dimeric sesquiterpenoids from *Sarcandra glabra*. J Nat Prod 73:45
132. Ni G, Zhang H, Liu H-C, Yang S-P, Geng M-Y, Yue J-M (2013) Cytotoxic sesquiterpenoids from *Sarcandra glabra*. Tetrahedron 69:564
133. He X-F, Zhang S, Zhu R-X, Yang S-P, Yuan T, Yue J-M (2011) Sarcanolides A and B: two sesquiterpenoid dimers with a nonacyclic scaffold from *Sarcandra hainanensis*. Tetrahedron 67:3170
134. Yang S, Chen H, Yue J (2012) Sesquiterpenoids from *Chloranthus spicatus*. Chin J Chem 30:1243
135. Kwon SW, Kim YK, Kim JY, Ryu HS, Lee HK, Kang JS, Kim HM, Hong JT, Kim Y, Han S-B (2011) Evaluation of immunotoxicity of shizukaol B isolated from *Chloranthus japonicus*. Biomol Ther 19:59

136. Kim S-Y, Kashiwada Y, Kawazoe K, Murakami K, Sun H-D, Li S-L, Takaishi Y (2011) Spicachlorantins G-J, new lindenane sesquiterpenoid dimers from the roots of *Chloranthus spicatus*. Chem Pharm Bull 59:1281
137. Yoyota M, Koyama H, Asakawa Y (1997) Sesquiterpenoids from the three Japanese liverworts *Lejeunea aquatica*, *L. flava* and *L. japonica*. Phytochemistry 46:145
138. Su B-N, Zhu Q-X, Jia Z-J (1999) Aquaticol, a novel bis-sesquiterpene from *Veronica anagallis-aquatica*. Tetrahedron Lett 40:357
139. Su B-N, Yang L, Gao K, Jia Z-J (2000) Aquaticol, a bis-sesquiterpene and iridoid glucosides from *Veronica anagallis-aquatica*. Planta Med 66:281
140. Shizuri Y, Yamada K (1985) Laurebiphenyl, a dimeric sesquiterpene of the cyclolaurane-type from the red alga *Laurencia nidifica*. Phytochemistry 24:1385
141. Irita H, Hashimoto T, Fukuyama Y, Asakawa T (2000) Herbertane-type sesquiterpenoids from the liverwort *Herbertus sakuraii*. Phytochemistry 55:247
142. Matsuo A, Yuki S, Nakayama M, Hayashi S (1982) Three new sesquiterpene phenols of the *ent*-herbertane class from the liverwort *Herberta adunca*. Chem Lett 11:463
143. Fukuyama Y, Asakawa Y (1991) Novel neurotrophic isocuparane-type sesquiterpene dimers, mastigophorenes A, B, C and D, isolated from the liverwort *Mastigophora diclados*. J Chem Soc Perkin Trans 1:2737
144. Fukuyama Y, Toyota M, Asakawa Y (1988) Mastigophorenes: novel dimeric isocuprane-type sesquiterpenoids from the liverwort *Mastigophora diclados*. J Chem Soc Chem Commun 1988:1341
145. Kladi M, Xenaki H, Vagias C, Papazafiri P, Roussis V (2006) New cytotoxic sesquiterpenes from the red algae *Laurencia obtusa* and *Laurencia microcladia*. Tetrahedron 62:182
146. Kladi M, Vagias C, Papazafiri P, Furnari G, Serio D, Roussis V (2007) New sesquiterpenes from the red alga *Laurencia microcladia*. Tetrahedron 63:7606
147. Bokesch HR, Blunt JW, Westergaard CK, Cardellina JH II, Johnson TR, Michael JA, McKee TC, Hollingshead MG, Boyd MR (1999) Alertenone, a dimer of suberosenone from *Alertigorgia* sp. J Nat Prod 62:633
148. Kanokmedhakul S, Lekphrom R, Kanokmedhakul K, Hahnvajanawong C, Bua-art S, Saksirirat W, Prabpai S, Kongsaeree P (2012) Cytotoxic sesquiterpenes from luminescent mushroom *Neonothopanus nambi*. Tetrahedron 68:8261
149. Hashimoto T, Irita H, Tanaka M, Takaoka S, Asakawa Y (1998) Two novel Diels-Alder reaction-type dimeric pinguisane sesquiterpenoids and related compounds from the liverwort *Porella acutifolia* subsp. *tosana*. Tetrahedron Lett 39:2977
150. Bohlmann F, Singh P, Jakupovic J (1982) Naturally occurring terpene derivatives. Part 429. Sesquiterpenes and a dimeric spiroketone from *Cineraria fruticulorum*. Phytochemistry 21:2531
151. Takaoka D, Kouyama N, Tani H, Matsuo A (1991) Structures of three novel dimeric sesquiterpenoids from the liverwort *Mylia taylorii*. J Chem Res (S):180
152. Rasser F, Anke T, Sterner O (2002) Terpenoids from *Bovista* sp. 96042. Tetrahedron 58:7785
153. Harinantenaina L, Asakawa Y, De Clercq E (2007) Cinnamacrins A-C, cinnafragrin D, and cytostatic metabolites with α-glucosidase inhibitory activity from *Cinnamosma macrocarpa*. J Nat Prod 70:277
154. Hori K, Satake T, Saiki Y, Murakami T, Chen C-M (1988) Chemical and chemotaxonomical studies of Fillices. LXXV. Chemical studies on the constituents of *Dennstaedtia distenta* MOORE. (1). Yakugaku Zasshi 108:422
155. Hori K, Satake T, Yabuuchi M, Saiki Y, Murakami T, Chen CM (1987) Chemical and chemotaxonomical studies of filices. LXVIII. The distribution of sesquiterpene dimers 'Monachosorins' and its chemotaxonomic implication. Yakugaku Zasshi 107:485
156. Satake T, Murakami T, Yokote N, Saiki Y, Chen C (1985) Chemical and chemotaxonomical studies on filices. LVIII. Chemical studies on the constituents of *Monachosorum arakii* TAGAWA (Pteridaceae). Chem. Pharm Bull 33:4175

157. Epstein WW, Sweat FW, VanLear G, Lovell FM, Gabe EJ (1979) Structure and stereochemistry of officinalic acid, a novel triterpene from *Fomes officinalis*. J Am Chem Soc 101:2748
158. Erb B, Borschberg H-J, Arigoni D (2000) The structure of laricinolic acid and its biomimetic transformation into officinalic acid. J Chem Soc Perkin Trans 1:2307
159. Guella G, Dini F, Pietra F (1999) Metaboliten aus marinen Ciliaten mit einem neuartigen C30-Rückgrat. Angew Chem 111:1217
160. Isaka M, Srisanoh U, Choowong W, Boonpratuang T (2011) Sterostreins A–E, new terpenoids from cultures of the Basidiomycete *Stereum ostrea* BCC 22955. Org Lett 13:4886
161. Jin HZ, Lee D, Lee JH, Lee K, Hong Y-S, Choung D-H, Kim YH, Lee JJ (2006) New sesquiterpene dimers from *Inula britannica* inhibit NF-κB activation and NO and TNF-α production in LPS-stimulated RAW264.7 cells. Planta Med 72:40
162. Zhu JX, Qin JJ, Jin HZ, Zhang WD (2013) Japonicones Q-T, four new dimeric sesquiterpene lactones from *Inula japonica* Thunb. Fitoterapia 84:40
163. Qin JJ, Jin HZ, Fu JJ, Hu XJ, Wang Y, Yan SK, Zhang WD (2009) Japonicones A-D, bioactive dimeric sesquiterpenes from *Inula japonica* Thunb. Bioorg Med Chem Lett 19:710
164. Qin JJ, Jin HZ, Zhu JX, Fu JJ, Hu XJ, Liu XH, Zhu Y, Yan SK, Zhang WD (2010) Japonicones E-L, dimeric sesquiterpene lactones from *Inula japonica* Thunb. Planta Med 76:278
165. Qin JJ, Wang LY, Zhu JX, Jin HZ, Fu JJ, Liu XF, Li HL, Zhang WD (2011) Neojaponicone A, a bioactive sesquiterpene lactone dimer with an unprecedented carbon skeleton from *Inula japonica*. Chem Commun 47:1222
166. Sun C-M, Syu W Jr, Don M-J, Lu J-J, Lee G-H (2003) Cytotoxic sesquiterpene lactones from the root of *Saussurea lappa*. J Nat Prod 66:1175
167. Qin J-J, Huang Y, Wang D, Cheng X-R, Zeng Q, Zhang S-D, Hu Z-L, Jin H-Z, Zhang W-D (2012) Lineariifolianoids A-D, rare unsymmetrical sesquiterpenoid dimers comprised of xanthane and guaiane framework units from *Inula lineariifolia*. RSC Adv 2:1307
168. Fu B, Su B-N, Takaishi Y, Honda G, Ito M, Takeda Y, Kodzhimatov OK, Ashurmetov O (2001) A bis-sesquiterpene and sesquiterpenolides from *Inula macrophylla*. Phytochemistry 58:1121
169. Su B-N, Takaishi Y, Tori M, Takaoka S, Honda G, Itoh M, Takeda Y, Kodzhimatov OK, Ashurmetov O (2000) Macrophyllidimers A and B, two novel sesquiterpene dimers from the bark of *Inula macrophylla*. Tetrahedron Lett 41:1475
170. Jakupovic J, Jia Y, King RM, Bohlmann F (1986) Rudbeckiolide, a dimeric sesquiterpene lactone from *Rudbeckia laciniata*. Liebigs Ann Chem 1986:1474
171. Yuan T, Zhu R-X, Yang S-P, Zhang H, Zhang C-R, Yue J-M (2012) Serratustones A and B representing a new dimerization pattern of two types of sesquiterpenoids from *Chloranthus serratus*. Org Lett 14:3198
172. Nagashima F, Murakami M, Takaoka S, Asakawa Y (2004) New sesquiterpenoids from the New Zealand liverwort *Chiloscyphus subporosus*. Chem Pharm Bull 52:949
173. Cherney EC, Baran PS (2011) Terpenoid-alkaloids: their biosynthetic twist of fate and total synthesis. Isr J Chem 51:391
174. Seigler D (1998) Alkaloids of terpenoid origin (excepting indole alkaloids and ergot alkaloids). In: Seigler D (ed) Plant secondary metabolism. Kluwer, New York, p 668
175. Gao HY, Wu LJ, Muto N, Fuchino H, Nakane T, Shirota O, Sano T, Kuroyanagi M (2008) Beyerane derivatives and a sesquiterpene dimer from *Japanese cypress* (*Chamaecyparis obtusa*). Chem Pharm Bull 56:1030
176. Liermann JC, Schüffler A, Wollinsky B, Birnbacher J, Kolshorn H, Anke T, Opatz T (2010) Hirsutane-type sesquiterpenes with uncommon modifications from three Basidiomycetes. J Org Chem 75:2955
177. Ksebati MB, Schmitz FJ (1988) Sesquiterpene furans and thiosesquiterpenes from the nudibranch *Ceratosoma brevicaudatum*. J Nat Prod 51:857

178. Opatz T, Kolshorn H, Birnbacher J, Schüffler A, Deininger F, Anke T (2007) The creolophins: a family of linear triquinanes from *Creolophus cirrhatus* (Basidiomycete). Eur J Org Chem 2007:5546
179. Fiorentino A, DellaGreca M, D'Abrosca B, Golino A, Pacifico S, Izzo A, Monaco P (2006) Unusual sesquiterpene glucosides from *Amaranthus retroflexus*. Tetrahedron 62:8952
180. Chen Y, Bean MF, Chambers C, Francis T, Huddleston MJ, Offen P, Westley JW, Carté, Timmermann BN (1991) Arrivacins, novel pseudoguaianolide esters with potent angiotensin II binding activity from *Ambrosia psilostachya*. Tetrahedron 47:4869
181. Wang S, Li J, Sun J, Zeng K-W, Cui J-R, Jiang Y, Tu P-F (2013) NO inhibitory guaianolide-derived terpenoids from *Artemisia argyi*. Fitoterapia 85:169
182. Sullivan BW, Faulkner DJ, Okamoto KT, Chen MHM, Clardy J (1986) (6*R*,7*S*)-7-Amino-7,8-dihydro-α-bisabolene, an antimicrobial metabolite from the marine sponge *Halichondria* sp. J Org Chem 51:5134
183. Satitpatipan V, Suwanborirux K (2004) New nitrogenous germacranes from a Thai marine sponge, *Axinyssa* n. sp. J Nat Prod 67:503
184. Ruangrungsi N, Likhitwitayawuid K, Kasiwong S, Lange GL, Decicco CP (1988) Constituents of *Michelia rajaniana*. Two new germacranolide amides. J Nat Prod 51:1220
185. Ruangrungsi N, Rivepiboon A, Lange GL, Lee M, Decicco CP, Picha P, Preechanukool K (1987) Constituents of *Paramichelia baillonii*: a new antitumor germacranolide alkaloid. J Nat Prod 50:891
186. Mahmoud II, Kinghorn AD, Cordell GA, Farnsworth NR (1980) Potential anticancer agents. XVI. Isolation of bicyclofarnesane sesquiterpenoids from *Capsicodendron dinisii*. J Nat Prod 43:365
187. Kuo Y-H, Chen W-C (1994) Chinensiol, a new dimeric himachalane-type sesquiterpene from the root of *Juniperus chinensis* Linn. Chem Pharm Bull 42:2187
188. Harinantenaina L, Takaoka S (2006) Cinnafragrins A-C, dimeric and trimeric drimane sesquiterpenoids from *Cinnamosma fragrans*, and structure revision of capsicodendrin. J Nat Prod 69:1193
189. Zdero C, Ahmed AA, Bohlmann F, Mungai GM (1990) Diterpenes and sesquiterpene xylosides from East African *Conyza* species. Phytochemistry 29:3167
190. Castro V, Murillo R, Klaas CA, Meunier C, Mora G, Pahl HL, Merfort I (2000) Inhibition of the transcription factor NF-κB by sesquiterpene lactones from *Podachaenium eminens*. Planta Med 66:591
191. Song Q, Gomez-Barrios ML, Fronczek FR, Vargas D, Thien LB, Fischer NH (1998) Sesquiterpenes from southern *Magnolia virginiana*. Phytochemistry 47:221
192. Takahashi H, Toyota M, Asakawa Y (1993) Drimane-type sesquiterpenoids from *Cryptoporus volvatus* infected by *Paecilomyces varioti*. Phytochemistry 33:1055
193. Hashimoto T, Tori M, Mizuno Y, Asakawa Y, Fukazawa Y (1989) The superoxide release inhibitors, cryptoporic acids C, D, and E; dimeric drimane sesquiterpenoid ethers of isocitric acid from the fungus *Cryptoporus volvatus*. J Chem Soc Chem Commun 1989:258
194. Asakawa Y, Hashimoto T, Mizuno Y, Tori M, Fukazawa Y (1992) Cryptoporic acids A-G, drimane-type sesquiterpenoid ethers of isocitric acid from the fungus *Cryptoporus volvatus*. Phytochemistry 31:579
195. Meng J, Li Y-Y, Ou Y-X, Song L-F, Lu C-H, Shen Y-M (2011) New sesquiterpenes from *Marasmius cladophyllus*. Mycology 2:30
196. Asakawa Y (2004) Chemosystematics of the Hepaticae. Phytochemistry 65:623
197. Bohlmann F, Gupta RK, Jakupovic J (1982) A further sesquiterpene lactone esterified with a sesquiterpenic acid. Phytochemistry 21:460
198. Zhang Q, Mandi A, Li S, Chen Y, Zhang W, Tian X, Zhang H, Li H, Zhang W, Zhang S, Ju J, Kurtan T, Zhang C (2012) N-N-coupled indolo-sesquiterpene atropo-diastereomers from a marine-derived actinomycete. Eur J Org Chem 2012:5256
199. Ishiyama H, Hashimoto A, Fromont J, Hoshino Y, Mikami Y, Ji K (2005) Halichonadins A-D, new sesquiterpenoids from a sponge *Halichondria* sp. Tetrahedron 61:1101

200. Kozawa S, Ishiyama H, Fromont J, Ji K (2008) Halichonadin E, a dimeric sesquiterpenoid from the sponge *Halichondria* sp. J Nat Prod 71:445
201. Suto S, Tanaka N, Fromont J, Ji K (2011) Halichonadins G-J, new sesquiterpenoids from a sponge *Halichondria* sp. Tetrahedron Lett 52:3470
202. Tanaka N, Suto S, Ishiyama H, Kubota T, Yamano A, Shiro M, Fromont J, Ji K (2012) Halichonadins K and L, new dimeric sesquiterpenoids from a sponge *Halichondria* sp. Org Lett 14:3498
203. Morita Y, Hayashi Y, Sumi Y, Kodaira A, Shibata H (1995) Studies on chemical components of mushrooms, Part VI. Haploporic acid A, a novel dimeric drimane sesquiterpenoid from the basidiomycete *Haploporus odorus*. Biosci Biotechnol Biochem 59:2008
204. Herz W, Gage D, Kumar N (1981) Damsinic acid and ambrosanolides from vegetative *Ambrosia hispida*. Phytochemistry 20:1601
205. Zidorn C, Ellmerer-Müller EP, Stuppner H (1998) Guaian-5,12-olides from *Leontodon hispidus*. Phytochemistry 49:797
206. Bohlmann F, Jakupovic J, Abraham W-R, Zdero C (1981) The first sesquiterpene lactones esterified with a sesquiterpenic acid. Phytochemistry 20:2371
207. Hou C-C, Lin S-J, Cheng J-T, Hsu F-L (2003) Antidiabetic dimeric guianolides and a lignan glycoside from *Lactuca indica*. J Nat Prod 66:625
208. Fu B, Zhu QX, Yang XP, Jia ZJ (2002) A new bisesquiterpene from *Ligularia macrophylla*. Chin Chem Lett 13:249
209. Ortega A, Maldonado RATE (1998) A costic acid guaianyl ester and other constituents of *Podachaenium eminens*. Phytochemistry 49:1085
210. Nozoe S, Agatsuma T, Takahashi A, Ohishi H, In Y, Kusano G (1993) Roseolide A, a novel dimeric drimane sesquiterpenoid from the basidiomycete *Roseoformes subflexibilis*. Tetrahedron Lett 34:2497
211. Liu Y, Nugroho AE, Hirasawa Y, Nakata A, Kaneda T, Uchiyama N, Goda Y, Shirota O, Morita H, Aisa HA (2010) Vernodalidimers A and B, novel orthoester elemanolide dimers from seeds of *Vernonia anthelmintica*. Tetrahedron Lett 51:6584
212. Zhang Z-X, Fei D-Q, Jia Z-J (2008) Virgaurols A-D: novel asymmetric eremophilane dimers from the roots of *Ligularia virgaurea*. Bull Chem Soc Jpn 81:1007
213. Kawabata J, Mizutani J (1992) Studies on the chemical constituents of Chloranthaceae plants. Part 8. Dimeric sesquiterpenoid esters from *Chloranthus serratus*. Phytochemistry 31:1293
214. Ye J, Qin J-J, Su J, Lin S, Huang Y, Jin H-Z, Zhang W-D (2013) Identification and structural characterization of dimeric sesquiterpene lactones in *Inula japonica* Thunb. by high-performance liquid chromatography/electrospray ionization with multi-stage mass spectrometry. Rapid Commun Mass Spectrom 27:2159
215. Alkorta I, Elguero J (2010) Computational NMR spectroscopy. In: Grunenberg J (ed) Computational spectroscopy. Wiley-VCH, Weinheim, p 37
216. Lodewyk MW, Siebert MR, Tantillo DJ (2011) Computational prediction of ^{1}H and ^{13}C chemical shifts: a useful tool for natural product, mechanistic, and synthetic organic chemistry. Chem Rev 112:1839
217. Nicolaou KC, Ortiz A, Zhang H, Dagneau P, Lanver A, Jennings MP, Arseniyadis S, Faraoni R, Lizos DE (2010) Total synthesis and structural revision of vannusals A and B: synthesis of the originally assigned structure of vannusal B. J Am Chem Soc 132:7138
218. Nicolaou KC, Zhang H, Ortiz A, Dagneau P (2008) Total synthesis of the originally assigned structure of vannusal B. Angew Chem Int Ed Engl 47:8605
219. Nicolaou KC, Ortiz A, Zhang H (2009) The true structures of the vannusals, Part 2: Total synthesis and revised structure of vannusal B. Angew Chem Int Ed Engl 48:5648
220. Saielli G, Nicolaou KC, Ortiz A, Zhang H, Bagno A (2011) Addressing the stereochemistry of complex organic molecules by density functional theory-NMR: vannusal B in retrospective. J Am Chem Soc 133:6072

221. Yang S-P, Gao Z-B, Wang F-D, Liao S-G, Chen H-D, Zhang C-R, Hu G-Y, Yue J-M (2007) Chlorahololides A and B, two potent and selective blockers of the potassium channel isolated from *Chloranthus holostegius*. Org Lett 9:903
222. Xu Y-J, Tang C-P, Ke C-Q, Zhang J-B, Weiss H-C, Gesing E-R, Ye Y (2007) Mono- and di-sesquiterpenoids from *Chloranthus spicatus*. J Nat Prod 70:1987
223. Hashimoto T, Irita H, Tanaka M, Takaoka S, Asakawa Y (2000) Pinguisane and dimeric pinguisane-type sesquiterpenoids from the Japanese liverwort *Porella acutifolia* subsp. *tosana*. Phytochemistry 53:593
224. Song Q, Gomez-Barrios ML, Fronczek FR, Vargas D, Thien LB, Fischer NH (1997) Sesquiterpenes from southern *Magnolia virginiana*. Phytochemistry 47:221
225. Ibragimov BT, Talipov SA, Dadabaev BN, Nazarov GB, Aripov TF (1988) X-ray structural investigation of gossypol and its derivatives. X. Unusual inclusion compounds based on gossypol. Khim Prir Soedin:675
226. Kim S-Y, Kashiwada Y, Kawazoe K, Murakami K, Sun H-D, Li S-L, Takaishi Y (2009) Spicachlorantins C-F, hydroperoxy dimeric sesquiterpenes from the roots of *Chloranthus spicatus*. Tetrahedron Lett 50:6032
227. Kim S-Y, Kashiwada Y, Kawazoe K, Murakami K, Sun H-D, Li S-L, Takaishi Y (2009) Spicachlorantins A and B, new dimeric sesquiterpenes from the roots of *Chloranthus spicatus*. Phytochem Lett 2:110
228. Yang S-P, Gao Z-B, Wu Y, Hu G-Y, Yue J-M (2008) Chlorahololides C-F: a new class of potent and selective potassium channel blockers from *Chloranthus holostegius*. Tetrahedron 64:2027
229. Di Bari L, Pescitelli G (2010) Electronic circular dichroism. In: Grunenberg J (ed) Computational spectroscopy. Wiley-VCH, Weinheim, p 241
230. Kong L-Y, Wang P (2013) Determination of the absolute configuration of natural products. Chin J Nat Med 11:193
231. Kusano G, Abe M, Koike Y, Uchida M, Nozoe S, Taira Z (1991) Studies on the constituents of *Chloranthus* spp. Further studies on the constituents of *Chloranthus japonicus*. Yakugaku Zasshi 111:756
232. Li C, Dian L, Zhang W, Lei X (2012) Biomimetic syntheses of (−)-gochnatiolides A-C and (−)-ainsliadimer B. J Am Chem Soc 134:12414
233. Chen QF, Liu ZP, Wang FP (2011) Natural sesquiterpenoids as cytotoxic anticancer agents. Mini Rev Med Chem 11:1153
234. Nie L-Y, Qin J-J, Huang Y, Yan L, Liu Y-B, Pan Y-X, Jin H-Z, Zhang W-D (2010) Sesquiterpenoids from *Inula lineariifolia* inhibit nitric oxide production. J Nat Prod 73:1117
235. Sultana N, Saeed Saify Z (2012) Naturally occurring and synthetic agents as potential anti-inflammatory and immunomodulants. Antiinflamm Antiallergy Agents Med Chem 11:3
236. Jung M, Lee K, Kim H, Park M (2004) Recent advances in artemisinin and its derivatives as antimalarial and antitumor agents. Curr Med Chem 11:1265
237. Lai H, Singh N, Sasaki T (2013) Development of artemisinin compounds for cancer treatment. Invest New Drugs 31:230
238. He R, Mott BT, Rosenthal AS, Genna DT, Posner GH, Arav-Boger R (2011) An artemisinin-derived dimer has highly potent anti-cytomegalovirus (CMV) and anti-cancer activities. PLoS One 6:e24334
239. Singh NP, Lai HC, Park JS, Gerhardt TE, Kim BJ, Wang S, Sasaki T (2011) Effects of artemisinin dimers on rat breast cancer cells *in vitro* and *in vivo*. Anticancer Res 31:4111
240. Chaturvedi D, Goswami A, Pratim Saikia P, Barua NC, Rao PG (2010) Artemisinin and its derivatives: a novel class of anti-malarial and anti-cancer agents. Chem Soc Rev 39:435
241. Njuguna NM, Ongarora DSB, Chibale K (2012) Artemisinin derivatives: a patent review (2006—present). Expert Opin Ther Pat 22:1179
242. Hall IH, Lee KH, Imakura Y, Sims D (1983) Antitumor agents. LXIII: The effects of microlenin on nucleic acid and protein syntheses of Ehrlich ascites cells. J Pharm Sci 72:1008
243. Tamaki M, Sadhu SK, Ohtsuki T, Toume K, Koyano T, Kowithayakorn T, Hayashi M, Komiyama K, Ishibashi M (2007) Parviflorene J, a cytotoxic sesquiterpene dimer with a new rearranged skeleton from *Curcuma parviflora*. Heterocycles 72:649

244. Ohtsuki T, Tamaki M, Toume K, Ishibashi M (2008) A novel sesquiterpenoid dimer parviflorene F induces apoptosis by up-regulating the expression of TRAIL-R2 and a caspase-dependent mechanism. Bioorg Med Chem 16:1756
245. Kwon OE, Lee HS, Lee SW, Bae K, Kim K, Hayashi M, Rho M-C, Kim Y-K (2006) Dimeric sesquiterpenoids isolated from *Chloranthus japonicus* inhibited the expression of cell adhesion molecules. J Ethnopharmacol 104:270
246. Hilmi F, Gertsch J, Bremner P, Valovic S, Heinrich M, Sticher O, Heilmann J (2003) Cytotoxic versus anti-inflammatory effects in HeLa, Jurkat T and human peripheral blood cells caused by guaianolide-type sesquiterpene lactones. Bioorg Med Chem 11:3659
247. Narisawa T, Fukaura Y, Kotanagi H, Asakawa Y (1992) Inhibitory effect of cryptoporic acid E, a product from fungus *Cryptoporus volvatus*, on colon carcinogenesis induced with *N*-methyl-*N*-nitrosourea in rats and with 1,2-dimethylhydrazine in mice. Jpn J Cancer Res 83:830
248. Hashimoto T, Asakawa Y (1998) Biologically active substances of Japanese inedible mushrooms. Heterocycles 47:1067
249. Matsuda H, Yoshida K, Miyagawa K, Nemoto Y, Asao Y, Yoshikawa M (2006) *Nuphar* alkaloids with immediately apoptosis-inducing activity from *Nuphar pumilum* and their structural requirements for the activity. Bioorg Med Chem Lett 16:1567
250. Matsuda H, Morikawa T, Oda M, Asao Y, Yoshikawa M (2003) Potent anti-metastatic activity of dimeric sesquiterpene thioalkaloids from the rhizome of *Nuphar pumilum*. Bioorg Med Chem Lett 13:4445
251. Wang Q, Kuang H, Su Y, Sun Y, Feng J, Guo R, Chan K (2013) Naturally derived anti-inflammatory compounds from Chinese medicinal plants. J Ethnopharmacol 146:9
252. Hu Z, Qin J, Zhang H, Wang D, Hua Y, Ding J, Shan L, Jin H, Zhang J, Zhang W (2012) Japonicone A antagonizes the activity of TNF-α by directly targeting this cytokine and selectively disrupting its interaction with TNF receptor-1. Biochem Pharmacol 84:1482
253. Yamahara J, Shimoda H, Matsuda H, Yoshikawa M (1996) Potent immunosuppressive principles, dimeric sesquiterpene thioalkaloids, isolated from *Nupharis Rhizoma*, the rhizome of *Nuphar pumilum* (Nymphaeaceae): structure-requirement of *Nuphar*-alkaloid for immunosuppressive activity. Biol Pharm Bull 19:1241
254. Matsuda H, Shimoda H, Yoshikawa M (2001) Dimeric sesquiterpene thioalkaloids with potent immunosuppressive activity from the rhizome of *Nuphar pumilum*. Structural requirements of *Nuphar* alkaloids for immunosuppressive activity. Bioorg Med Chem 9:1031
255. Shieh C-C, Coghlan M, Sullivan JP, Gopalakrishnan M (2000) Potassium channels: molecular defects, diseases, and therapeutic opportunities. Pharmacol Rev 52:557
256. Wulff H, Castle NA, Pardo LA (2009) Voltage-gated potassium channels as therapeutic targets. Nat Rev Drug Discov 8:982
257. Pu H-L, Wang Z-L, Huang Q-J, Xu S-B, Lin Y-C, Wu X-Y (2000) Effects of biatractylolide on isolated guinea pig myocardium. Zhongguo Yaolixue Tongbao (Chin Pharm Bull) 16:60
258. Lee YM, Moon JS, Yun B-S, Park KD, Choi GJ, Kim J-C, Lee SH, Kim SU (2009) Antifungal activity of CHE-23C, a dimeric sesquiterpene from *Chloranthus henryi*. J Agric Food Chem 57:5750
259. Yang Y, Cao Y-L, Liu H-Y, Yan H, Guo Y (2012) Shizukaol F: a new structural type inhibitor of HIV-1 reverse transcriptase RNase H. Yaoxue Xuebao 47:1011
260. Gille G, Pabst T, Janetzky B, Bringmann G, Reichmann H, Rausch WD (2000) Neurotrophic effects of simplified mastigophorene analogs on mesencephalic dopaminergic cells in primary culture. Drug Dev Res 50:153
261. Xing F, Wang Z-L, Lin Y-C, Zhou Y, Liu Y-Z, Yang H-Z (2009) Effects of biatractylolide on the AD rats induced by $A\beta_{1-40}$. Zhongguo Yaolixue Tongbao (Chin Pharm Bull) 25:949
262. Liu Y, Liao C-M (2006) Effects of biatractylolide on the AD rats induced by $AlCl_3$. J Hunan Norm Univ 3:25
263. Oikawa H, Tokiwano T (2004) Enzymatic catalysis of the Diels-Alder reaction in the biosynthesis of natural products. Nat Prod Rep 21:321
264. Vrettou M, Gray AA, Brewer ARE, Barrett AGM (2007) Strategies for the synthesis of C_2 symmetric natural products—a review. Tetrahedron 63:1487

265. Kawabata J, Fukushi E, Mizutani J (1993) Symmetric sesquiterpene dimer from *Chloranthus serratus*. Phytochemistry 32:1347
266. Shiina J, Oikawa M, Nakamura K, Obata R, Nishiyama S (2007) Synthesis of pinguisane-type sesquiterpenoids acutifolone A, pinguisenol, and bisacutifolones by a Diels-Alder dimerization reaction. Eur J Org Chem 2007:5190
267. Lu Y-S, Peng X-S (2011) A concise construction of the chlorahololide heptacyclic core. Org Lett 13:2940
268. Li C, Yu X, Lei X (2010) A biomimetic total synthesis of (+)-ainsliadimer A. Org Lett 12:4284
269. Berube A, Drutu I, Wood JL (2006) Progress toward the total synthesis of bacchopetiolone: application of a tandem aromatic oxidation/Diels-Alder reaction. Org Lett 8:5421
270. Gagnepain J, Castet F, Quideau S (2007) Total synthesis of (+)-aquaticol by biomimetic phenol dearomatization: double diastereofacial differentiation in the Diels–Alder dimerization of orthoquinols with a C_2-symmetric transition state. Angew Chem Int Ed Engl 46:1533
271. Bringmann G, Pabst T, Rycroft DS, Connolly JD (1999) First synthesis of mastigophorenes A and B, by biomimetic oxidative coupling of herbertenediol. Tetrahedron Lett 40:483
272. Degnan AP, Meyers AI (1999) Total syntheses of (−)-herbertenediol, (−)-mastigophorene A, and (−)-mastigophorene B. Combined utility of chiral bicyclic lactams and chiral aryl oxazolines. J Am Chem Soc 121:2762
273. Bringmann G, Pabst T, Henschel P, Kraus J, Peters K, Peters E-M, Rycroft DS, Connolly JD (2000) Nondynamic and dynamic kinetic resolution of lactones with stereogenic centers and axes: stereoselective total synthesis of herbertenediol and mastigophorenes A and B. J Am Chem Soc 122:9127
274. Bringmann G, Hinrichs J, Pabst T, Henschel P, Peters K, Peters E-M (2001) From dynamic to non-dynamic kinetic resolution of lactone-bridged biaryls: Synthesis of mastigophorene B. Synthesis 2001:155
275. Srikrishna A, Rao MS (2001) Formal total synthesis of (±)-herbertenediol and (±)-mastigophorenes A and B. Tetrahedron Lett 42:5781
276. Bringmann G, Pabst T, Henschel P, Michel M (2001) First total synthesis of the mastigophorenes C and D and of simplified unnatural analogs. Tetrahedron 57:1269
277. Yue G, Yang L, Yuan C, Du B, Liu B (2012) Total syntheses of lindenane-type sesquiterpenoids: (±)-chloranthalactones A, B, F, (±)-9-hydroxy heterogorgiolide, and (±)-shizukanolide E. Tetrahedron 68:9624
278. Bagal SK, Adlington RM, Baldwin JE, Marquez R, Cowley A (2003) Biomimetic synthesis of biatractylolide and biepiasterolide. Org Lett 5:3049
279. Bagal SK, Adlington RM, Baldwin JE, Marquez R (2004) Dimerization of butenolide structures. A biomimetic approach to the dimeric sesquiterpene lactones (±)-biatractylolide and (±)-biepiasterolide. J Org Chem 69:9100
280. Bagal SK, Adlington RM, Marquez R, Cowley AR, Baldwin JE (2003) Studies towards the biomimetic synthesis of bisesquiterpene lactones. Tetrahedron Lett 44:4993
281. Goswami A, Saikia PP, Saikia B, Barua NC (2011) Dinitroaliphatics as linkers: application in the synthesis of novel artemisinin carba-dimer. Mol Divers 15:707
282. Harinantenaina L, Noma Y, Asakawa Y (2005) *Penicillium sclerotiorum* catalyzes the conversion of herbertenediol into its dimers: mastigophorenes A and B. Chem Pharm Bull 53:256

Shang-Gao Liao was born in Guizhou, People's Republic of China, in 1972. After receiving his B.S. degree in Chemistry from Beijing Normal University in 1995, he moved to Guiyang Medical College (presently Guizhou Medical University) as an instructor in Chemistry. In 2002, he joined Professor Jian-Min Yue's group as a postgraduate and graduated in 2007 with a Ph.D. degree in Medicinal Chemistry from Shanghai Institute of Materia Medica, CAS. He then returned to the School of Pharmacy, Guiyang Medical College and worked as a Lecturer (2007–2008), Associate Professor (2008–2013), and Professor (2013–present). He is also a Senior Research Scientist at Guizhou Provincial Key Laboratory of Pharmaceutics, and Engineering Research Center for the Development and Application of Ethnic Medicine and TCM (Ministry of Education) (2011–present). He has published over 50 scientific papers in the field of natural products chemistry, pharmaceutical analysis, and pharmacology. His research interests include the isolation and structure elucidation of secondary metabolites of plants used in the Traditional Chinese Medicine system and local Ethnic Medicines, and the investigation of their biological significance.

Jian-Min Yue is a full-time Professor in Medicinal Chemistry at SIMM, CAS. He received his B.Sc. degree in Chemistry in 1984 from Lanzhou University, where he completed his Ph.D. degree in Organic Chemistry in 1990. After his postdoctoral research at Kunming Institute of Botany (KIB), CAS (1991–1993), and at the University of Bristol (1993–1994), he returned to KIB as an Associate Professor (1994–1996). He joined the staff of the Joint Laboratory of Unilever Research and Shanghai Institute of Organic Chemistry, CAS, as a Senior Research Scientist (1996–1999) in Natural Products Chemistry. He then moved to SIMM where he has remained until the present. During his research career, he has published over 220 scientific papers and book chapters, and has been named as co-inventor on several patents. He currently serves as an associate editor of the journal *Natural Products and Bioprospecting*, and is an editorial board member for the *Journal of Asian Natural Products Research* and the *Journal of Chinese Pharmaceutical Sciences*. His group is actively involved in the isolation, structure determination, and synthesis optimization of natural products, and is currently focusing on natural products with significant bioactivities against infectious diseases, cancer, and neurodegenerative disorders.

Acetogenins from Annonaceae

Chih-Chuang Liaw, Jing-Ru Liou, Tung-Ying Wu, Fang-Rong Chang, and Yang-Chang Wu

Contents

C.-C. Liaw
Department of Marine Biotechnology and Resources, National Sun Yat-sen University, Kaohsiung 804, Taiwan
e-mail: ccliaw@mail.nsysu.edu.tw

J.-R. Liou • T.-Y. Wu • F.-R. Chang
Graduate Institute of Natural Products, Kaohsiung Medical University, Kaohsiung 807, Taiwan
e-mail: liu740122@gmail.com; u96831003@kmu.edu.tw; aaronfrc@kmu.edu.tw

Y.-C. Wu (✉)
School of Pharmacy, College of Pharmacy, China Medical University, Taichung 404, Taiwan
e-mail: yachwu@mail.cmu.edu.tw

A.D. Kinghorn, H. Falk, S. Gibbons, J. Kobayashi (eds.), *Progress in the Chemistry of Organic Natural Products, Vol. 101*, DOI 10.1007/978-3-319-22692-7_2

1 Introduction

Annonaceous acetogenins (AGEs) constitute a series of polyketides found almost exclusively from plants in the family Annonaceae, with some of their species of origin being important economic crops in Asia and North and South America. The study of AGEs was initiated as a result of the first report on the bioactive uvaricin (**1**) in 1982, from the roots of *Uvaria accuminata* Oliv. by Jolad et al., which exhibited excellent bioactivity in the P-388 lymphocytic leukemia system in mice [1]. Since then, numerous AGEs have been isolated and identified from various parts of annonaceous plants, especially the seeds, by virtue of advances in separation technology. These initial results have promoted further work on the structural elucidation and classification of AGEs, as well as their biogenetic hypotheses. Annonaceous AGEs are a unique compound class of C_{35} or C_{37} secondary metabolites, derived from the polyketide pathway, which include structurally a γ-lactone ring along with several oxygenated functionalities. For example, hydroxy group, ketone, epoxide, tetrahydrofuran (THF), and tetahydropyran (THP) moieties, and even double and triple bonds are structural features encountered among the AGEs. Annonaceous acetogenins have been found to exhibit a broad range of biological properties, such as antineoplastic, antiparasitic, cytotoxic, immunosuppressive, neurotoxic, and pesticidal effects. Among the broad array of biological properties documented in the biomedical literature for the AGEs, their cytotoxic and antitumor effects and the underlying mechanisms for such effects have received the most attention.

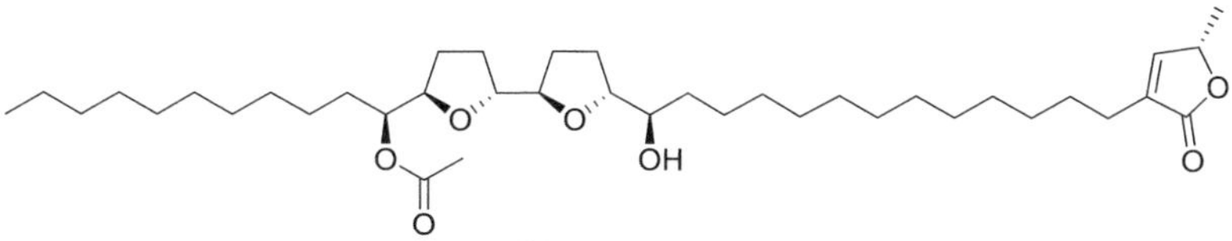

1 (uvaricin)

In 1997, Cavé et al. wrote a thorough contribution on AGEs in this book series, covering the classification, extraction, isolation, structure elucidation, biogenetic hypotheses, syntheses, and biological activities of these type of compounds. Altogether, they collaboratively reported 242 AGEs and described relevant studies on their synthesis and bioactivities [2].

An overall advancement in experimental techniques has led to many worldwide efforts focusing on the isolation and structural identification of new bioactive AGEs. Most importantly, organic chemists have overcome the challenges of meeting the total and rapid synthesis of AGEs with multiple stereocenters during the past 15 years. Moreover, growing interest in investigating the mechanisms of biochemical action of AGEs has been triggered by recent advances in understanding the processes involved in tumor cell death. Members of this class of natural compounds are considered as possible candidates for future anticancer drugs. Bioactivity and mechanism of action studies on annonaceous AGEs have both focused on their potent cytotoxicity against cancer cells and the inhibition of mitochondrial respiratory chain complex I. However, recent studies have reported the relation between this type of compound and sporadic neurodegenerative tau pathologies in those humans who have ingested annonaceous plants containing AGEs [3]. The purpose of this contribution is to give a short historical introduction as well as a description of current studies on AGEs and their analogues. As a result, 203 AGEs dating from 1997 to 2014 were isolated and characterized. The bioactivities of each AGE are highlighted as well. Recent investigations on the mechanisms of action of the pesticidal and antitumor effects of AGEs are reviewed within each individual section. In addition, modified analogues of AGEs and synthesized AGE mimics that were mentioned to play an important role in verifying hypotheses on the modes of action of AGEs are covered.

Annonaceous acetogenins, derivatives of the polyketide pathway, contain 35 or 37 carbons. Biogenetically, they appear to originate from polyhydroxy C_{32} or C_{34} fatty acids to which a 2-propanol unit is added to form a methylated α,β-unsaturated γ-lactone. During the past two decades, advances in chromatographic methodology, such as repeated open column chromatography and high-performance liquid chromatography (HPLC) [4–7], have made the ready and efficient separation of AGEs with only minor structural differences possible. Based on the isolation and characterization of different acetogenins, their general structural features can be divided into various classes dependent on the nature of the γ-lactone rings, such as an α,β-unsaturated γ-lactone ring (normal form) or a ketolactone (isoform), in addition to the oxygen-bearing moieties evident [2, 8]. However, Cavé et al. suspected that acetogenins with terminal ketolactones (isoforms) are artifacts of the translactonization of 4-hydroxy-AGEs. To validate this suspicion, they performed the extraction and characterization of the initial AGEs from fresh crude materials under the effects of alkalis, other basic media, and alcohols. These reagents affected the kinetics of the translactonization [9], which was later supported by the work of Figadère and colleagues describing how 4-hydroxylated AGEs led to iso-derivatives under basic conditions [10].

In summary, the common features on the structures of AGEs are a terminal γ-lactone ring and a terminal aliphatic side chain connecting to some hydrophilic functional groups, such as one to three THF rings and several hydroxy groups. In 1998, Cavé et al. discussed the previously mentioned features in terms of two major structural factors, the terminal γ-lactone ring and the substituents on the long aliphatic chain [8]. Under such a classification system, AGEs were divided into ten subtypes: (1) AGEs without THF rings: linear AGEs; (2) AGEs without THF rings: epoxy-AGEs; (3) mono-THF α,α′-dihydroxylated γ-lactone AGEs; (4) mono-THF α-hydroxylated γ-lactone AGEs; (5) mono-THF AGEs with various lactone moieties; (6) adjacent bis-THF α,α′-dihydroxylated γ-lactone AGEs; (7) adjacent bis-THF α-hydroxylated γ-lactone AGEs; (8) non-adjacent bis-THF γ-lactone AGEs; (9) saturated lactone bis-THF AGEs; (10) miscellaneous AGEs.

2 Annonaceous Acetogenins in the Plant Kingdom

Since their first dicovery over 30 years ago, phytochemists have focused their studies on AGEs in the plant family Annonaceae. The isolation of AGEs has been reported from 15 genera of this family, including *Annona* (19 species), *Asimina* (3), *Artabotrys* (2), *Cananga* (2), *Dasymaschalon* (1), *Disepalum* (1), *Fissistigma* (2), *Goniothalamus* (5), *Mitrephora* (2), *Ophrypetalum* (1), *Polyalthia* (6), *Rollinia* (7), *Saccopetalum* (1), *Uvaria* (10), and *Xylopia* (1), up to the present (see Table 1). Among all the reported isolations of AGEs, a series of linear acetylenic C_{25}-AGEs has been found from species of four genera (i.e. *Cananga*, *Polyalthia*, *Mitrephora*, and *Saccopetalum*). Surprisingly, a paper in 2008 reported that a new AGE was isolated from the root of an *Ampelocissus* species collected in the Philippines, which belongs to the family Vitaceae [34]. However, this remains the first and only example suggesting that this type of compound could be found from plants other than in the family Annonaceae (see Table 1).

Table 1 Distribution of acetogenins in the plant kingdom

Plant source	Genus	Species	Refs.
Annonaceae	*Annona*	*A. atemoya (A. cherimola* x *A. squamosa)*	
		A. bullata	
		A. cherimola Mill.	
		A. coriacea	
		A. crassiflora	
		A. densicoma	
		A. glabra L.	
		A. glauca	
		A. jahnii	[11, 12]
		A. montana Macf.	
		A. muricata L.	
		A. nutans	[13]
		A. purpurea	
		A. reticulata L.	
		A. salzmanii	
		A. senegalensis	
		A. spinescens	
		A. spraguei	
		A. squamosa L.	
	Asimina	*A. longifolia*	
		A. parviflora	
		A. triloba	
	Artabotrys	*A. hexaptalus* (L.f.) Bhandari	[14]
		A. uncinatus (Lam.) Merr.	
	Cananga	*C. odorata* (Lam.) Hook.f & Thomas	
		C. latifolia (Hook.f. & Thomson) Finet & Gagnep	[15]
	Dasymaschalon	*D. sootepense*	[16]
	Disepalum	*D. anomalum*	[17]
		D. plagioneurum	[18]
	Fissistigma	*F. glaucescens* (Hance) Merr.[a]	
		F. oldhami (Hemsl.) Merr.[a]	
	Goniothalamus	*G. amuyon* (Blanco) Merr.[a]	
		G. donnaiensis	[19–21]
		G. giganteus	
		G. howii	
		G. undulatus Ridl.	[22]
	Mitrephora	*M. glabra*	[23]
		M. maingayi	[24]
	Ophrypetalum	*O. odoratum*	[25]
	Polyalthia	*P. debilis*	[26]
		P. longifolia Benth. et Hook.f	
		P. longifolia Benth. et Hook.f 'pendula'	
		P. plagioneura	
		P. suberosa Hook.f	
		P. liukiuensis Hatusima[a]	

(continued)

Table 1 (continued)

Plant source	Genus	Species	Refs.
	Rollinia	*R. emarginata*	[27]
		R. membranacea	
		R. mucosa Baill.	
		R. papilionella	
		R. sericea	
		R. sylvatica	
		R. ulei	
	Saccopetalum	*S. prolificum*	[28]
	Uvaria	*U. acuminata*	
		U. boniana	[29]
		U. calamistrata	[30]
		U. grandiflora	
		U. hookeri	
		U. microcarpa	
		U. narum	
		U. pauci-ovulata	[31]
		U. rufa Bl.	
		U. tonkinensis	[32]
	Xylopia	*X. aromatica*	
		X. emarginata	[33]
Vitaceae	*Ampelocissus*	*A.* sp.	[34]

[a]Annonaceous plants native to Taiwan

3 Classification of Annonaceous Acetogenins (Since 1997 Until the End of 2014)

Studies on annonaceous acetogenins (AGEs) after 1997 have focused on efficient compound identification by hyphenated chromatographic techniques and other spectroscopic methods. Although normal- and reversed-phase HPLC are powerful tools for the isolation of natural products, limitations such as the allowed amount of sample to be purified per unit time, the solvent cost, and the size of the columns, still remain. Therefore, searching for new approaches to facilitate chromatographic work is crucial. McLaughlin et al. used countercurrent chromatography (CCC) to isolate four AGEs, (2,4-*cis* and *trans*)-9-hydroxyasimicinone (**2**), (2,4-*cis* and *trans*)-squamoxinone B (**3**), (2,4-*cis* and *trans*) squamoxinone C (**4**), and isoannoreticuin (**5**), from the bark of *A. squamosa* [35]. Cavé et al. also applied high-speed countercurrent chromatography (HSCCC) to the separation of AGEs from *A. atemoya* to give two major AGEs, squamocin (**6**), bullatacin (rolliniastatin-2) (**7**), asimicin (**8**), and isodesacetyluvaricin (**9**), as well as four other known analogues **10–13** [36]. Moreover, our group introduced the recycle-HPLC system for isolating of gigantetronenin (**14**) and montalicin J (**15**) [37] in an attempt to separate mixtures of AGEs that cannot be easily purified by regular HPLC methods (unpublished data, see Fig. 1).

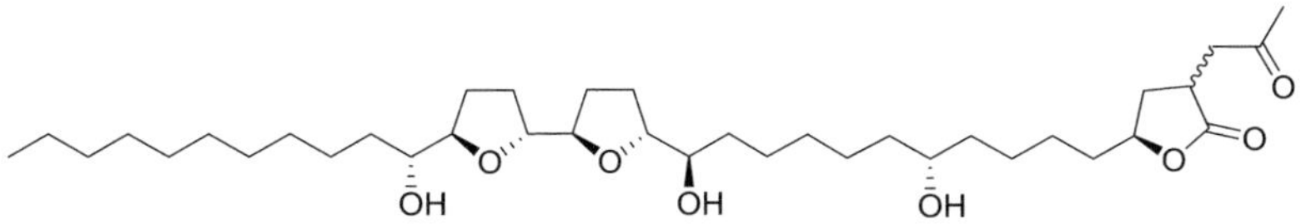

2 ((2,4-*cis* and *trans*)-9-hydroxyasimicinone)

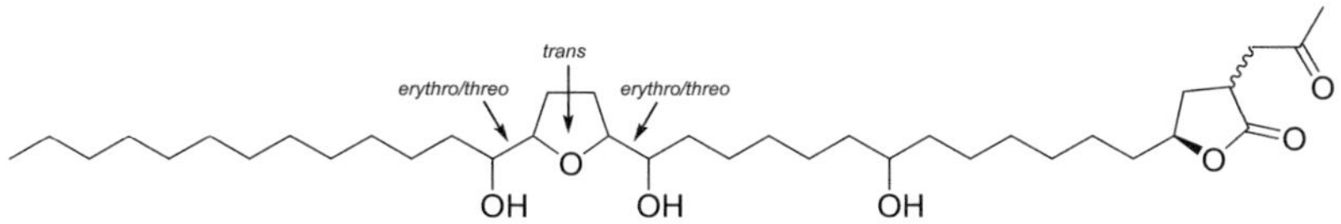

3 ((2,4-*cis* and *trans*)-squamoxinone B)

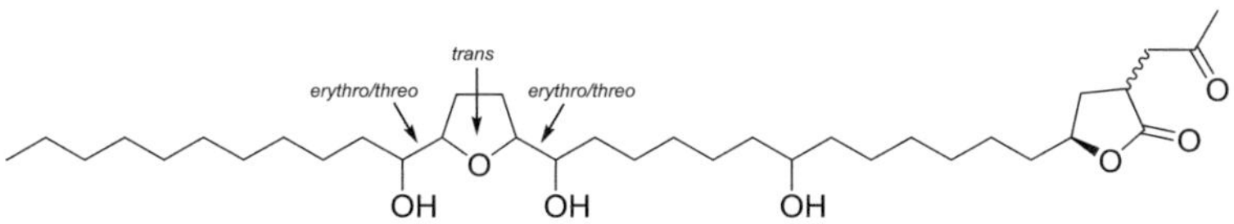

4 ((2,4-*cis* and *trans*)-squamoxinone C)

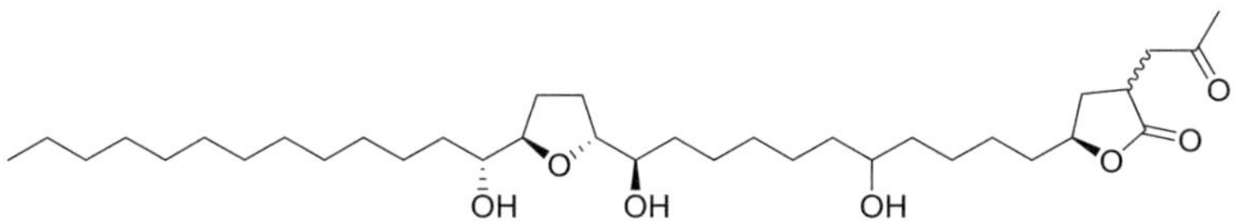

5 (isoannoreticuin)

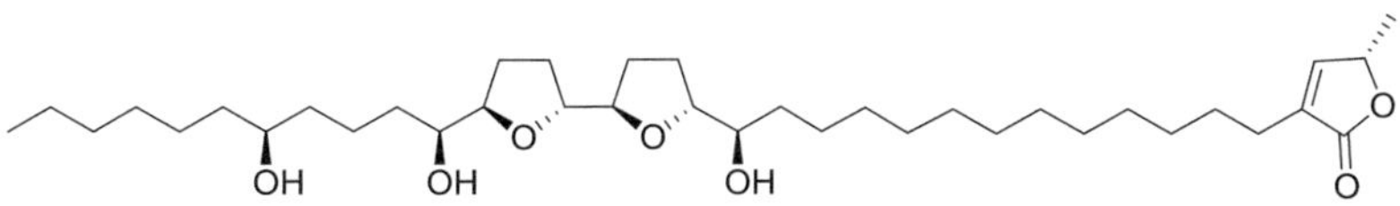

6 (squamocin)

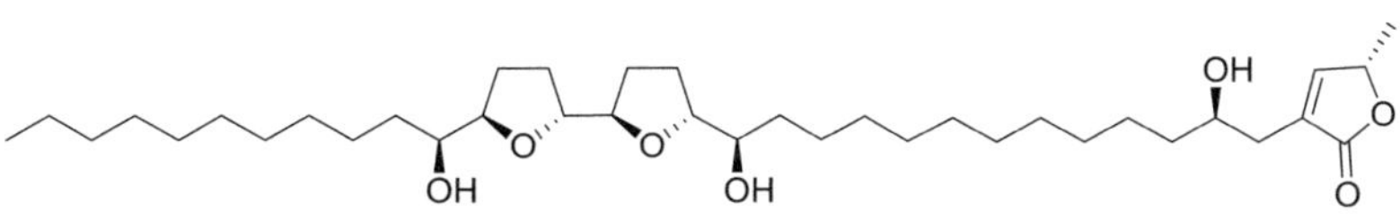

7 (bullatacin = rolliniastatin-2)

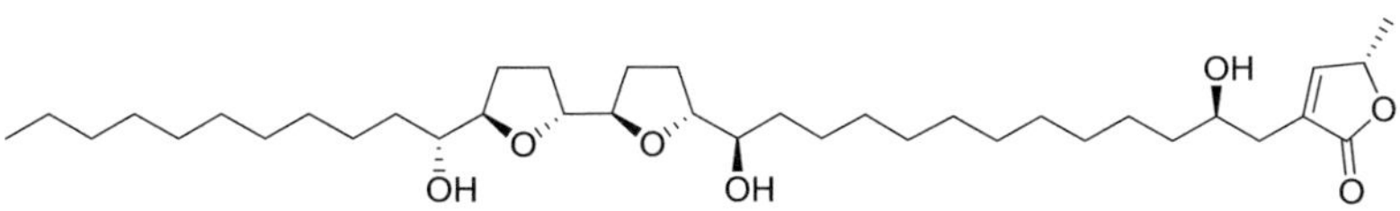

8 (asimicin)

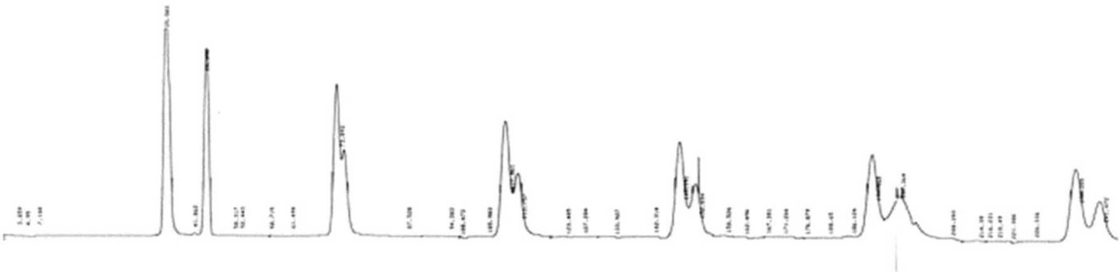

Fig. 1 HPLC separation profiles of gigantetronenin (**14**) and montalicin J (**15**) from the JAI recycling HPLC system

With the idea of developing a convenient yet reliable spectroscopic methodology for determining the stereochemistry of AGEs, Gawronski et al. established the absolute configuration of the γ-lactone ring moiety by analyzing the CD spectra of butenolides [38]. Cavé et al. not only modified the Mosher ester method, but they also determined the stereochemistry of asimicin (**8**) using the long-range anisotropic effect of 2-NMA (naphthylmethoxyacetic acid) [39].

Another key tool for determining the structures of AGEs is mass spectrometry (MS). Electron-impact mass spectrometry (EI-MS) has been a preferred technique for determining the location of AGE tetrahydrofuran rings and functional groups (i.e. hydroxy, ketone, acetoxy, and double bond) along the hydrocarbon chain. Derivatized AGEs, such as TMS- and acetyl derivatives, are helpful in the elucidation of these structures. In addition, the direct-inlet-probe technique (DIP) and a lower volatile energy setting (e.g. 30 eV) have been suggested for use with EI-MS scanning, as AGEs can decompose readily under thermal conditions. Such EI-MS fragmentations are quite useful to determine the planar structures of AGEs despite being a seemingly old-fashioned method. The structure of squamocin (**6**) from *A. squamosa* was characterized by a combination of chemical derivatization and precursor-ion scanning mass spectrometry. The lactone portion of squamocin (**6**) was modified with *N*,*N*-dimethylethylene-diamine in the vapor phase to afford a strong positive charge at one end of the skeleton [40]. In 1997, Wu et al. (Xeno-Biotic Laboratories, Inc.) in cooperation with the McLaughlin group, analyzed AGEs from *R. mucosa* that were subjected to liquid chromatography/mass spectrometry (LC/MS) with ionization source-atmospheric pressure in-source collision-induced dissociation (APICID). They were able to detect the presence of 40 known AGEs along with four new AGEs of diverse structures, from the bioactive crude methanol-soluble fraction of this plant extract [41]. They also observed a unique fragmentation profile for AGEs with a hydroxy group at C-4, which gave a characteristic loss of a terminal γ-lactone (*m*/*z* 112) during ESI-MS scanning [41]. This rapid and straightforward selective ionization procedure also provided a convenient and useful method for identifying AGEs with or without a hydroxy group at C-4 [41].

From a detailed literature survey, it is apparent that the structures of more than 200 AGEs have been published since 1996 until the end of 2014. Among those published, 113 AGEs were included in McLaughlin's review from the year of 1999 [42] yet excluded from chapter written by Cavé's group in 1997 [2]. In addition, AGEs with many previously unprecedented skeletons were isolated and elucidated from plants in the Annonaceae. In view of this, the criteria of Cavé et al. have been modified and the structural characteristics to classify the AGEs have been simplified [8]. Depending on the substituents along the aliphatic chain, AGEs have been classified into four groups, namely, (1) linear and epoxy AGEs, including those without any THF ring but with substitution by an epoxide ring and/or a double bond; (2) mono-THF AGEs, including AGEs with a mono-THF ring; (3) bis-THF AGEs, including those with adjacent or non-adjacent bis-THF rings; and (4) miscellaneous (see Fig. 2 and formulas **9–15**).

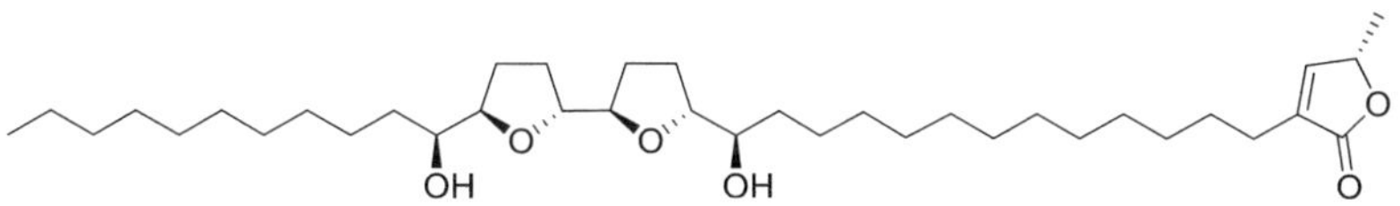

9 (isodesacetyluvaricin)

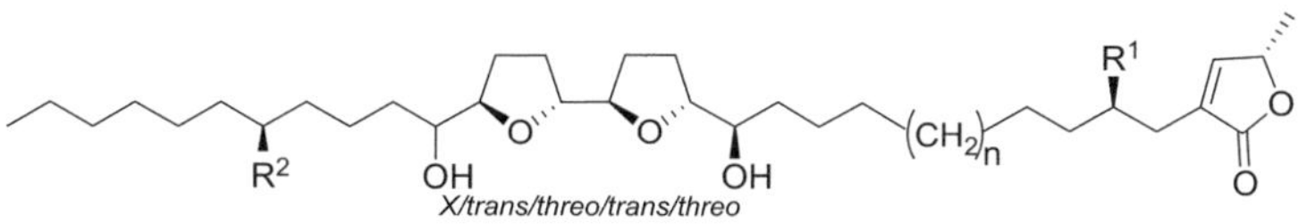

10 (molvizarin) R^1 = OH R^2 = H n = 4 *X* = *erythro*
11 (neoannonin) R^1 = H R^2 = H n = 4 *X* = *erythro*
12 (atemoyin) R^1 = H R^2 = H n = 4 *X* = *threo*
13 (desacetyluvaricin) R^1 = OH R^2 = H n = 6 *X* = *erythro*

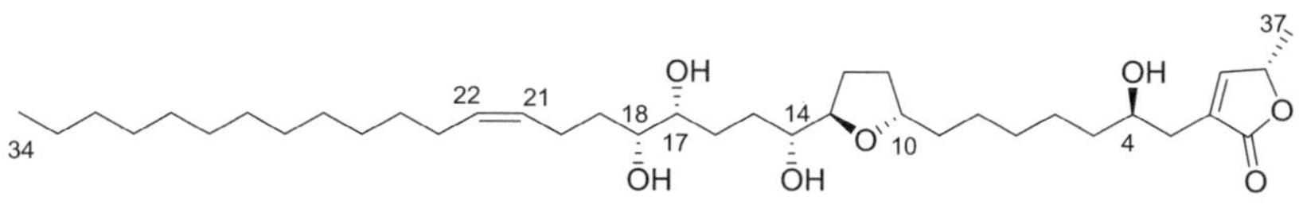

14 (gigantetronenin)

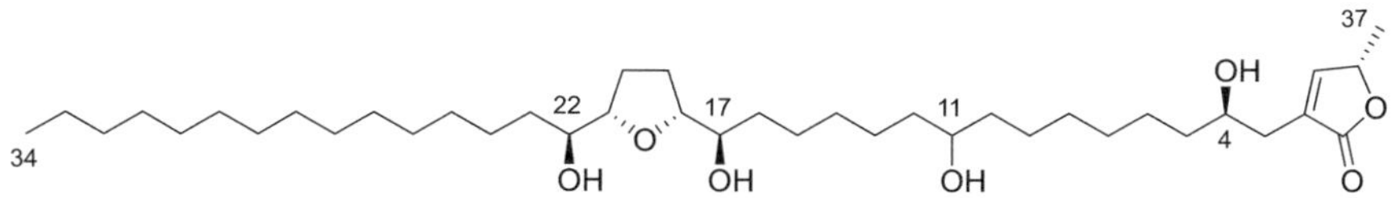

15 (montalicin J)

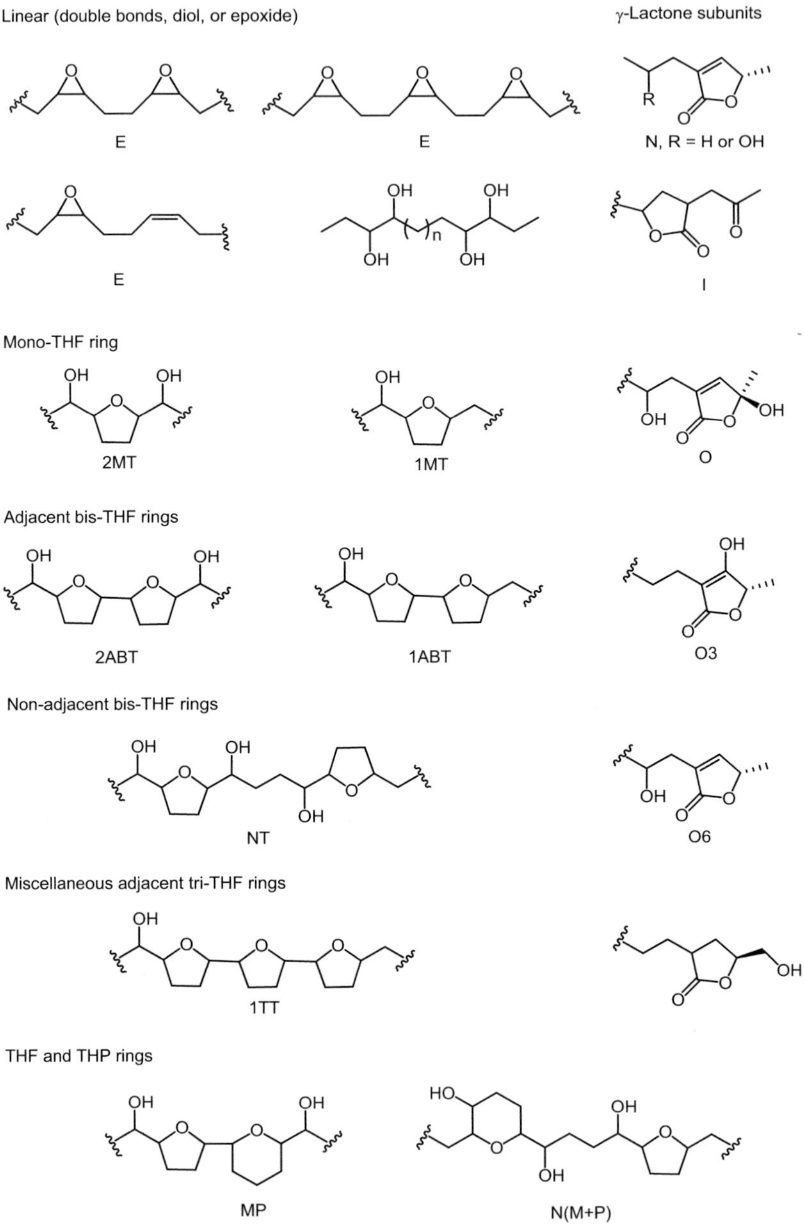

Fig. 2 γ-Lactone and tetrahydrofuran (THF), tetrahydropyran (THP), and other oxygen-bearing subunits in annonaceous acetogenins (AGEs)

3.1 Linear and Epoxy Annonaceous Acetogenins

In general, this structural sub-type of AGEs includes compounds with no THF ring but that have a substituent like an epoxide ring and/or a double bond. Fifty-two new linear and epoxide ring-containing AGEs were isolated from 15 species in nine genera, including *Annona* (22), *Cananga* (9), *Goniothalamus* (7), *Mitrephora* (3), *Polyalthia* (6), *Rollinia* (2), *Saccopetalum* (2), *Uvaria* (2), and *Xylopia* since 1997 (1). The features of this group are double bonds, a diol group, and/or an epoxide ring in place of the THF ring (see Table 2). Notably, AGEs with an epoxide ring of this structural sub-type usually have the terminal γ-lactone ring without a C-4 hydroxy group often suggested by some specific biosynthesis pathways for these compounds. In 2002, the first linear acetylenic C_{25}-AGEs (**16** and **17**) with a new type of saturated γ-hydroxymethyl-γ-lactone terminal ring, but with no oxygenated substituents on its alkyl chain were found in *Saccopetalum prolificum* (Annonaceae) [28]. However, such compounds were later found in *Goniothalamus gardneri* (Annonaceae) [52], *Mitrephora glabra* [23], *Mitrephora maingayi* [24], *Polyalthia debilis* [26], and *Cananga latifolia* [15].

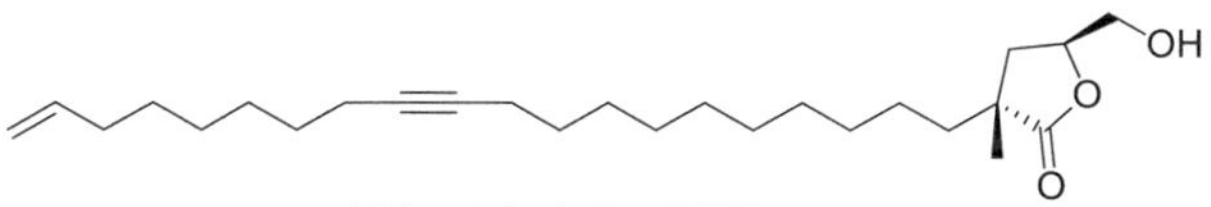

16 (acetylenic C_{25}-AGE)

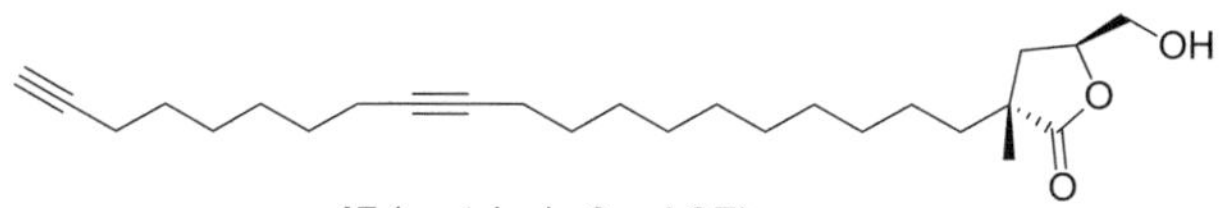

17 (acetylenic C_{25}-AGE)

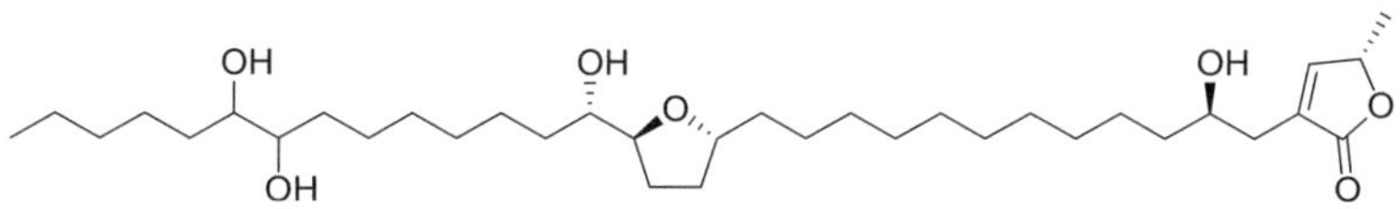

18 (muricin A)

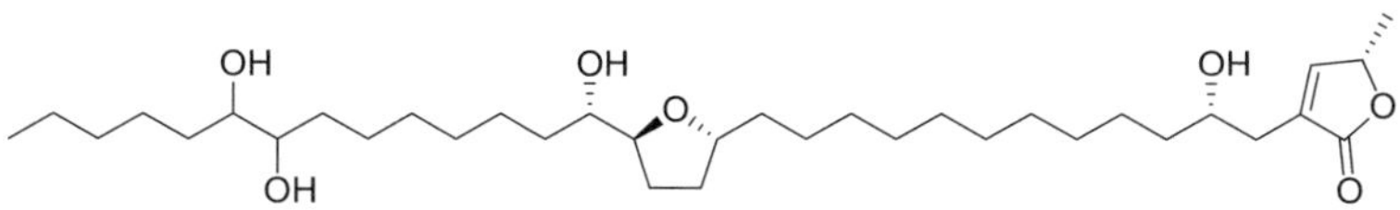

19 (muricin B)

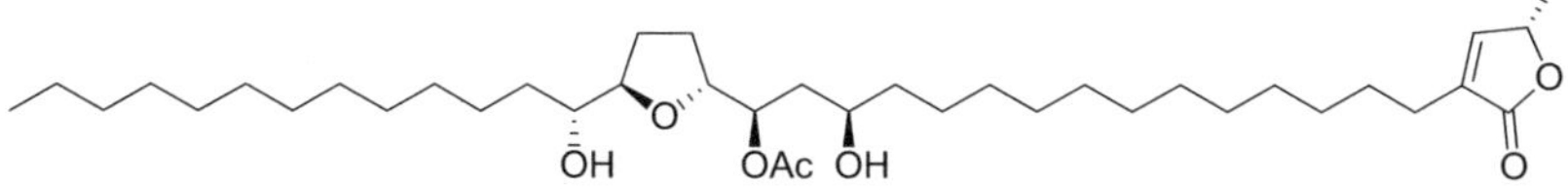

20 (22-epicalimistrin)

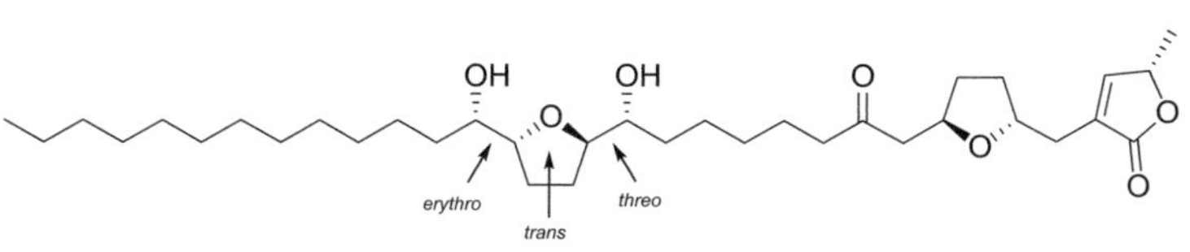

21 (aromin-A)

Table 2 Linear and epoxy AGEs isolated since 1997 until the end of 2014

	Name	No. of C	-OH/epoxide/double bond	Plant	Ref.
1	Annojahnin	37	10=O, (17*R*,18*R*)	*A. jahnii*	[11]
2	Artemoin A	35	23,24	*A. atemoya*	[43]
3	Artemoin B	35	21,22	*A. atemoya*	[43]
4	Artemoin C	35	19,20	*A. atemoya*	[43]
5	Artemoin D	35	17,18	*A. atemoya*	[43]
6	Cananginone A	23	11T 13T 15D 19D	*C. latifolia*	[15]
7	Cananginone B	23	11T 13T 15D	*C. latifolia*	[15]
8	Cananginone C	23	11T 13T 19D	*C. latifolia*	[15]
9	Cananginone D	23	11T 13T	*C. latifolia*	[15]
10	Cananginone E	23	11T 13D 19D	*C. latifolia*	[15]
11	Cananginone F	23	11T 19D	*C. latifolia*	[15]
12	Cananginone G	21	9T 11D 17D	*C. latifolia*	[15]
13	Cananginone H	21	9T 11D	*C. latifolia*	[15]
14	Cananginone I	25	13T 15T 17D 21D	*C. latifolia*	[15]
15	Cohibin A	35	15,16	*A. muricata*	[44]
16	Cohibin B	35	13,14	*A. muricata*	[44]
17	Cohibin C	37	17,18,21	*A. muricata* *A. nutans*	[45]
18	Cohibin D	37	15,16,19	*A. muricata* *A. nutans*	[45]
19	Coronin	37	(13*E*,17*E*),21	*A. muricata*	[46]
20	Debilisone A	25	13T 15T	*P. debilis*	[26]
21	Debilisone B	25	13T 15T	*P. debilis*	[26]
22	Debilisone C	25	13T 15T 17D	*P. debilis*	[26]
23	Debilisone D	25	13T 15T 17D 21D	*P. debilis*	[26]
24	Debilisone E	25	13T 15T 17D 21D	*P. debilis*	[26]
25	Debilisone F	27	15T 17T 19D	*P. debilis*	[26]
26	Diepomuricanin B	35	(15*E*,19*E*)	*R. membranacea*	[47]
27	Dieposabadelin	35	(13*E*,17*E*)	*A. squamosa*	[48]
28	Diepoxyrollin	37	(17*E*,21*E*)	*R. membranacea*	[47]
29	Donbutocin	35	(4*R*),10,15,16	*G. donnaiensis*	[49]
30	Donhepocin	35	(4*R*),10,15,16,19,20	*G. donnaiensis*	[49]
31	34-*epi*-Donhepocin	35	(4*R*),10,15,16,19,20	*G. donnaiensis*	[49]
32	Donhexocin	35	(4*R*),10,15,16,19,20	*G. donnaiensis*	[49]
33	Epomurinin-A	35	(15*E*)	*A. muricata*	[50]
34	Epomurinin-B	35	(13*E*)	*A. muricata*	[50]
35	Gardnerilin A	35	(4*R*),8,15,16,19,20	*G. gardneri*	[51]
36	Gardnerilin B	35	4*R*,10,17,18	*G. gardneri*	[51]
37	Goniothalamusin	25	13T	*G. gardneri*	[52]
38	9,10-Dihydrooropheolide	19	9D 11T 13T 17D	*M. glabra*	[23]
39	Mitregenin	21		*M. maingayi*	[24]
40	(+)-Monhexocin	35	4,9,15,16,19,20	*A. montana*	[53]

(continued)

Table 2 (continued)

	Name	No. of C	-OH/epoxide/double bond	Plant	Ref.
41	(−)-Monhexocin	35	4,9,15,16,19,20	*A. montana*	[53]
42	Montecristin	37	13,14	*A. muricata*	[54]
43	Muricadienin	35	15,19	*A. muricata*	[13]
44	Muridienin-3	37	13,17	*A. muricata*	[13]
45	Muridienin-4	37	17,21	*A. muricata*	[13]
46	Murihexol	35	4,10,15,16,19,20	*A. muricata*	[55]
47	Oropheolide	19	9T 11T 13T 17D	*M. glabra*	[23]
48	Sabadelin	35	13,(17*E*)	*A. muricata*	[56]
49	Saccopetrin A	25	13T 21D	*S. prolificum*	[28]
50	Saccopetrin B	25	13T 21D	*S. prolificum*	[28]
51	Squamocenin[a]	35	(15*E*,19*E*),23	*A. squamosa*	[48]
52	Xymarginatin	35	10=O 15D 19D	*X. emarginata*	[33]

[a]This compound has the same name as one of the mono-THF AGEs [57]

3.2 *Mono-THF Annonaceous Acetogenins, Including Derivatives with a Mono-THF Ring*

The structural features of this type of AGE are one THF ring with one or two flanking hydroxy groups in the long aliphatic chain (see Table 3). There are two subtypes based on the number of flanking hydroxy groups including: (1) the THF ring flanking one hydroxy group and (2) the THF ring flanking two hydroxy groups. Mono-THF acetogenins are indeed the largest single group of these plant secondary metabolites. One hundred and nineteen new mono-THF compounds were isolated from 15 species in seven genera, including *Ampelocissus* (1), *Annona* (91), *Asimina* (6), *Disepalum* (8), *Goniothalamus* (5), *Rollinia* (3), and *Uvaria*, since 1997 (5). In particular, two epimeric AGEs, muricins A (**18**) and B (**19**), were isolated from *A. muricata*, of which the absolute configurations were determined by the modified Mosher ester method [86]. Muricin B (**19**) is the first AGE to possess a hydroxy group with the (*S*)-configuration at C-4 where the typical configuration of this hydroxy group is (*R*). Moreover, 22-epicalimistrin B (**20**) is the first AGE that was isolated from the genus *Ampelocissus* (Vitaceae), which does not belong to the family Annonaceae [34].

Table 3 Mono-THF AGEs isolated since 1997 until the end of 2014

	Name	No. of C	OH	THF/epo	Plant	Ref.
1	Anmontanin A	35	(4*R*),8=O	15,20	*A. montana*	[58]
2	Anmontanin B	35	(4*R*,8*R*),10=O	(17*S*,22*R*)	*A. montana*	[58]
3	Anmontanin C	35	(4*R*),10=O,34	17–22	*A. montana*	[58]
4	Annocherimolin	37	(4*R*),9	(13*R*,18*R*)	*A. cherimola*	[59]
5	Annocatalin	35	4,28,29	19	*A. muricata*	[60]
6	Annocherin	35	(4*R*),7=O	(15*R*,20*R*)	*A. cherimola*	[61]
7	(2,4)-*cis*- and *trans*-Annocherinones	35	7=O,34=O	(15*R*,20*R*)	*A. cherimola*	[61]
8	Annoglacin A	37	(4*R*,12*R*)	(17*R*,22*S*)	*A. glabra*	[62]
9	Annoglacin B	37	(4*R*,12*R*)	(17*R*,22*R*)	*A. glabra*	[62]
10	Annoglaxin	35	(8*R*),12=O, (22*S*)	(15*R*,20*R*)	*A. glabra*	[63]
11	Annomocherin	35	(4*R*),10,23-	(15*R*,20*R*)	*A. cherimola*	[64]
12	Annomolin	35	(4*R*,7*R*,8*R*)	(18*S*)	*A. cherimola*	[59]
13	Annomolon A	35	11=O,34	(15*R*,20*R*)	*A. cherimola*	[65]
14	34-*epi*-Annomolon A	35	11=O,34	(15*R*,20*R*)	*A. cherimola*	[65]
15	Annomolon B	35	(4*R*),11=O,34	(15*R*,20*R*)	*A. cherimola*	[65]
16	34-*epi*-Annomolon B	35	(4*R*),11=O,34	(15*R*,20*R*)	*A. cherimola*	[65]
17	*cis*-Annomontacin	37	4,10	17–22	*A. muricata*	[60]
18	4-Deoxyannomontacin	37	(10*R*)	(17*R*,22*R*)	*G. giganteus*	[66]
19	(2,4-*cis* and *trans*)-Annomontacinone	37	(10*S*)36=O	(17*R*,22*R*)	*G. giganteus*	[66]
20	Annomuricin E	35	4,10,11	15–20	*A. muricata*	[67]
21	*cis*-Annoreticuin	35	4,9	15,20	*A. montana*	[68]
22	4-Deoxyannoreticuin	35	9	15–20	*A. squamosa*	[69]
23	*cis*-4-Deoxyannoreticuin	35	9	15–20	*A. squamosa*	[69]
24	*cis*-Annotemoyin-1	35		17–22	*A. squamosa*	[37]
25	Asitrilobin A	37	(4*R*),10	(17*R* or *S*,22*R* or *S*)	*A. triloba*	[70]
26	Asitrilobin B	35	(4*R*),10	(17*R* or *S*,22*R* or *S*)	*A. triloba*	[70]
27	Asitrilobin C	37	(4*R*,10*R*,15*S*)	(17*R*,22*R*)	*A. triloba*	[71]
28	Asitrilobin D	37	(4*R*,10*R*,17*S*)	(19*R*,24*R*)	*A. triloba*	[71]
29	Asitrocin	35	(4*R*,12*R*)	(15*S*,20*R*)	*A. triloba*	[72]
30	(2,4)-*cis*- and *trans*-Asitrocinones	35	(12*R*),34=O	(15*S*,20*R*)	*A. triloba*	[72]
31	Calamistrin A	37	(15*S*)	(17*R*,22*S*)	*U. calamistrata*	[30]
32	Calamistrin B	39	(15*R*)	17*R*OAc, (22*S*)	*U. calamistrata*	[30]
33	22-*epi*-Calamistrin B	39	15,17-OAc	(22*R*)	*Ampelocissus* sp.	[34]

(continued)

Table 3 (continued)

	Name	No. of C	OH	THF/epo	Plant	Ref.
34	Calamistrin C	37	13	(19*R*,24*R*)	*U. calamistrata*	[73]
35	Calamistrin D	37	13	(19*R*,24*S*)	*U. calamistrata*	[73]
36	Calamistrin E	37	(5*R*),23-	(15*R*,20*R*)	*U. calamistrata*	[73]
37	Coriaheptocin A	35	4,14,16,19,20	7,12	*A. coriacea*	[74]
38	Coriaheptocin B	35	4,14,16,19,20	7,12	*A. coriacea*	[74]
39	*cis*-Corossolone	35	10=O	15,20	*A. muricata*	[60]
40	Dotistenin	35	23	15,20	*A. squamosa*	[48]
41	Glabrencin A	37		13,18	*A. glabra*	[75]
42	Glabrencin B	37		17,22	*A. glabra*	[75]
43	Glacin A	35	(4*R*,12*R*)	(17*R*,22*R*)	*A. glabra*	[76]
44	Glacin B	35	(4*R*,12*R*)	(15*R*,20*S*)	*A. glabra*	[76]
45	Glaucabellin	37	4	17,22	*A. glauca*	[77]
46	Glaucaflorin	37	4,19,20	16	*A. glauca*	[77]
47	Mixture of (2,4-*cis* and *trans*)-gonioneninone	37	(10*R*)	(13*R*,18*R*)	*G. giganteus*	[78]
48	Goniotetracin	37	(4*R*),10	(13*R*,18*R*)	*G. giganteus*	[78]
49	*cis-/trans*-Isomurisolin	35	34=O	15,20	*A. reticulata*	[79]
50	Jimenezin	37	4,23	15,20	*R. mucosa*	[80]
51	Lepirenin	35	23	15,20	*A. squamosa*	[48]
52	Monlicin A	35	(4*R*,7*R*,8*R*)	(14*R*,18*S*)	*A. montana*	[53]
53	Monlicin B	35	(4*R*,7*S*,8*S*)	(14*S*,18*R*)	*A. montana*	[53]
54	Montacin	35	(4*R*),7=O,(9*S*)	(20*S*,25*S*)	*A. montana*	[81]
55	*cis*-Montacin	35	(4*R*),7=O,(9*R*)	(20*S* or *R*),25(*R* or *S*)	*A. montana*	[81]
56	Montalicin A	33	4	13,18	*A. montana*	[68]
57	Montalicin B	35	4	13,18	*A. montana*	[68]
58	Montalicin C	35	4,7	13,18	*A. montana*	[68]
59	Montalicin D	35	4,11	13,18	*A. montana*	[68]
60	Montalicin E	37	4,7	13,18	*A. montana*	[68]
61	Montalicin F	35	4,9	15,20	*A. montana*	[68]
62	Montalicin G	35	(4*R*),7,9	(15*R*,20*R*)	*A. montana*	[53]
63	Montalicin H	35	4,7,9	15,20	*A. montana*	[53]
64	Montalicin I	37	4,9	15,20	*A. montana*	[68]
65	Montalicin J	37	4,11	17,22	*A. montana*	[68]
66	Montanacin B	35	(4*R*,8*R*),10=O	(15*R*,20*R*)	*A. montana*	[82]
67	Montanacin C	35	4,8,10=O	15,(20*S*)	*A. montana*	[82]
68	Montanacin D	35	10=O	15*R*-20	*A. montana*	[82]
69	Montanacin E	35	10=O	15,20	*A. montana*	[82]
70	Montanacin F	35	(29*S*)	(15*R*,20*S*)	*A. montana*	[83]
71	Montanacin G	35	(8*R*),10=O	(15*R*,20*S*)	*A. montana*	[84]

(continued)

Table 3 (continued)

	Name	No. of C	OH	THF/epo	Plant	Ref.
72	(34-*epi*)-Montanacin H	35	4,8,10=O	(15*R*,20*S*)	*A. montana*	[84]
73	(34-*epi*)-Montanacin I	35	(4*R*,29*S*)	(15*R*,20*R*)	*A. montana*	[84]
74	(34-*epi*)-Montanacin J	35	(4*R*,29*S*)	(15*R*,20*S*)	*A. montana*	[84]
75	Mosin B	35	(4*R*)	(15*R* or *S*,20*S* or *R*)	*A. squamosa*	[85]
76	Mosin C	35	(4*R*)	(15*R*,20*S*)	*A. squamosa*	[85]
77	(2,4-*cis* and *trans*)-Mosinone A	37	9=O	(15*R*,20*R*)	*A. squamosa*	[85]
78	Muricapentocin	35	4,8,12	15,20	*A. muricata*	[67]
79	Muricin A	35	(4*R*),26,27	(19*R*)	*A. muricata*	[86]
80	Muricin B	35	(4*S*),26,27	(19*R*)	*A. muricata*	[86]
81	Muricin C	35	4,24,25	21	*A. muricata*	[86]
82	Muricin D	33	4,22,23	19	*A. muricata*	[86]
83	Muricin E	33	4,22,23	16	*A. muricata*	[86]
84	Muricin F	35	4,27,28	21	*A. muricata*	[86]
85	Muricin G	35	4,10	15,20	*A. muricata*	[86]
86	Muricin H	35	24,25	(19*R*)	*A. muricata*	[60]
87	Muricin I	37	24,25	19	*A. muricata*	[60]
88	Muricoreacin	35	4,8,10,19,20	16	*A. muricata*	[87]
89	Murihexocin C	35	4,7,8,19,20	16	*A. muricata*	[87]
90	*cis*-Panatellin	35		13,18	*A. muricata*	[88]
91	Parisin	37	4,23,24	15,20	*A. salzmanii*	[89]
92	Plagioneurin A	39	10,15-OAc	(17*R*,22*R*)	*D. plagioneurum*	[18]
93	Plagioneurin B	39	10,15-OAc	17–22	*D. plagioneurum*	[18]
94	Plagioneurin C	39	10=O,15-OAc	(17*R*,22*R*)	*D. plagioneurum*	[18]
95	Plagioneurin D	39	(5*R*),10=O,15-OAc	(17*R*,22*R*)	*D. plagioneurum*	[18]
96	Plagioneurin E	39	5,10=O,15-OAc	17,22	*D. plagioneurum*	[18]
97	Plagionicin B	35	5,10=O	15,20	*D. plagioneurum*	[18]
98	Plagionicin C	35	4,5,11	15,20	*D. plagioneurum*	[18]
99	Plagionicin D	35	5=O,10,11	15,20	*D. plagioneurum*	[18]
100	*cis*-Reticulatacin	37		17,22	*A. muricata*	[88]
101	*cis*-Reticulatacin-10-one	37	10=O	17,22	*A. muricata*	[88]
102	Rolliacocin	35	(4*R*),11	(15*S*,20*S*)	*R. mucosa*	[90]
103	Rollicosin	22	(4*R*),19=O	(15*R*)	*R. mucosa*	[91]
104	*cis*-Solamin	35		15,20	*A. muricata*	[88]
105	Squadiolin A	37	15,16,28	19,24	*A. squamosa*	[37]
106	Squadiolin B	37	19,20,23,24	16	*A. squamosa*	[37]
107	Squadiolin C	37	4,21,22	16	*A. squamosa*	[37]

(continued)

Table 3 (continued)

	Name	No. of C	OH	THF/epo	Plant	Ref.
108	Squafosacin B	37		15,20	*A. squamosa*	[37]
109	Squafosacin C	35		17,22	*A. squamosa*	[37]
110	Squafosacin F	35		(15*S*,20*S*)	*A. squamosa*	[37]
111	Squafosacin G	37		(19*S*,24*S*)	*A. squamosa*	[37]
112	Squamocenin[a]	37	25B	17–22	*A. squamosa*	[57]
113	(2,4-*cis* and *trans*)-Squamoxinone	37	(11*S*),36=O	(17*R*,22*R*)	*A. squamosa*	[69]
114	(2,4-*cis* and *trans*)-Squamoxinone B	37	11,36=O	17,22	*A. squamosa*	[35]
115	(2,4-*cis* and *trans*) Squamoxinone C	35	11,34=O	17,22	*A. squamosa*	[35]
116	Tucupentol	35	4,8,19,20	15	*A. montana*	[92]
117	*cis*-Uvariamicin I	37		15,20	*A. muricata*	[88]
118	*cis*-Uvariamicin IV	37		13,18	*A. muricata*	[88]
119	Mixture of (2,4-*cis* and -*trans*)-xylomaticinones	37	(10*R*),36=O	(15*R*,20*R*)	*G. giganteus*	[93]

[a]This compound has the same name as one of the linear AGEs [48]

3.3 *Bis-THF Annonaceous Acetogenins, Including Derivatives with Adjacent or Non-adjacent Bis-THF Rings*

The primary structural feature of bis-THF AGEs is two THF rings flanking one or two hydroxy groups. Two subtypes of this group can be identified based on the nature of the bis-THF moieties, namely, (1) compounds with an adjacent bis-THF moiety and (2) those with a non-adjacent bis-THF moiety, in which the latter possesses a four-carbon aliphatic chain between the THF rings. Since 1997, 63 new bis-THF AGEs were found, including 48 adjacent bis-THF AGEs isolated from 15 species in the four genera, *Annona* (35 species), *Asimina* (four species), *Rollinia* (five species), and *Uvaria* (four species), along with 15 non-adjacent bis-THF AGEs from six species in the four genera, *Annona* (11 species), *Asimina* (one species), *Goniothalamus* (two species), and *Rollinia* (one species) (see Table 4). Interestingly, aromin-A (**21**) from *A. cherimola* possesses one THF ring that is cyclized between C-4 and C-7 by an ether linkage [116], which shares many similarities with the analogues found from *Xylopia aromatica* [120]. As a result, the bis-THF ring-containing compounds have been organized into groups of adjacent and non-adjacent compounds.

Table 4 Bis-THF AGEs isolated since 1997 until the end of 2014

	Name	No. of C	OH	THF/epo	Plant	Ref.
Adjacent Bis-THF AGEs						
1	Annocatacin A	35	(4*S*)	(23*S*)	*A. muricata*	[94]
2	Annocatacin B	35	(4*S*)	23	*A. muricata*	[94]
3	Annonisin	35	4,8	13,22	*A. atemoya*	[95]
4	Annosquacin A	35		11,20	*A. squamosa*	[96]
5	Annosquacin B	37		13,22	*A. squamosa*	[96]
6	Annosquacin C	37	25	13,22	*A. squamosa*	[96]
7	Annosquacin D	37		13,22	*A. squamosa*	[96]
8	Annosquacin-I	37	23	10,19	*A. squamosa*	[97]
9	(2,4-*cis* and *trans*)-9-Hydroxy-asimicinone	37	(9*S*),36=O	(15*R*,24*R*)	*A. squamosa*	[35]
10	(2,4-*cis* and *trans*)-9-Oxo-asimicinone	35	9=O,36=O	(15*R*,24*R*)	*A. squamosa*	[98]
11	Asimitrin	37	(4*R*,17*R*)	(15*R*,24*R*)	*A. triloba*	[99]
12	Atemotetrolin	37	28,29	15,24	*A. atemoya*	[100]
13	Bullacin B	37	(6*R*)	(15*R*,24*R*)	*A. squamosa*	[98]
14	Bulladecin	37	4,23,24	11,20	*A. atemoya*	[100]
15	27-Hydroxybullatacin	37	(4*R*,27*S*)	(15*R*,24*S*)	*A. glabra*	[63]
16	Calamistrin F	37	(5*R*)	17,(26*R*)	*U. calamistrata*	[73]
17	Calamistrin G	37	(5*S*)	(17*R*,26*S*)	*U. calamistrata*	[73]
18	Carolin A	37	28	15,24	*A. spinescens*	[101]
19	Carolin B	37	29	15,24	*A. spinescens*	[101]
20	Carolin C	35	26	1322	*A. spinescens*	[101]
21	Chamuvarinin	37		15	*U. chamae*	[102]
22	Cornifolin	37	7	17,26	*A. cornifolia*	[103]
23	9-Hydroxyfolianin	37	9	(12*R*,21*S*)	*A. cornifolia*	[104]
24	Folianin B	37		12,21	*A. cornifolia*	[104]
25	Glabracin A	37	(4*R*),23,24	(10*S*)	*A. glabra*	[105]
26	Glabracin B	37	(4*R*),23,24	(10*S*)	*A. glabra*	[105]
27	Guanaconetin-1	41	24,30=$OCOCH_3$	15	*A.* aff. *spraguei*	[106]
28	Guanaconetin-2	41	15,30=$OCOCH_3$	24	*A.* aff. *spraguei*	[106]
29	Guanaconetin-3	39	24=$OCOCH_3$,30	15	*A.* aff. *spraguei*	[106]
30	Guanaconetin-4	39	30=$OCOCH_3$	15,24	*A.* aff. *spraguei*	[106]
31	Joolanin	37	5=O	15,24	*U. chamae.*	[107]
32	Purpuracenin	37	(4*R*)	(15*R*,24*S*)	*A. purpurea*	[108]
33	Purpurediolin	37	28,29*S*	(5*R*,24*S*)	*A. purpurea*	[109]
34	Purpurenin	37	(10*R*),28,29*S*	(15*R*,24*S*)	*A. purpurea*	[109]
35	Rollidecin C	35		20	*R. mucosa*	[110]
36	Rollidecin D	37		22	*R. mucosa*	[110]
37	Rollimusin	37	10,28	15,24	*R. mucosa*	[90]
38	Rollinacin	35	4,10	20	*R. mucosa*	[111]
39	Rollitacin	37	28,29	15,24	*R. mucosa*	[111]

(continued)

Table 4 (continued)

	Name	No. of C	OH	THF/epo	Plant	Ref.
40	Salzmanolin	37	15,17,28,29	24	*A. salzmanii*	[89]
41	Squamocin-O_1	37	(12*R*,28*S*)	(15*R*,24*S*)	*A. squamosa*	[112]
42	Squamocin-O_2	37	(12*S*,28*S*)	(15*R*,24*S*)	*A. squamosa*	[112]
43	(2,4-*cis* and *trans*)-Squamolinone	35	36=O	(15*R*,24*S*)	*A. squamosa*	[98]
44	2,4-*cis*-Trilobacinone	37	36=O	(15*R*,24*R*)	*A. triloba*	[113]
45	2,4-*trans*-Trilobacinone	37	36=O	(15*R*,24*R*)	*A. triloba*	[113]
46	4-Hydroxytrilobin	37	(4*R*),10	(15*R*,24*R*)	*A. triloba*	[99]
47	Tucumanin	37		15,24	*A. cherimola*	[114]
48	Compound 1	37	5	15,24	*A. squamosa*	[115]
Non-adjacent-bis-THF AGEs						
1	Aromin-A	35	9=O	15,20	*A. cherimola*	[116]
2	Annosquatin A	37		13,16,21	*A. squamosa*	[96]
3	Annosquatin B	37	29	13,16,21	*A. squamosa*	[96]
4	Annosquatin-I	37	30	17,24,29	*A. squamosa*	[97]
5	Annosquatin-II	37	4	17,24,29	*A. squamosa*	[97]
6	Mixture of 20,23-*cis*-2,4-*cis* and *trans*-bullatalicinone	37		16,19,24	*R. mucosa*	[90]
7	Mixture of (2,4-*cis* and *trans*)-gigantecinone	37	36=O	(14*S*,17*R*,22*R*)	*G. giganteus*	[117]
8	Goniotricin	37	4,18	10	*G. giganteus*	[93]
9	Squamostanin-A	37	(27*R*)	(15*R*,18*R*,23*R*)	*A. squamosa*	[118]
10	12,15-*cis*-Squamostatin-A	37	28	16,19,24	*A. atemoya*	[43]
11	Squamostanin-B	37	(27*R*)	(15*R*,18*R*,23*S*)	*A. squamosa*	[118]
12	Squamostanin-C	37	(5*R*)	(16*R*,19*R*,24*R*)	*A. squamosa*	[119]
13	Squamostanin-D	37	(5*R*)	(16*R*,19*R*,24*S*)	*A. squamosa*	[119]
14	12,15-*cis*-Squamostatin-D	37		16,19,24	*A. atemoya*	[43]
15	Trilobalicin	35	(4*R*)	14,17,22	*A. triloba*	[113]

3.4 Miscellaneous

This group includes AGEs with an atypical substituted alkyl chain (see Table 5). For instance, muricatacin (**22**) appears like a normal AGE without the γ-lactone ring moiety and the aliphatic chain between the THF ring and lactone ring [122]. In 2003, our group reported a novel skeleton of an abridged AGE, rollicosin (**23**), from the unripe fruits of *Rollinia mucosa*, the first identified AGE containing lactone moieties

Table 5 Miscellaneous types of AGEs isolated since 1997 until the end of 2014

	Name	No. of C	OH	THF/epo	Plant	Ref.
1	Montanacin D	35	10=O	(15*R*),20	*A. montana*	[82]
2	Montanacin E	35	10=O	(15*R*),20	*A. montana*	[82]
3	Chamuvarinin	37		15,28	*U. chamae*	[102]
4	Rollicosin	22	(4*R*),19=O	(15*R*)	*R. mucosa*	[91]
5	Squamostolide	22	19=O	(15*R*)	*A. squamosa*	[121]

on both sides of an aliphatic chain [91]. Soon after this report, a Chinese group communicated the second AGE of this type, squamostolide (**24**), from *A. squamosa* [121].

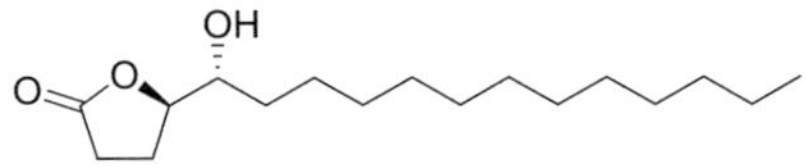

22 (muricatacin)

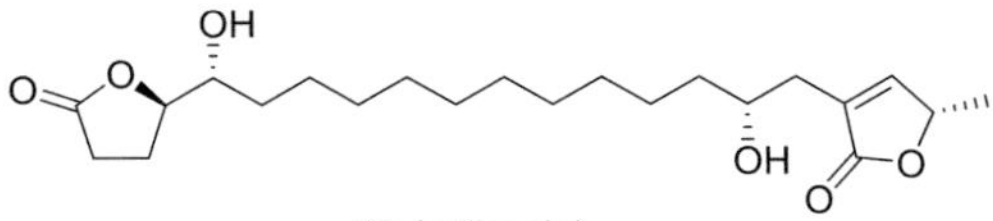

23 (rollicosin)

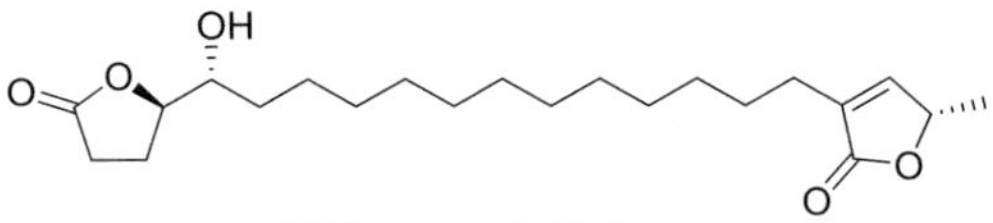

24 (squamostolide)

4 Chemotaxonomy of the Annonaceae Family

The family Annonaceae is composed of 2000 species including 129 genera found worldwide. Plants of this family are recognized sources of alkaloids, diterpenes, flavonoids, and polyketide compounds. Among them, acetogenins are regarded as characteristic secondary metabolites of this family. More than half of the annonaceous AGEs have been isolated from the genus *Annona*. A plausible biosynthesis pathway of AGEs should be related to various polyketide synthases deduced by different species of plants. Three types of AGEs were selected, including mono-THF AGEs (MT), adjacent bis-THF AGEs (ABT) and non-adjacent bis-THF AGEs (NBT). They were all isolated from the plants of the genus *Annona* according to the reported literature. By employing qualitative analysis, three patterns of radar charts were observed for deducing a cladistics chart and the phylogenetic inference of selected species in the genus *Annona*. These seem to correlate with the appearance of the fruits of these plants (see Fig. 3). For instance, the species examined may have fruits with the skin covered with many short fleshy spines (e.g. *A. montana* and

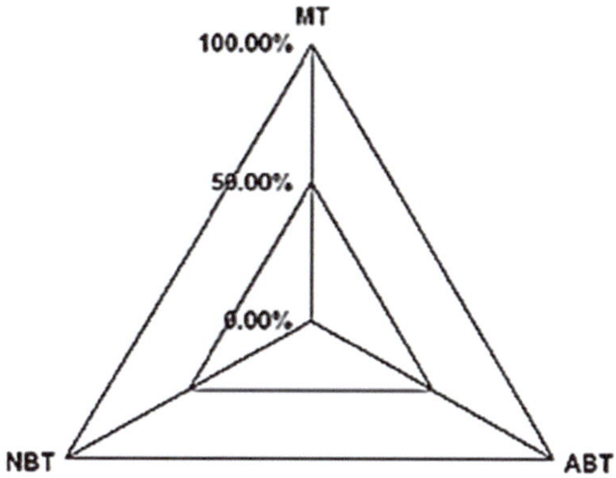

Fig. 3 Radar chart of AGEs. *MT* mono-THF AGEs, *ABT* adjacent bis-THF AGEs, *NBT* non-adjacent bis-THF AGEs

Table 6 Radar charts of AGEs of the plants in the genus *Annona* in qualitative analysis and the appearance of their fruits

A. montana	*A. muricata*	*A. cherimola*	*A. glabra*
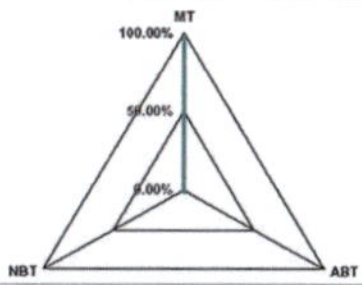	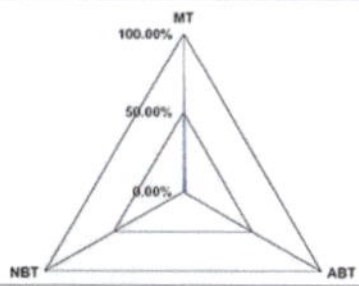	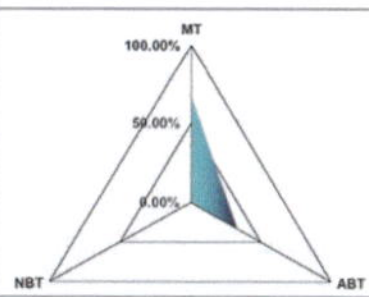	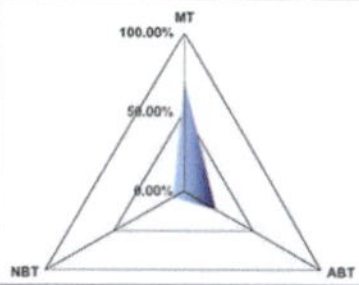

A. reticulata	*A. purpurea*	*A. squamosa*	*A. atemoya*
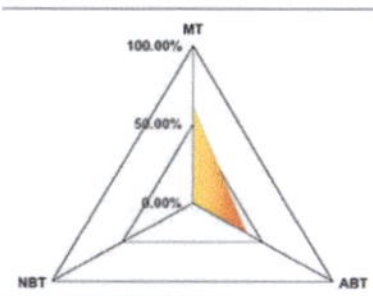	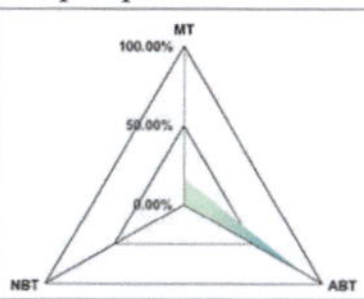	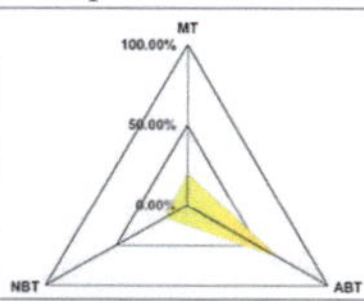	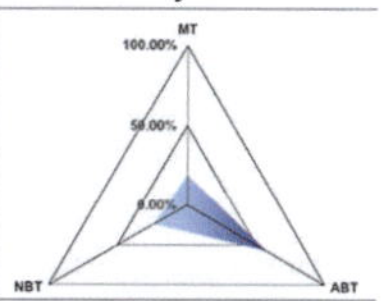

Photographs by the authors: *A. montana, A. glabra, A. squamosa*; giardinaggi.it: *A. muricata*; Ken Lowe: *A. cherimola*; photomazza.com: *A. reticulata*; *A. purpurea, A. atemoya*: wikipedia.org

A. muricata), fruits with the skin overlapping scales or knob-like warts (e.g. *A. cherimola*, *A. glabra*, and *A. reticulata*), and fruits with many round protuberances (e.g. *A. purpurea*, *A. squamosa*, and *A. atemoya*) (see Table 6).

Based on the radar chart analysis of their AGEs, plants of the genus *Annona* can be classified into three sub-genera. Sub-genus I with mono-THF (MT) AGEs (one hydroxy group) as the major type includes *A. montana* and *A. muricata*.

Sub-genus II with mono-THF and bis-THF AGEs can be further separated into two groups, including one with mono-THF (MT) AGEs and the other with adjacent bis-THF (ABT) AGEs as the major ones. The former includes *A. reticulata*, *A. cherimola*, and *A. glabra*, whereas the latter includes *A. purpurea*, *A. atemoya* and *A. squamosa* (see Fig. 4).

The evalution of the patterns of AGEs was extended to other plants of the Annonaceae. It was found that *Rollina mucosa* seems to be phylogenetically related to *A. squamosa* based on the analysis of AGEs isolated from this particular plant (see Fig. 5).

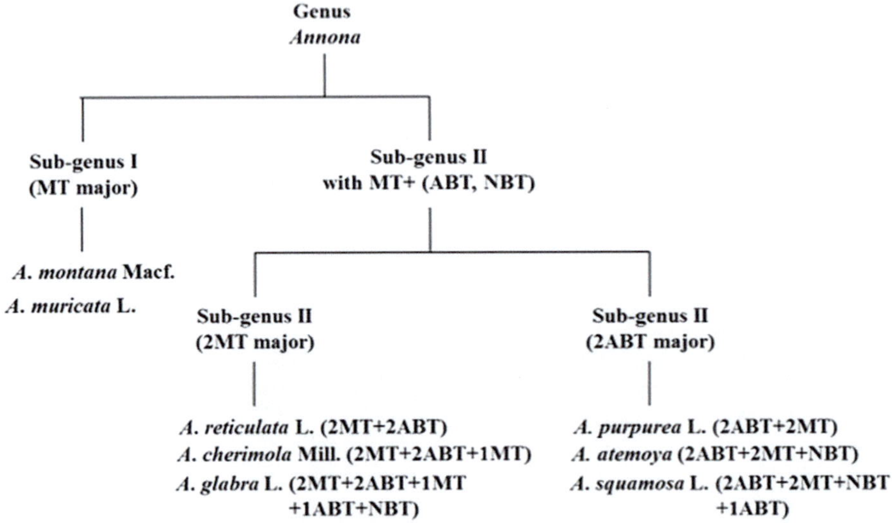

Fig. 4 Plausible interrelationships of plants of the genus *Annona*. 1MT: AGEs with a mono-THF moiety flanked by one hydroxy group, 2MT: AGEs with a mono-THF moiety flanked by two hydroxy groups, 1ABT: AGEs with an adjacent bis-THF moiety flanked by one hydroxy group, 2ABT: AGEs with an adjacent bis-THF moiety flanked by two hydroxy groups, NBT: AGEs with a non-adjacent bis-THF moiety

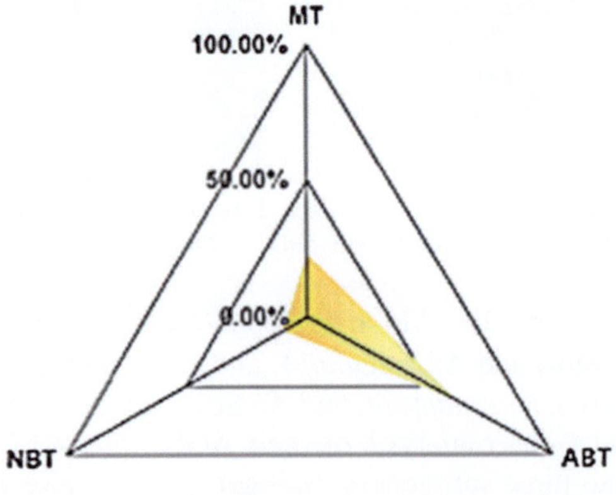

Fig. 5 Radar chart of AGEs from *R. mucosa*

5 Synthesis of Annonaceous Acetogenins

Owing to the great scientific interest in AGEs and their abundant structural types and interesting biological activities, these components have attracted the attention of many researchers working on chemical synthesis. However, the structural variations of AGEs, such as the presence of a γ-lactone ring moiety, and 1–3 THF/THP rings with multiple chiral centers, and an alkyl chain have been considered as challenging targets to utilize synthesis methods. A number of total syntheses for pure AGEs samples have also been reported in the literature since the 1990s. Since 1997, the total syntheses of AGEs achieved include those for the mono-THF AGEs: murisolin (**25**) [123, 124], longicin (**26**) [125, 126], and *cis*-solamin (**27**) [88, 127], adjacent bis-THF AGEs: bullatacin (**7**) [128], rolliniastatin 1 (**28**) [129, 130], rollimembrin (**29**) [130, 131], 10-hydroxyasimicin (**30**) [132, 133], membranacin (**31**) [130, 134], asimicin (**8**) [135, 136], longimicin D (**32**) [137, 138], and mucoxin (**33**) [139, 140], and non-adjacent bis-THF AGEs: *cis*-sylvaticin (**34**) [141, 142] and gigantecin (**35**) [143, 144], and others, jimenezin (**36**) [80, 145], mucocin (**37**) [146, 147], pyranicin (**38**) [148, 149], pyragonicin (**39**) [148, 150, 151], rollicosin (**23**) [91, 152], and squamostolide (**24**) [121, 153]. Indeed, more than 100 investigations on the synthesis of AGEs have been published during the last 15 years (see Fig. 6).

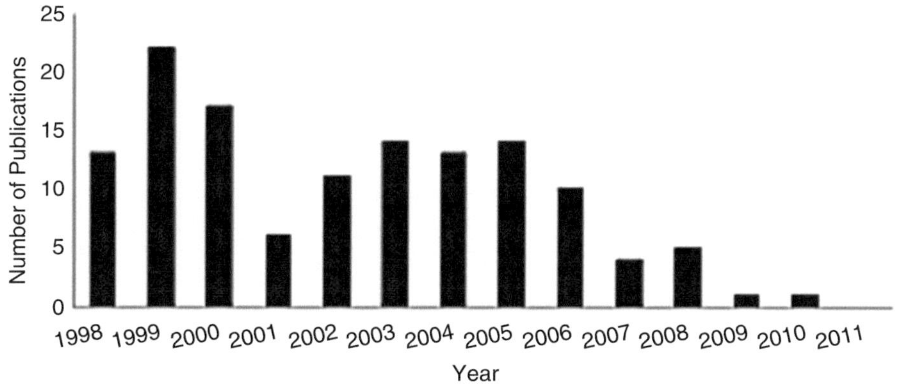

Fig. 6 Number of publications per year on the investigation of AGE synthesis from 1998 to 2011

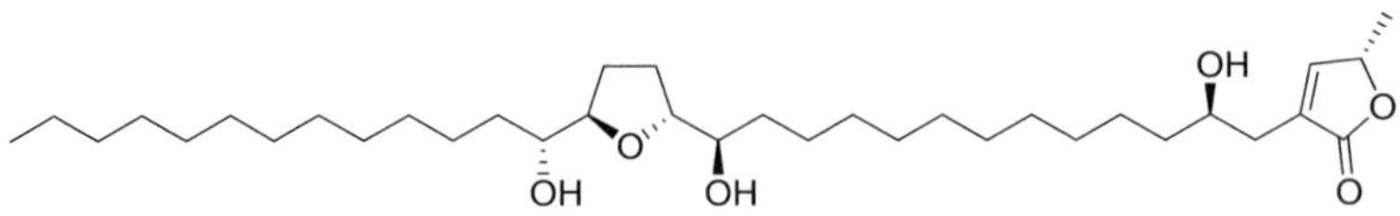

25 (murisolin)

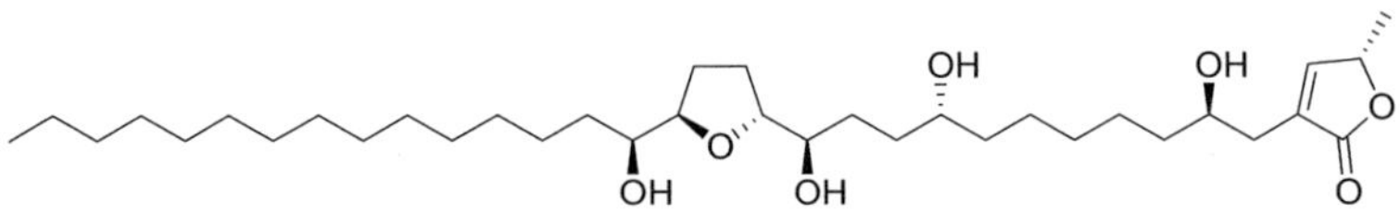

26 (longicin)

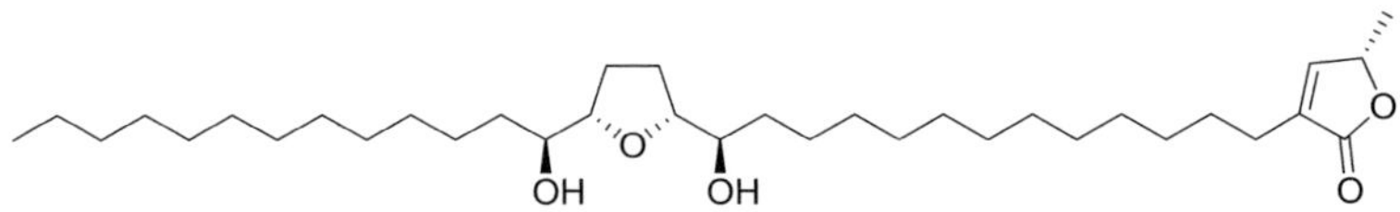

27 (*cis*-solamin)

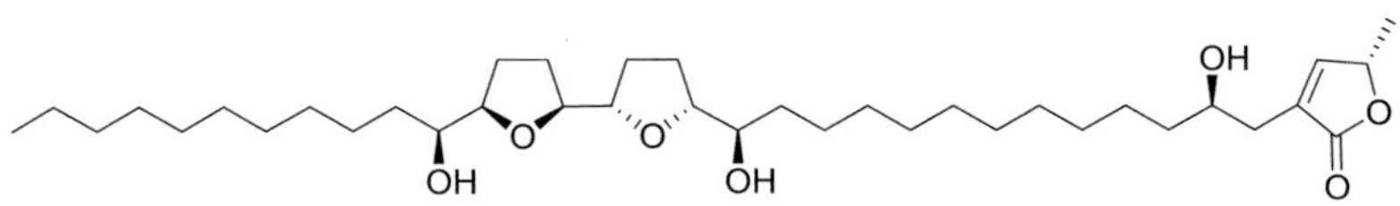

28 (rolliniastatin; also named as rolliniastatin 1)

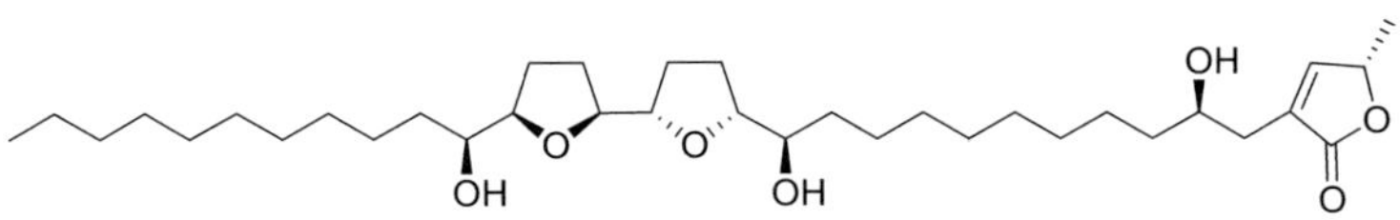

29 (rollimembrin)

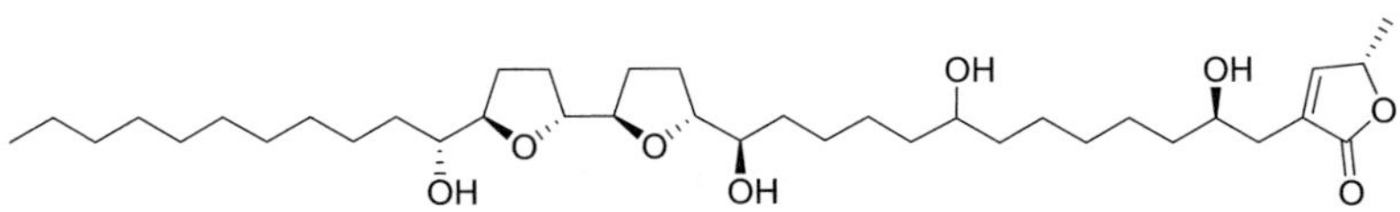

30 (10-hydroxyasimicin)

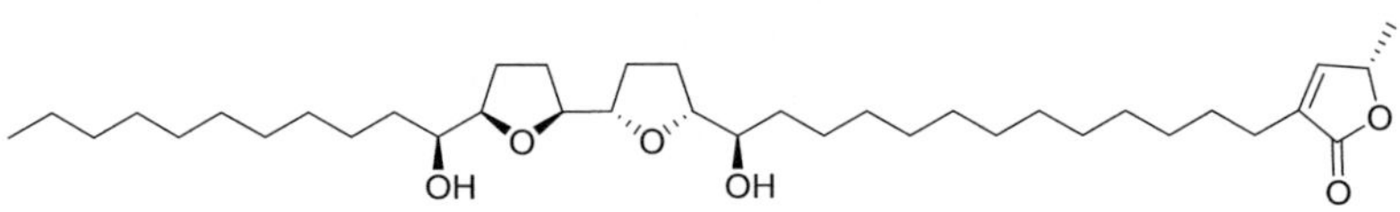

31 (membranacin)

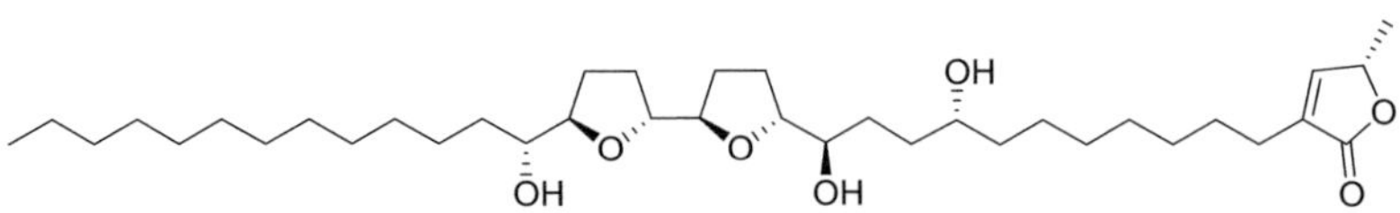

32 (longimicin D)

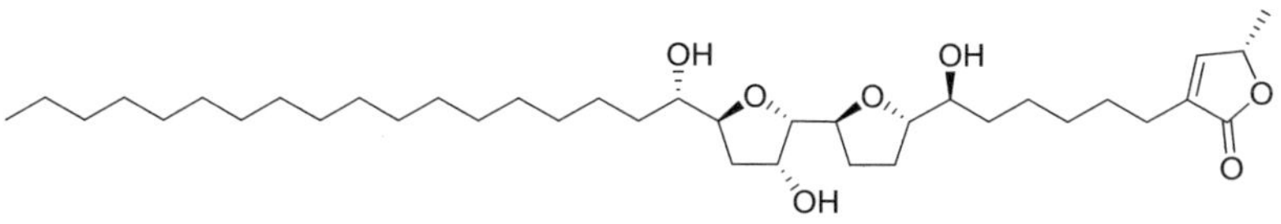

33 (mucoxin)

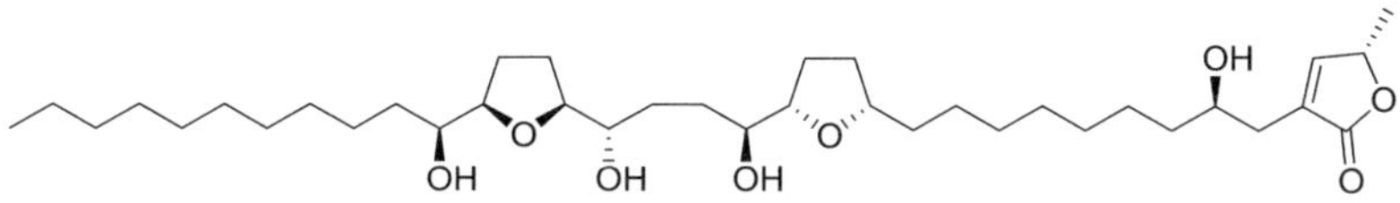

34 (*cis*-sylvaticin)

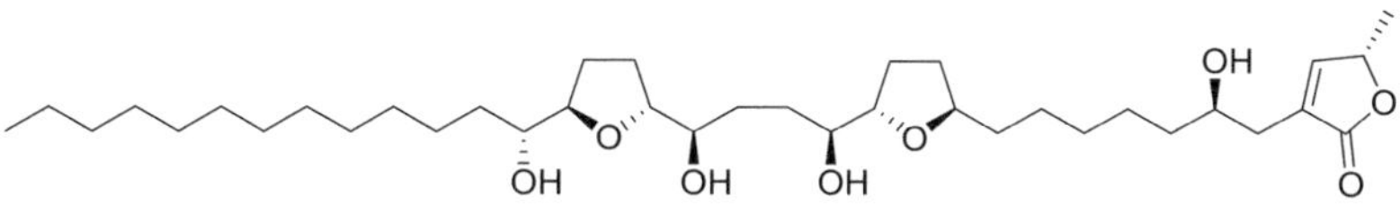

35 (gigantecin)

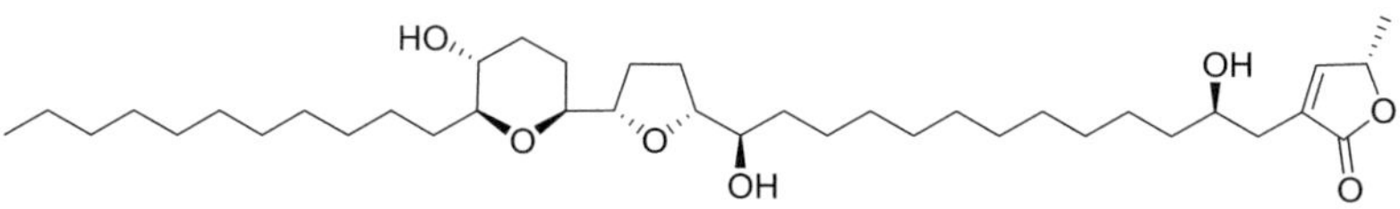

36 (jimenezin)

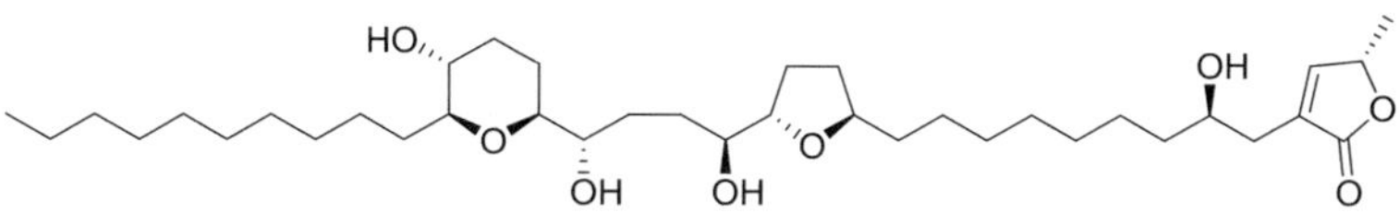

37 (mucocin)

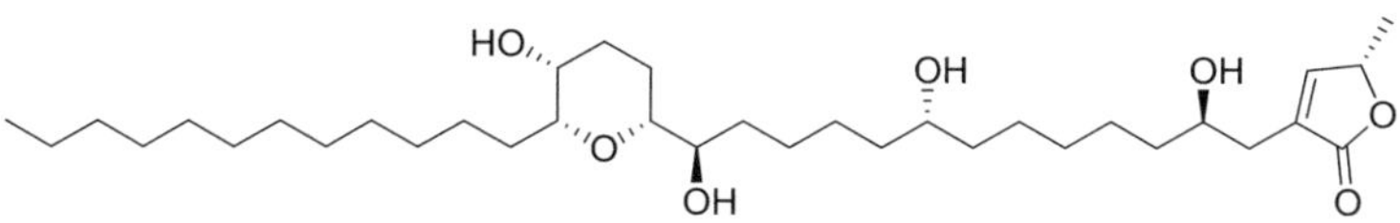

38 (pyranicin)

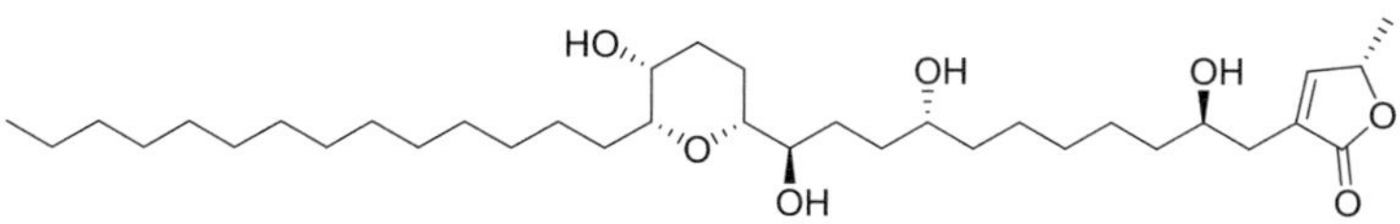

39 (pyragonicin)

Most information on the synthesis of AGEs is evident for the bis-THF AGEs (BT), followed by AGEs with a THP moiety (O), then the mono-THF AGEs (MT), and the linear AGEs (L) (see Fig. 7). Moreover, the methodologies described for the synthesis of the adjacent type (ABT) are about four times more numerous than

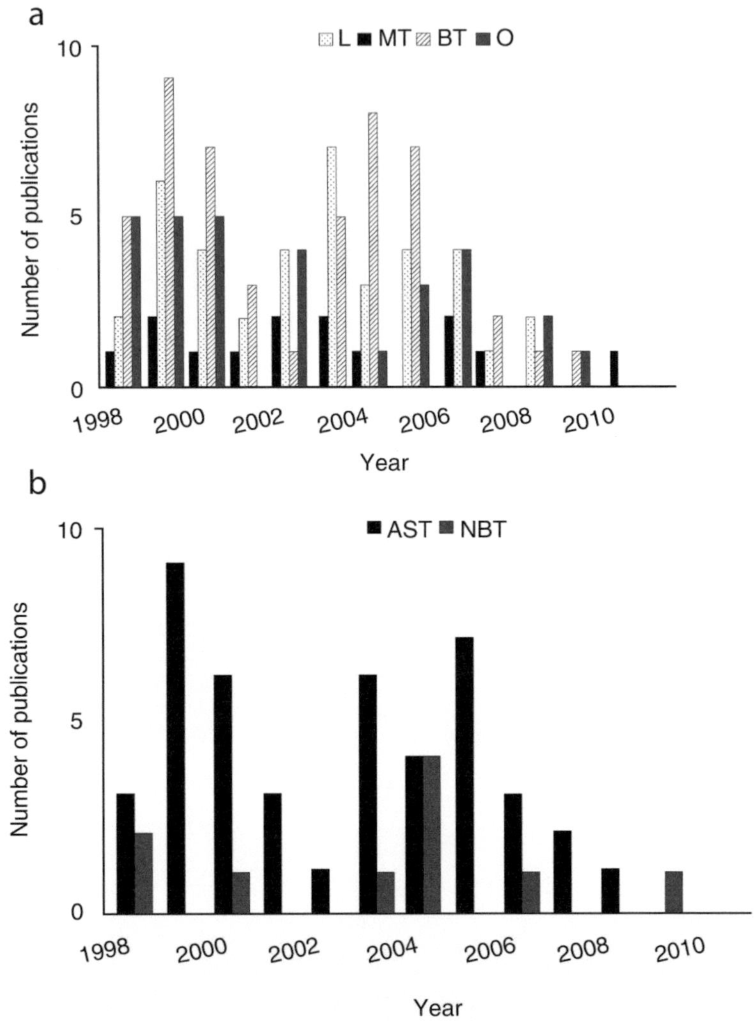

Fig. 7 Publications on AGEs divided into four major types from 1988 to 2010. (**a**) L linear AGEs, MT mono-THF ring of AGEs, BT bis-THF ring of AGEs, O other AGEs. (**b**) ABT adjacent bis-THF ring of AGEss, NBT nonadjacent bis-THF ring of AGEs

those on the non-adjacent type (NBT). Summarized below are some concepts of synthesis for AGEs presented according to the previously described classification method for these natural products.

5.1 *Linear Annonaceous Acetogenins*

Methods for the synthesis of three characteristic linear AGEs, including montecristin (**40**), muricatacin (**22**), and tonkinelin (**41**), have been proposed. In particular, 72% of the articles published on synthesis methods have had a special focus on muricatacin (**22**), see Chart 1.

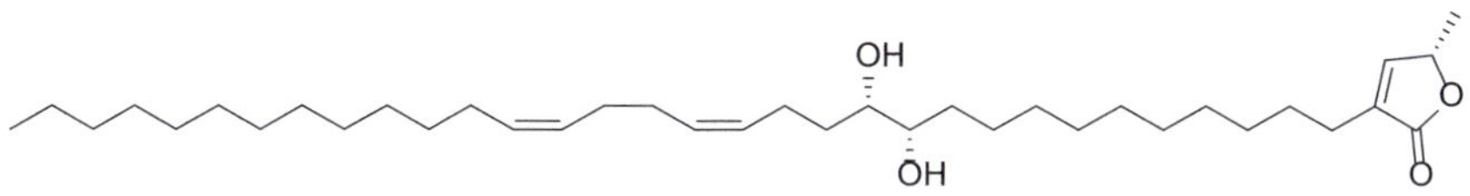

40 (montecristin)

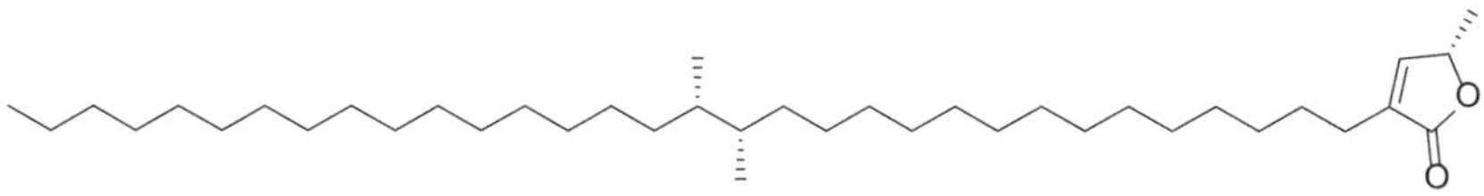

41 (tonkinelin)

Chart 1 Analysis of linear and epoxy ACGs synthesis from 1998 to 2012

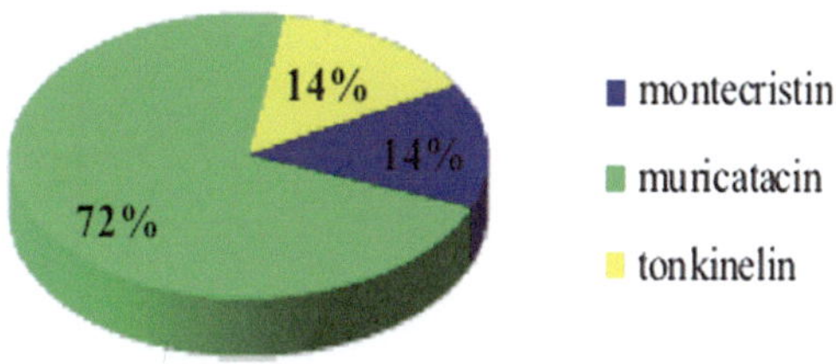

5.1.1 Montecristin

Montecristin (**40**) is a linear C_{32}-AGE with an α,β-unsaturated γ-lactone and a *syn*-form glycol moiety along with two *cis*-form C=C bonds in its side chain. A rapid but effective synthesis methodology was developed by the Brückner group, which was also used to determine the stereostructure of montecristin (**40**). These investigators prepared (*S*)- and (*R*)-4-hydroxy-5-methyl-dihydrofuran-2(3*H*)-one (the lactone substructure) from a commercially available pentenoic ester under asymmetric dihydroxylation conditions and further performed an asymmetric dihydroxylation of an appropriate (*E*)-olefin to obtain an acetonide iodide, which was conjugated to the lactone group while deprotecting this group simultaneously (see Table 7) [154].

Table 7 Long-chain AGE synthesis from 1998 to 2011

Compound name	Total synthesis	Synthesis method	Starting materials	Steps and yields	Other points	Ref.
5-*epi*-Montecristin and (−)-montecristin	First total synthesis	1. Asymmetric dihydroxylation 2. Nucleophile and electrophile	Pentenoic ester	5-*epi*-Montecristin (13 steps, 88%) and (−)-montecristin (13 steps, 83%)	(+)-Montecristin was established to have (11′*R*,12′*R*) configuration.	[154]
(−)-(4*R*,5*R*)-Muricatacin and (−)-(4*R*,5*S*)-*epi*-muricatacin		1. New chiral Witting reagent: a β-keto-γ-(*S*)-hydroxy-δ-(*R*)-*p*-tolysufinyl phosphonate 2. DIBAL-H reduction of *β*-hydroxy-*γ*-ketosulfoxides 3. Pummerer rearrangement 4. Swern oxidation	Ethyl oxalate and sulfoxide	(−)-(4*R*,5*R*)-Muricatacin (13 steps, 27%) and (−)-(4*R*,5*S*)-*epi*-muricatacin (13 steps, 25%)		[155]
(−)-(4*R*,5*R*)-Muricatacin	Total synthesis	1. Sharpless epoxidation on (*Z*) or (*E*) allylic alcohol using (+)-DET or (−)-DET 2. A regio- and stereospecific ring-opening of a substituted vinyl epoxide under Lewis acid catalysis using catalytic amounts of BF_3-Et_2O	Propargylic alcohol	11 Steps, 96%	1. The spectroscopic data (^{1}H and ^{13}C NMR, IR, MS) and optical rotation were in close agreement with reported values. 2. Condensation of alcohol on vinyl epoxide using equal amounts of both units has not previously been reported in the literature.	[156]

(4*R*,5*S*)- and (4*S*,5*R*)-Muricatacins, and (4*S*,5*R*)-aza-muricatacin, unnatural analogues		1. Sharpless asymmetric dihydroxylation 2. Grignard reaction 3. Johnson-Claisen rearrangement 4. Mitsunobu inversion or Weinreb amidation	Lauryl bromide	(4*R*,5*R*)-Muricatacins (5 steps, 87%, >98% *ee*) (4*R*,5*S*)-muricatacins (8 steps, 30%) (4*S*,5*R*)-muricatacins (10 steps, 90%) (4*S*,5*R*)-aza-muricatacins (11 steps, 84%)		[157]
(−)-Muricatacin	Total synthesis	Wittig olefination	D-Mannitol	8 Steps, 92–95%		[158]
(−)-(4*R*,5*R*)-Muricatacin and the pheromone (*R*)-japonilure		Lithium salt catalyzed ring expansion of nonracenmic oxaspiroprntenes		6 Steps	1. As far as we know this is the first report of a ring expansion of an oxaspiropentane into cyclobutanone occurring preferentially with retention of configuration. 2. Dioxolane ring protected.	[159]
(17*S*,18*S*)-Tonkinelin and (17*R*,18*R*)-Tonkinelin		1. Asymmetric dihydroxylation by the Sharpless procedure using AD mix α and spontaneous epoxidation afforded an epoxy alcohol 2. Grignard reaction 3. Sonogashira cross-coupling reaction	$C_{16}H_{33}I$	(17*S*,18*S*)-Tonkinelin (9 steps, 56%)	1. Another reason for the weak activity may be due to the length of the spacer. The spacer of tonkinelin is longer (15 carbon atoms) than the optimal length (13 carbon atoms). 2. Natural compound is (17*S*,18*S*)-tonkinelin. 3. MOM ether protected.	[160, 161]

5.1.2 (−)-Muricatacin

(−)-Muricatacin (**22**), an AGE with only one terminal γ-lactone unit and a long alkyl chain moiety, possesses a broad spectrum of biological activities. This particular AGE has greatly attracted many research groups to further explore its synthesis. Solladié et al. synthesized the *syn*- and *anti*-1,2-diol units of (−)-muricatacin by a highly stereoselective method, DIBAL-H reduction of β-hydroxy-γ-ketosulfoxides [155]. The Mioskowski group later produced a high yield of 96% of natural muricatacin (**22**) [156] through development of a new method, an epoxide ring-opening reaction under Lewis acid catalysis. In addition, Singh's group synthesized muricatacin (**22**) efficiently and highly stereoselectively using a 5-hydroxy-alkylbutan-4-olide obtained from D-mannitol [158]. The Konno group successfully prepared enantiomerically pure muricatacin (**22**) (87% yield, >98% *ee*) using the Sharpless asymmetric dihydroxylation method [157]. Moreover, Bernard et al. unexpectedly discovered a simple ring expansion procedure that led to the synthesis of **22**. The step involves the use of LiI to catalyze ring expansion of non-racemic oxaspiropentanes into cyclobutanones (see Table 7) [159].

5.1.3 Tonkinelin

Tonkinelin (**41**) is a nonclassical linear C_{37} AGE with the lactone ring in a (*S*)-configuration and with an undefined diol moiety [160]. In 2007, Makabe et al. first synthesized this compound using a Sonogashira cross-coupling reaction to link the dihydroxy and γ-lactone parts of the molecule together (see Table 7) [161].

5.2 *Mono-THF Annonaceous Acetogenins*

Mono-THF AGEs are one of the major types of AGEs, and have at least four stereogenic centers that have made the organic synthesis of these compounds challenging (Table 8). Several mono-THF AGEs, such as annonacin (**42**), corossolin (**43**), 4-deoxyannoreticuin (**44**), 4-deoxyannomontacin (**45**), *cis*-gigantrionenin (**46**), gigantetrocin A (**47**), (+)-longicin (**26**), longifolicin (**48**), mosin B (**49**), muricatetrocin C (**50**), murisolin (**25**), pseudoannonacin A (**51**), reticulatain (**52**), solamin (**53**), and tonkinecin (**54**) have been chosen as targets for total synthesis. Among them, solamin (**53**) (29%) was the most popular synthesis target (see Fig. 8).

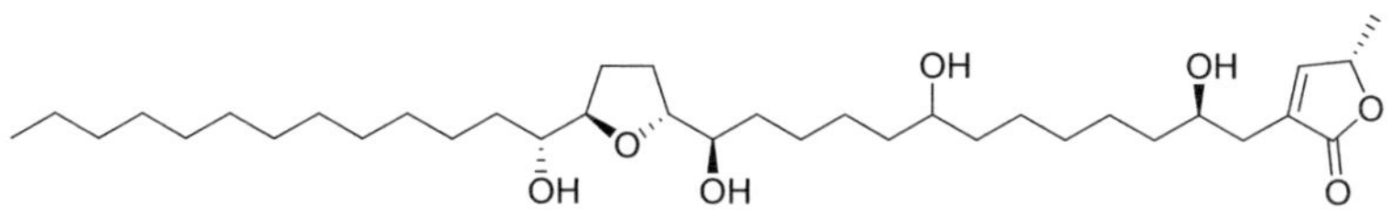

42 (annonacin)

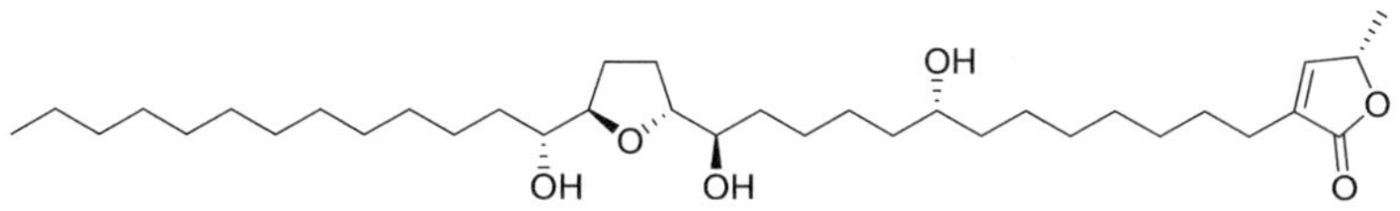

43 ((10*R*)-corossolin)

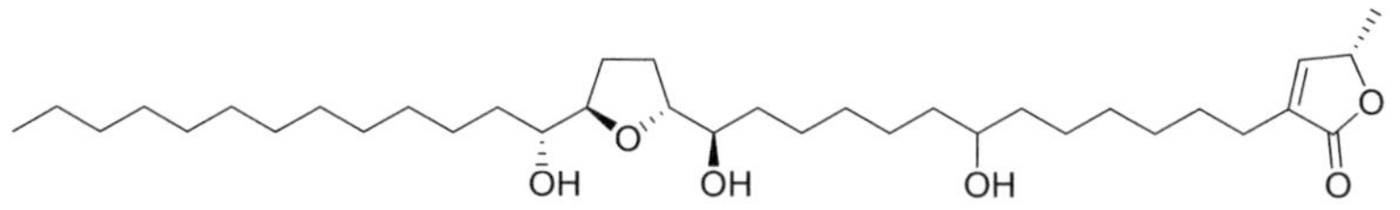

44 (4-deoxyannoreticuin)

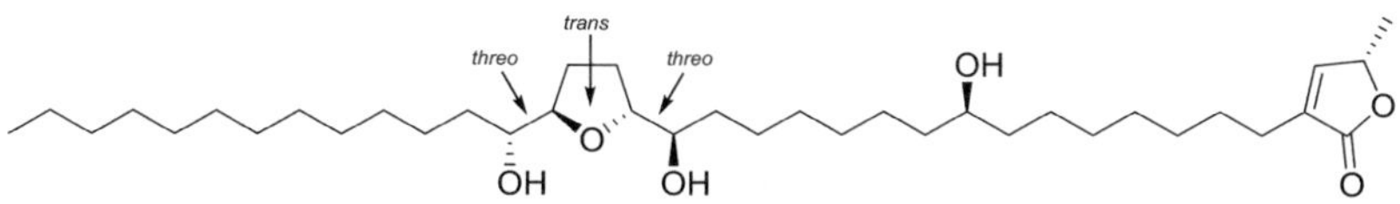

45 (4-deoxyannomontacin)

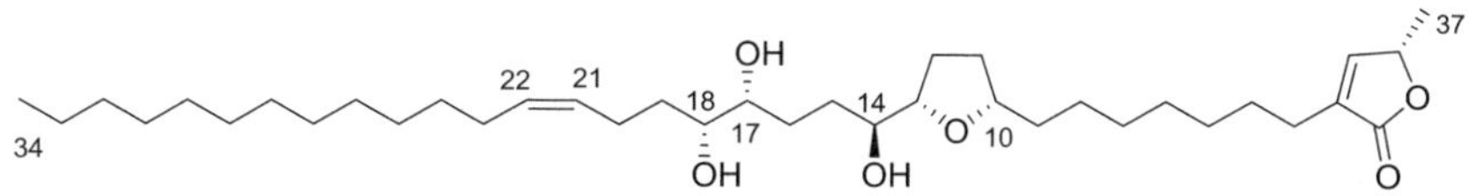

46 (*cis*-gigantrionenin)

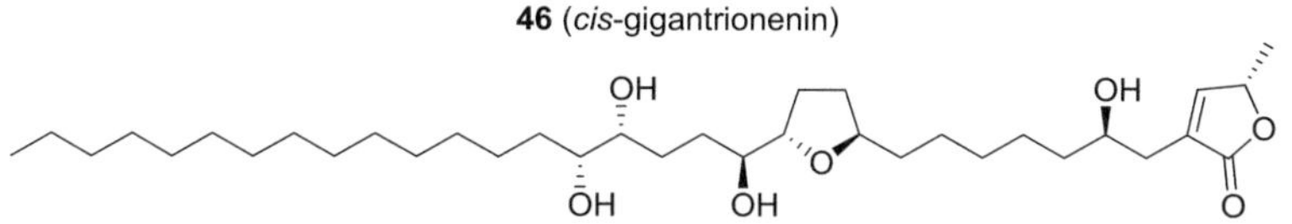

47 (gigantetrocin A)

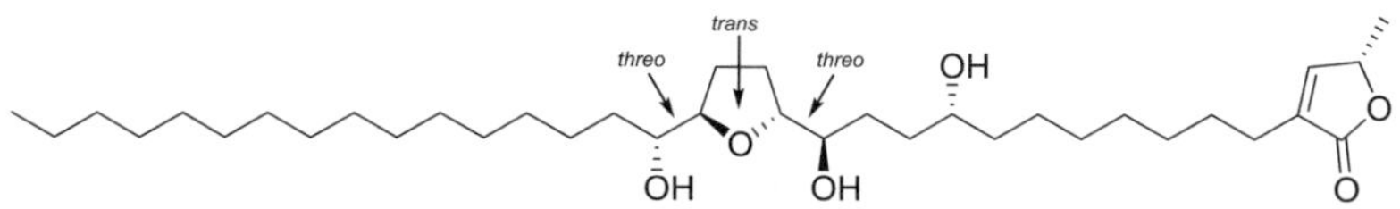

48 (longifolicin)

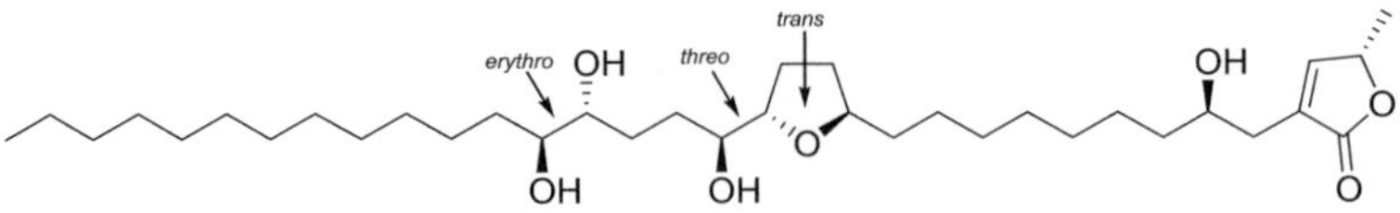

50 (muricatetrocin C)

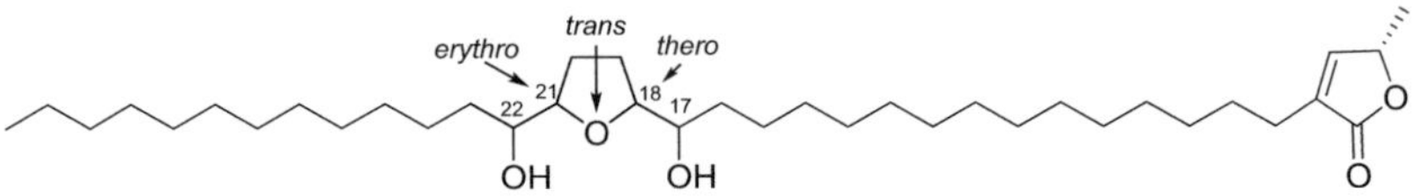

52 (reticulatain)

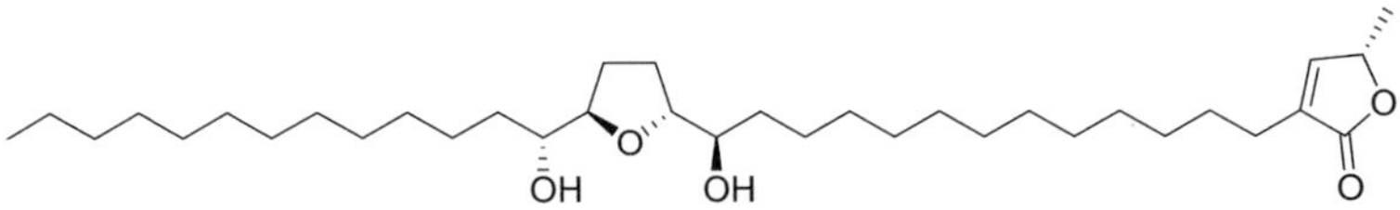

53 (solamin)

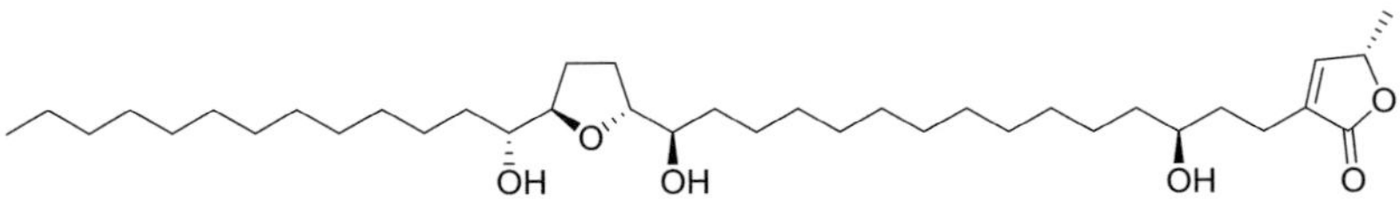

54 (tonkinecin)

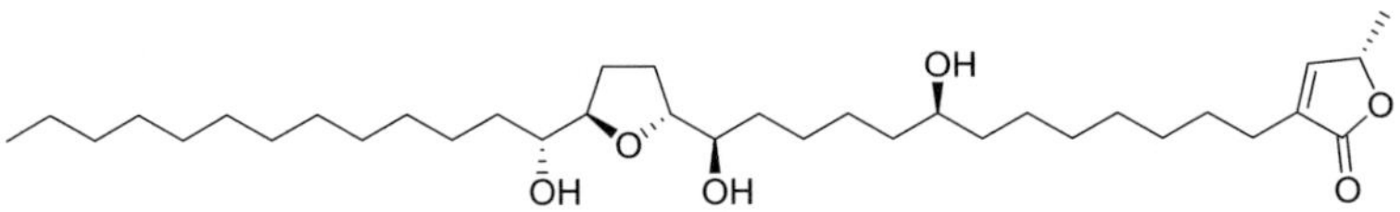

55 ((10*S*)-corossolin)

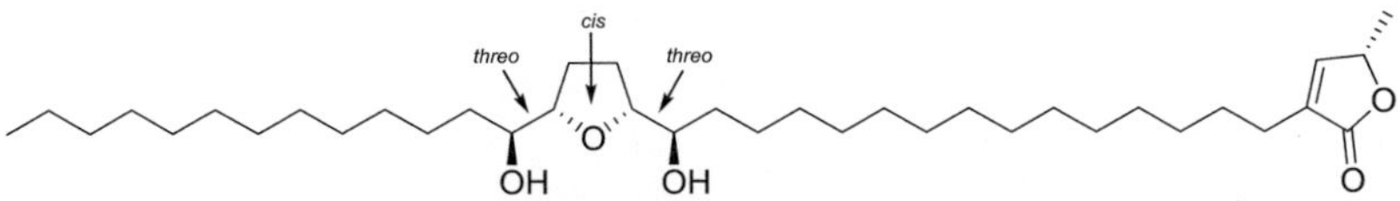

56 (*cis*-reticulatain)

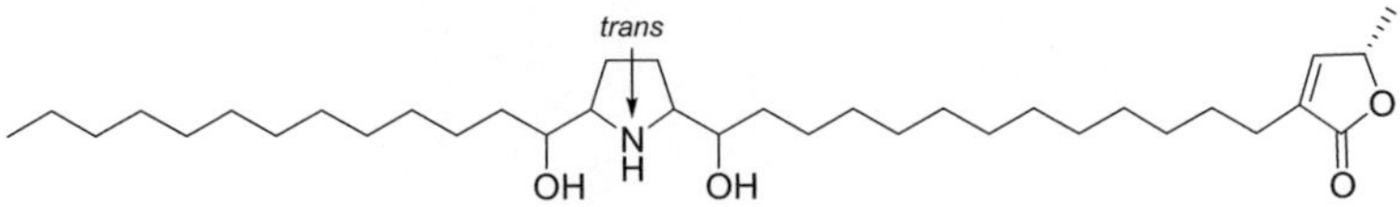

57a-**57d** (aza-solamines)

57a: *erythro-trans-erythro* **57b**: *erythro-trans-threo*

57c: *threo-trans-erythro* **57d**: *threo-trans-threo*

In 2000, Wu et al. synthesized annoacin (**42**), a typical mono-THF AGE with seven chiral centers, via a Sharpless asymmetric dihydroxylation reaction method with the incorporation of three natural hydroxy acids, achieving a high yield of 85% [162, 163]. Wu et al. have also developed a strategy to couple the alkyne

Table 8 Mono-THF AGE synthesis from 1998 to 2011

Compound	Total synthesis	Synthesis method	Starting materials	Steps and yields	Other points	Ref.
Annonacin	First total synthesis	Sharpless AD reaction	L-Ascorbic acid; D-glucono-1,5-lactone and ethyl L-lactate	31 Steps, 85%	1. MOM protection. 2. *Rf* value and spectroscopic data are identical to those reported for the natural product.	[162]
Annonacin and tonkinecin	Total synthesis	1. Asymmetric dihydroxylation 2. Pd(0)-catalyzed coupling reaction with vinyliodides 3. Introduction of the butenolide moiety by aldol condensation of protected S-lactal followed by cleavage of all MOM ethers	D-Glucono-1,5-lactone; D-xylose; (−)-(*S*)-ethyl lactate and L-ascorbic acid	Tonkinecin (18 steps, 86%) and annonacin (33 steps, 85%)	MOM protection.	[163]
(10*R*)- and (10*S*)-Corossolin	First total synthesis	1. Witting reaction 2. Honer-Emmons reaction 3. Swern oxidation 4. Hunsdiecker reaction 5. Wilkinson reaction	D-Gluconolactone and azelic acid monoethyl ester	(10*R*)-Corossolin (24 steps, 62%) and (10*S*)-corossolin (24 steps, 86%)	1. Natural compound is (10*R*)-corossolin. 2. Both compounds can cause suppression of the proliferation of the tumor cell; (*R*) > 18x (*S*) against B16BL6 cell line. 3. (10*S*)-Corossolin was first synthesized. 4. TBS and MOM ether protection.	[164]

(continued)

Table 8 (continued)

Compound	Total synthesis	Synthesis method	Starting materials	Steps and yields	Other points	Ref.
(9*R*)- and (9*S*)-4-Deoxyannoreticuin		Olefin cross-metathesis (CM) coupling reaction		(*R*)-4-Deoxyannoreticuin (6 steps, 70–72%) and (*S*)-4-deoxyannoreticuin (6 steps, 90%)	1. TBDPS ether protection. 2. Unfortunately, identification of one or the other epimeric structures with the natural product was not possible because of the closeness of the physical data for all three compounds. 3. Both C-9 epimeric analogues showed similar toxicity in the low micromolar range, against two human tumor cell lines PC-3 (prostate) and Jurkat (T-cell leukemia).	[165]
4-Deoxyannomontacin	First total synthesis	Sharpless asymmetric dihydroxylation and salen Co^{III}-catalyzed hydrolytic kinetic resolution	D-Glucose and ethyl (*S*)-lactate	17 Steps, 88%	MOM ether protection.	[166]
cis-Gigantrionenin	First asymmetric total synthesis	1. An enzyme-catalyzed epoxide hydrolysis and an enzyme-triggered double cyclization 2. Sonogashira coupling with a γ-lactone segment	Sulfanyl acetic acid; 1,5-hexanediyne and pentynol	14 Steps		[167]

Gigantetrocin A	Total synthesis	1. Wittig reaction 2. Sharpless asymmetric dihydroxylation 3. Honer-Emmons reaction 4. Swern oxidation	*trans*-1,4-Dichloro-2-butene	19 Steps, 59%	MOM protection.	[168]
Gigantetrocin A	First total synthesis	1. Wittig reaction 2. Sharpless asymmetric dihydroxylation 3. Honer-Emmons reaction 4. Swern oxidation	*trans*-1,4-Dichloro-2-butene	19 Steps, 59%	MOM protection.	[169]
(+)-Longicin	First total synthesis	1. Grubbs RCM reaction 2. The butenolide subunit was constructed via an aldol reaction with a macrocyclic lactone precursor	D- and L-Glutamic acids	19 Steps, 62% [internal translactonization strategy (18 steps, 50%)]	These data were identical to the natural product on the basis of the reported physical constants.	[126]
Longifolicin	Total synthesis	1. Asymmetric dihydroxylation 2. Allenyl Pd hydrocarbonylation	Chiral long-chain α- and γ-OMOM allylic stannanes and (*E*)-ethyl 3-formyl-2-propenoate	29 Steps, 97%	TBS ether and MOM ether protection.	[170]
Mosin B and a diastereomer	First total synthesis	Asymmetric desymmetrization of the γ-symmetric diol and the Nozaki-Hiyama-Kishi reaction	A common intermediate, 4-cyclohexene-1,2-diol; l; (*R*)-malic acid and (*S*)-propylenoxide	20 Steps, 72%	Natural compound type is **1a**.	[171]

(continued)

Table 8 (continued)

Compound	Total synthesis	Synthesis method	Starting materials	Steps and yields	Other points	Ref.
Mosin B and one of its diastereomers	Total synthesis	Asymmetric desymmetrization of the γ-symmetric diol and the Nozaki-Hiyama-Kishi reaction	4-Cyclohexene-1,2-diol; (*R*)-malic acid and (*S*)-propylenoxide	Mosin B (20 steps, 72%) and one of its diastereomers (20 steps, 78%)	1. Natural type is **1a**. 2. Diastereomer of mosin B (**1b**) exhibited a higher antiproliferative effect than Adriamycin and had a similar profile of growth inhibition as natural (**1a**) against the cancer cells used.	[172]
Muricatetrocin C	First total synthesis	1. Sonogashira coupling and chelated addition 2. Anomeric O–C rearrangement and HAD reaction	(*R*,*R*)-Dimethyltartrate, butane-1,4-diol and (*R*)-oxiran-2-yl methanol	22 Steps, 82%		[173]
Muricatetrocin C	First total synthesis	1. Sonogashira coupling and chelated addition 2. Anomeric O–C rearrangement and hetero-Diels-Alder (HAD) reaction	(*R*,*R*)-Dimethyltartrate, butane-1,4-diol and (*R*)-oxiran-2-yl methanol	22 Steps, 82%		[174]
Murisolin, natural 16,19-*cis*-murisolin and unnatural 16,19-*cis*-murisolin	Total synthesis	Asymmetric alkynylation of α-tetrahydrofuranic aldehyde with a diyne and Sonogashira coupling with a γ-lactone segment	1,6-Heptadiyne and α-oxyaldehyde	Murisolin (4 steps, 91%), natural 16,19-*cis*-murisolin (11 steps, 85%), and unnatural 16,19-*cis*-murisolin (11 steps, 82%)	1. Three compounds showed strong inhibitory activity against lung cancer cells (DMS114). 2. Natural 16,19-*cis*-murisolin exhibited potent activity against stomach cancer cells.	[175]

Pseudo annonacin A	First total synthesis	Chiron approach	L-Glutamic acid; D-glutamic acid and L-lactic acid	36 Steps, 83%	1. Pseudo annonacin A (15*R*,16*S*,19*S*,20*S*), annonacin A (15*R*,16*R*,19*R*,20*S*). 2. The synthetic product (a mixture of epimers at C-10) had spectroscopic data identical to that of the natural product, but a different optical rotation.	[176]
(17*R*,18*R*,21*R*,22*S*)- and (17*S*,18*S*,21*S*,22*R*)-Reticulatain-1		Mitsunobu inversion and hydrolysis	Acrolein and lauryl magnesium bromide	(17*R*,18*R*,21*R*,22*S*)-Reticulatain-1 (17 steps, 87%)	1. Comparison of the specific optical rotations of **1a** and **1b** did not allow for the strict determination of the absolute configuration. However, the bis-(*R*)-MTPA esters of **1a** and **1b** showed a clear difference in chemical shifts in the ^{1}H NMR spectra. 2. Both compounds showed inhibitory activity against the bovine heart mitochondrial complex I.	[177]
Solamin	Total synthesis	Direct coupling between γ-lactone and mono-THF unit	D-Glutamic acid	19 steps, 97%	TBDPS ether and MOM protection.	[178]
(15*R*,16*R*,19*S*,20*S*)- and (15*S*,16*S*,19*R*,20*R*)-*cis*-Solamin	Concise total syntheses	Permanganate-promoted oxidative cyclization	Tridecanal and (*S*)-ethyl 4-hydroxypent-2-ynoate	*cis*-Solamin (13 steps, 94%)	1. No protecting groups. 2. Relative stereochemical relationship of the THF-diol portion is *threo*/*cis*/*threo*, but the absolute stereochemistry could not be defined.	[179]

(continued)

Table 8 (continued)

Compound	Total synthesis	Synthesis method	Starting materials	Steps and yields	Other points	Ref.
(15*R*,16*R*,19*S*,20*S*)- and (15*S*,16*S*,19*R*,20*R*)-*cis*-Solamin	Total synthesis	TBHP-VO(acac)$_2$ diastereoselective epoxidation and cyclization	1,8-Diiodooct-1-ene and 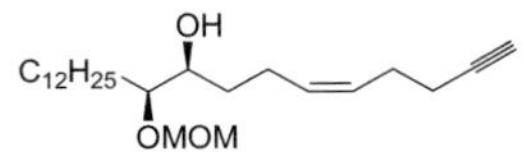 (9*S*,10*S*,*Z*)-10-(methoxymethoxy)docos-5-en-1-yn-9-ol 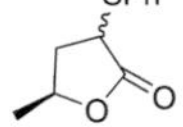 (*S*)-5-methyl-3-(phenylthio)-dihydrofuran-2(3*H*)-one	13 Steps, 60%	Naturural *cis*-solamin is of configuration (15*S*,16*S*,19*R*,20*R*) (**1a**).	[180]
(15*R*,16*R*,19*S*,20-*S*,34*S*)- and (15*S*,16*S*,19*R*,20-*R*,34*S*)-*cis*-Solamin	First total synthesis	VO(acac)$_2$-catalyzed diastereoselective epoxidation of (*Z*)-bis-homoallylic alcohol followed by spontaneous cyclization for the *cis*-THF ring formation.	1,8-Diiodooct-1-ene and 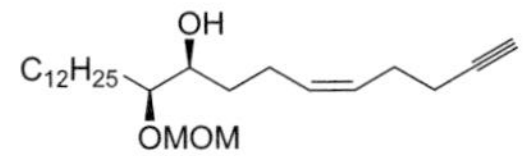 (9*S*,10*S*,*Z*)-10-(methoxymethoxy)docos-5-en-1-yn-9-ol 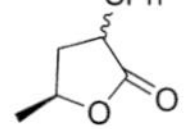 (*S*)-5-methyl-3-(phenylthio)-dihydrofuran-2(3*H*)-one	(15*R*,16*R*,19*S*,20-*S*,34*S*)-*cis*-Solamin (13 steps, 60%)	1. Natural *cis*-solamin is of the configuration (15*R*,16*R*,19*S*,20*S*,34*S*). 2. Both compounds showed inhibitory activity against the bovine heart mitochondrial complex I.	[181]
Solamin	Total synthesis	1. A ring-closing metathesis (RCM) reaction using a ruthenium imidazolylidene complex 2. Asymmetric epoxidation	Propargylic alcohol	24 Steps, 85%	The first application of RCM using the ruthenium catalyst for the total synthesis of solamin.	[182]

cis-Solamin A [(15*R*,16*R*,19*S*,20*S*,34*S*)-*cis*-solamin]	Total synthesis	1. Olefin crossmetathesis between the tetrahydrofuran moiety and γ-lactone moiety 2. An enzymatic kinetic transesterification procedure was successfully applied to the synthesis of an optically pure γ-lactone moiety	(−)-Muricatacin	11 Steps, 95%	No protection/ deprotection steps.	[183]
cis-Solamin A, *cis*-solamin B, and reticulatacin	Total synthesis	1. Olefin crossmetathesis between the tetrahydrofuran moiety and γ-lactone moiety 2. An enzymatic kinetic transesterification procedure was successfully applied to the synthesis of an optically pure γ-lactone moiety	Muricatacin	*cis*-Solamin A (10 steps, 95%) and *cis*-solamin B (10 steps, 93%) and reticulatacin (9 steps, 91%)	1. No protection/ deprotection steps. 2. *cis*-Solamin A and *cis*-solamin B are of the (15*R*,16*R*,19*S*,20*S*,34*S*) and *cis*-solamin of (15*S*,16*S*,19*R*,20*R*,34*S*) configurations.	[184]
aza-Solamin isomers		Coupling with vinyl iodide segment and *trans*-pyrrolidine segment	2,5-*trans*-Bis(methoxycarbonyl)pyrrolidine; ethyl (*S*)-lactate and undecylenic acid	aza-Solamin (9 steps, 40%)	1. Solamin analogs active against several tumor cell lines were also observed. 2. Being first synthesized with a facile route.	[185]
Tonkinecin	First synthesis	Palladium-catalyzed cross-coupling raction and Sharpless asymmetric dihydroxylation	D-Glucose; L-lactate; D-xylose	16 Steps, 81%	1. The physical data are the same as those of the natural one. 2. MOM protection.	[186]

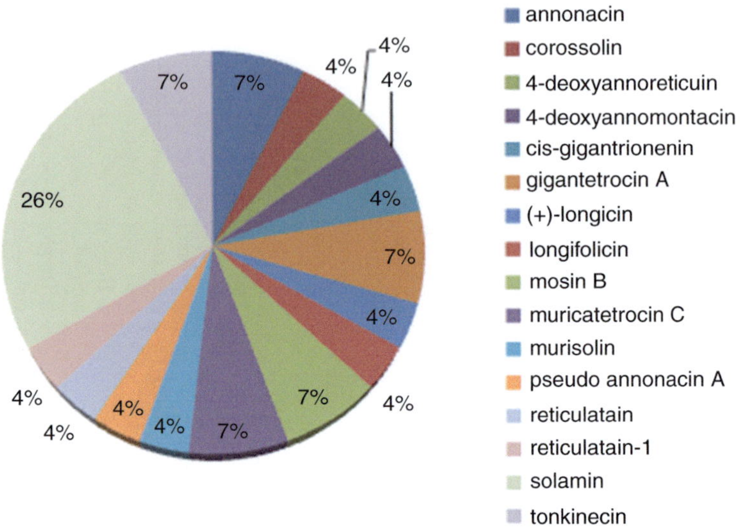

Fig. 8 Analysis of linear and epoxy AGEs synthesis from 1998 to 2011

intermediate and the terminal epoxide to produce (10*R*)-(**43**) and (10*S*)-corossolin (**55**) [164]. The comparison of the optical rotation, the ^{13}C NMR spectrum and the in vitro activity of the synthesized compound with the natural form allowed further postulation of the absolute configuration of the natural form at C-10 as (*R*).

Mootoo et al. employed an olefin cross metathesis method with THF and butenolide alkene to convergently synthesize the C-9 epimers of 4-deoxy-annoreticuin (**45**) [165] since both stereoisomers exhibit similar cytotoxicity against prostate tumor and T-cell leukemia cells. However, the identification of the unassigned configuration at C-9 was unsuccessful owing to the similar physical data of the two synthetic products.

Orru et al. reported an asymmetric total synthesis of *cis*-gigantrionenin (**46**) by utilizing a rapid and efficient enzyme-catalyzed epoxide hydrolysis method and an enzyme-triggered double cyclization method for the construction of the THF ring and the alkyl chain with a double bond. In addition, the Sonogashira coupling method was employed to overcome a synthesis problem inherent from the linkage between the THF fragment and γ-lactone segment [167]. Various other methods were reported also for synthesizing gigantetrocin A (**47**), such as the Sharpless asymmetric dihydroxylation, the Horner-Emmons reaction, Swern oxidation, and the Wittig reaction [169].

Hanessian et al. used D- and L-glutamic acid as chirons corresponding to two five-carbon segments, harboring stereogenic centers at C-4 and at C-17 of longicin (**26**) using the Grubbs RCM reaction as the "chain elongation" strategy, which underwent coupling and assembling of the complete aliphatic chain. The butenolide unit was subjected to an aldol reaction with a macrocylic lactone precursor.

The synthetic product was identical to the natural product according to the reported physical constants and spectroscopic data obtained [126].

Marshall et al. designed a bidirectional synthesis strategy with minor modifications toward longifolicin (**48**), a C_{35} mono-THF AGE with a *threo/trans/threo* configuration. The group successfully achieved a high yield. The allylic stannanes used in the experiment demonstrated the potential for converging mono-THF AGEs efficiently [170].

Tanaka et al. attempted to determine the absolute configuration of mosin B (**49**). Despite the ^{1}H and ^{13}C NMR spectroscopic data at hand, Mosher ester methodology and X-ray analysis were not conclusive [171, 172]. An efficient asymmetric desymmetrization method of cyclic *meso*-1,2-diols using C_2-symmetric bis-sulfoxide synthesis has provided two possible candidates for determing the absolute configuration (see Fig. 9). The THF ring fragment was introduced stereoselectively by a stereodivergent synthesis starting from 4-cyclohexene-1,2-diol based on a desymmetrization strategy. The γ-lactone fragment was synthesized via coupling a triflate and a chiral α-sulfenyl γ-lactone. By plotting the difference between the chemical shifts of the ^{13}C NMR spectroscopic data of natural mosin B (**49**) and both candidate structures, the THF moiety of mosin B (**49**) was even more closely matched to the **49a** configuration (see Fig. 10) [172].

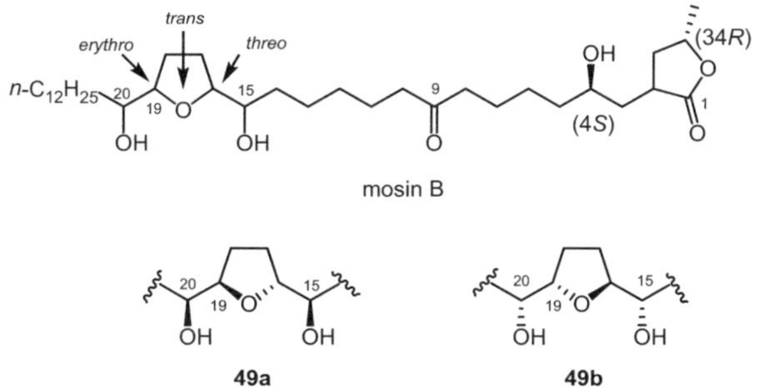

Fig. 9 Possible structures of mosin B (**49**)

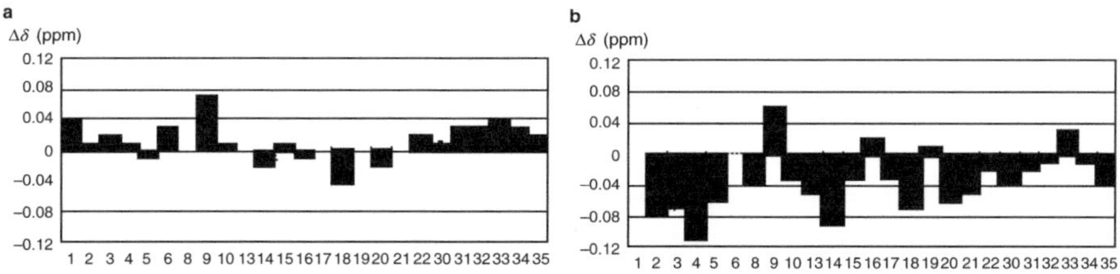

Fig. 10 Differences between the characteristic chemical shifts of the carbon atoms of natural mosin B (**49**) and those of each candidate **49a** (*left*) and **49b** (*right*) (75 MHz, $CDCl_3$). The x and y axes represent the carbon number and $\Delta\delta$ ($= \delta_{\mathbf{a,b}} - \delta_{\text{mosin B}}$)

Fig. 11 Structures of analogues of murisolin (**25**)

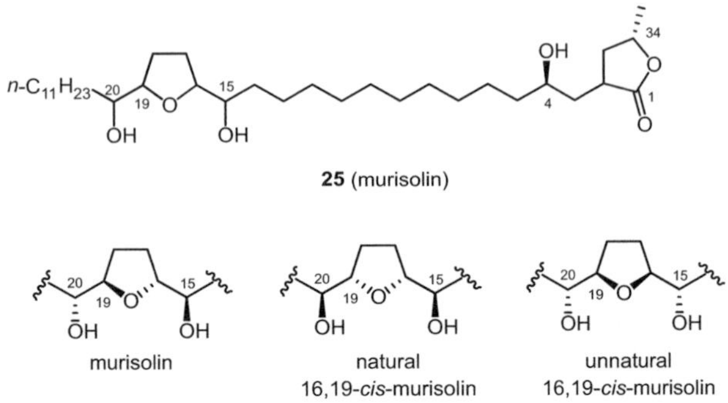

Muricatetrocin C (**50**) exhibited excellent cytotoxic activities against three human cell lines, including PC-3, PACA-2, and A-549; thus it may be regarded as a potential antitumor agent. Ley et al. provided a stereoselective strategy with a linear sequence of 22 steps and achieved a 82% yield. They applied 2,3-butanediacetal (BDA)-protected butane tetrol as a building block for the *anti*-1,2-diol units. The use of the anomeric oxygen to carbon rearrangement of alkynyl stannanes for the stereoselective construction of the 2,5-*trans*-disubstituted THF ring component, and finally the implementation of a hetero-Diels-Alder (HDA) reaction allowed construction of the hydroxy-butenolide terminus [173, 174].

Tanaka et al. provided a procedure for the total synthesis of murisolin (**25**): asymmetric alkynylation of α-tetrahydrofuranic aldehyde with a diyne and Sonogashira coupling method using a γ-lactone segment as key steps [175]. The approach achieved a good yield and high diastereoselectivity. Based on the interesting stereodivergent differences in biological activity of these compounds, unnatural murisolin with an opposiste configuration from that of the natural ones was synthesized (**25a**). Using the COMPARE analysis, a tool to examine a pair of compounds in terms of their mean graphs, both compounds showed inhibitory activity against DMS114 lung cancer cells, and indicated that they share the same mode of action (see Fig. 11).

Hanessian et al. used three natural acids and took a chiron approach for the total synthesis of the *erythro,trans,threo*-(15*R*,16*S*,19*S*,20*S*)-diastereomer of annonacin A (**51a**) [176]. The inavailability of this compound made the comparison of the synthetic sample with the natural product difficult and only referred to the physical data, specifically the optical rotation (see Fig. 12). However, the undefined configuration of positions C-4, C-10, and C-34 of annonacin A (**51**) still remains a challenge.

Although Makabe et al. successfully synthesized two diastereomers of reticulatain-1 (**52**) using the Mitsunobu inversion method in 2004 (see Fig. 13) [177], the absolute configuration of natural reticulatain-1 (**52**) has not yet been resolved. However, the use of the Mosher ester method showed a clear difference between the two synthetic products in terms of their ^{1}H NMR chemical shifts. This demonstrated that if Mosher esters were to be prepared, the absolute configuration of naturally occuring reticulatain-1 (**52**) could be determined. Moreover, both synthetic epimers displayed very similar reactivity to bovine heart mitochondrial complex I.

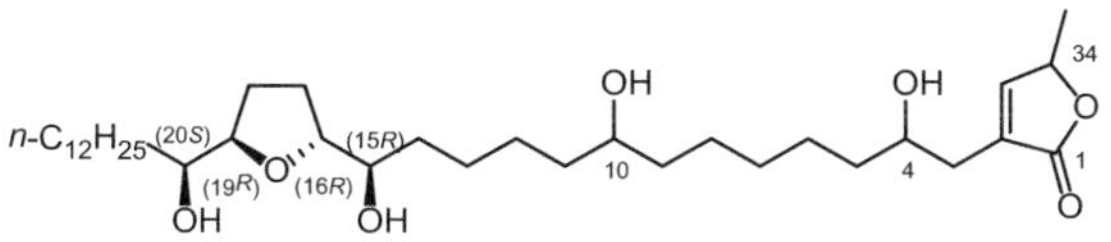

51 (annonacin A)
threo,trans,erythro
((15*R*) → (20*S*))

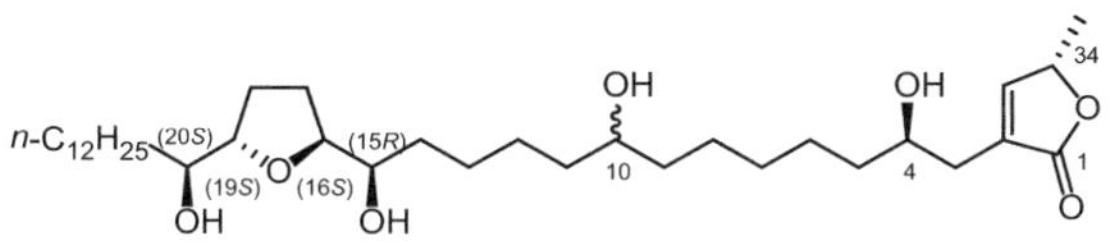

51a (pseudoannonacin A)
(4*S*,10*R*/*S*,15*R*,16*S*,19*S*,20*S*)

Fig. 12 Structures of analogues of annonacin A (**51** and **51a**)

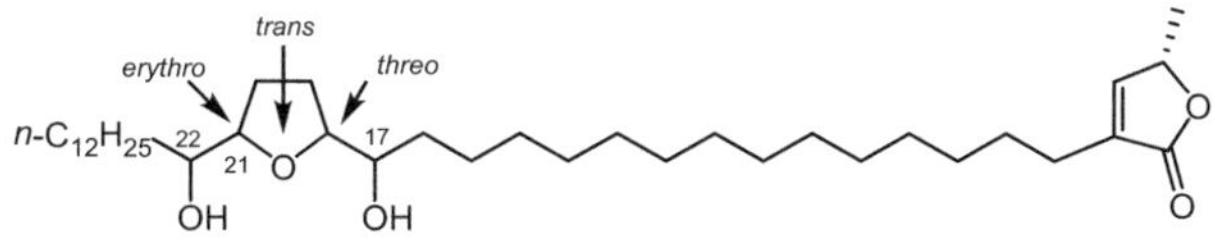

52 (reticulatacin-1)

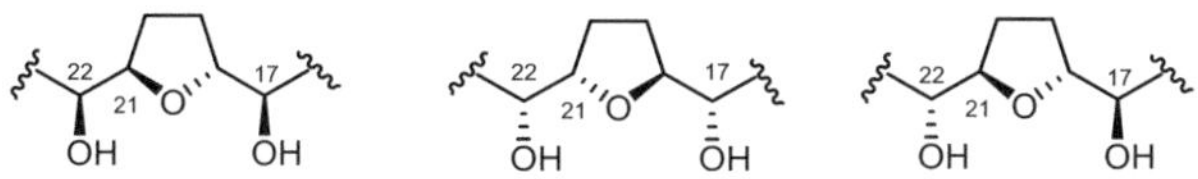

1a **1b** reticulatacin

Fig. 13 Structures of analogues of reticulatacin-1 (**52**)

Kitahara et al. completed the synthesis of solamin (**53**) via direct coupling between a long chain iodide and γ-lactone, which resulted in an excellent yield of 97% from D-glutamic acid in 19 steps [178]. Heck et al. further described a convergent total synthesis method for solamin (**53**) in 2004 [182]. The central THF core was obtained by means of a ring-closing metathesis (RCM) reaction using a ruthenium imidazolylidene complex. All the stereoisomers of solamin (**53**) could be obtained readily by employing this strategy using (*E*)- or (*Z*)-allyl alcohol and (+)- or (−)-DET for the asymmetric epoxidations. Brown et al. synthesized *cis*-solamin (**27**) and its diastereoisomer, 15,16-di-*epi*-solamin (also named as *cis*-solamin B, **27a**) with potassium permanganate under phase-transfer conditions with an excellent yield (94%) (see Fig. 14) [179]. No hydroxy group protection was required during this approach. The absolute configuration of *cis*-solamin (**27**) based only on the optical rotation data has remained uncertain. Makabe et al. reported the synthesis of two possible variants of *cis*-solamin (**27**) using VO(acac)$_2$-catalyzed diastereoselective epoxidation followed by cyclization of bis-homoallylic alcohol, in which the specific rotation ($[\alpha]_D^{21}$ +26°cm^2/g) of the product with the (15*R*,16*R*,19*S*,20*S*) configuration matched with the natural

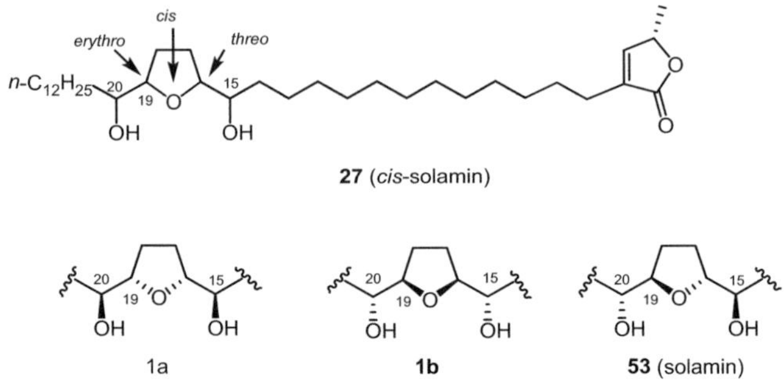

Fig. 14 Structures of solamin (**53**) and *cis*-solamin (**27**)

cis-solamin (**27**) ([α]$_D$ +22°cm^2/g) [180, 181]. Konno et al. developed a synthesis strategy for *cis*-solamin (**27**) via the enzymatic kinetic transesterification method producing an optically pure γ-lactone moiety and the olefin cross-metathesis reaction constructing two distinct motifs between the tetrahydrofuran moiety and γ-lactone moiety [183]. This approach also favored the synthesis of solamin-type AGEs, like the isomer of *cis*-solamin (**27a**) and *cis*-reticulatacin (**56**) [184]. In addition, Shen et al. were able to couple vinyl iodide and *trans*-pyrrolidine segments to prepare the four possible relative configurations of aza-solamin (**57a–57d**), in which the THF core unit is replaced by pyrrolidine [185]. The stereochemistry proposed was supported by ^{1}H NMR spectroscopic analysis.

Tonkinecin (**54**) is a mono-THF AGE with an unusual C-5 carbinol center. Wu et al. reported the first total synthesis of tonkinecin (**54**) by an efficient and highly stereoseclective palladium-catalyzed cross-coupling reaction [163, 186]. Four stereogenic centers were derived from three carbohydrates, D-glucose, L-lactate, and D-xylose, and two stereogenic centers were produced via Sharpless asymmetric dihydroxylation. The data of the synthetic product matched with those of natural product.

5.3 Adjacent Bis-THF Annonaceous Acetogenins

Evolution of synthesis methodology after 1998 has allowed higher yields of adjacent bis-THF AGEs (Fig. 15; Table 9).

In 1999, Mootoo et al. reported a two-directional strategy for synthesizing the bis-THF core of asimicin (**8**). This method relied on the iodoetherification reaction for desymmetrization of a C_2-symmetric precursor in a relatively easy and inexpensive reaction through large-scale preparations [205].

Sinha et al. divided asimicin (**8**) into two major fragments, the butenolide moiety and the bis-THF component [187]. Three strategies were employed comprising: the naked carbon skeleton strategy, the convergent strategy, and the hybrid synthesis strategy to construct the bis-THF component. The advantages of the naked carbon skeleton strategy relied on the production of all stereogenic centers via specific

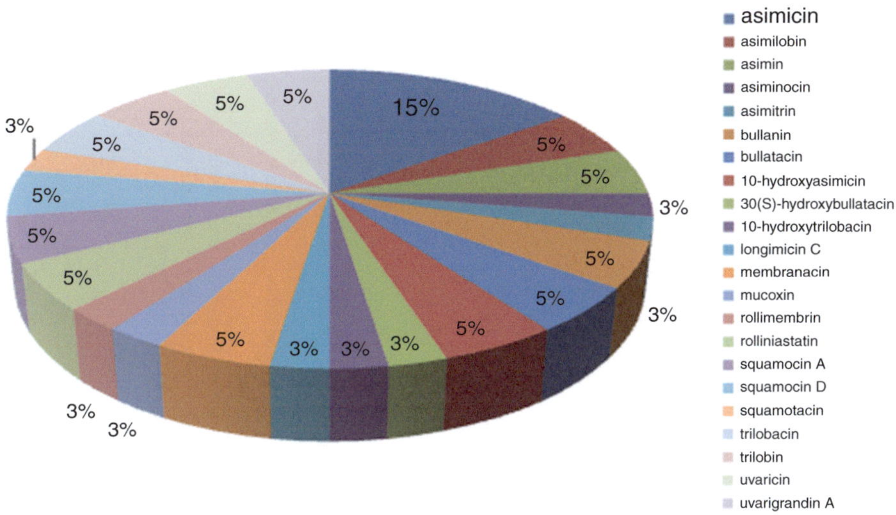

Fig. 15 Analysis of bis-THF AGEs (adjacent type) synthesis from 1998 to 2011

positioning of the oxygen functions onto the unsaturated, nonfunctionalized carbon skeleton. The convergent strategy can couple two series of diastereomeric fragments while taking into account their efficiency and versatility. The hybrid approach involves partially functionalized intermediates integrating the advantages of the linear and the convergent strategies that further improves synthesis efficiency and diversity. The butenolide fragment was prepared from deca-1,9-diene via an eight-step procedure that achieved a good yields. As a result, asimicin (**8**) was produced successfully with its spectroscopic data identical to those of the naturally occurring compound.

Roush et al. reported a double asymmetric [3 + 2]-annulation reaction toward the synthesis of asimicin (**8**), which used a bis-THF core with a *trans-threo-trans* configuration [188]. The key step with this approach is that the chiral (*E*)-γ-(dimethylphenylsilyl)allylborane reagent reacts with aldehydes to afford chiral, nonracemic *anti*-β-silyloxy allysilanes with high enantioselectivity. In order to develop a highly stereoselective synthesis of asimicin (**8**), the coupling of a 2,5-*trans*-tetrahydrofuryl aldehyde with the chiral allysilane was demonstrated (see Fig. 16). This approach provided the general synthesis of six diastereomeric bis-THF structures and the derived six diastereomeric desilylated bis-THF structures were related to the core sub-structure of several bis-THF type of AGEs.

Marshall et al. constructed the bis-THF core of asimicin (**8**) using a bidirectional outside-in hydroxy mesylate cascade cyclization pathway from 4-penten-1-ol, and then a subsequent Grubbs cross-metathesis reaction, which resulted in a good yield [136]. It is worth noting that almost 100% of the material was recovered from this study. They further synthesized three terminal hydroxy analogs **58a–58c** and a truncated analog **59** of asimicin (**8**) via the key Grubbs cross-metathesis reaction method [189]. Cytotoxicity testing of these analogs against HCT-116 human colon cancer cells revealed that the cytotoxic effect was not attributed to the terminal hydroxy group.

Table 9 Synthesis topics on adjacent bis-THF AGEs from 1998 to 2011

Compound name	Total synthesis	Synthesis method	Starting materials	Steps and yields	Other points	Ref.
Asimicin and bullatacin	Total synthesis	1. The naked carbon skeleton strategy 2. The convergent strategy 3. The hybrid synthetic strategy	Undecanal; deca-1,9-diene	Asimicin (8 steps, 81%) and bullatacin (9 steps, 79%)	MOM protection.	[187]
Asimicin and 12 diastereomeric bis-THF structures		Double stereodifferentiating [3 + 2]-annulation reactions of chiral allylsilanes with 2-tetrahydrofuryl aldehydes		Asimicin (7 steps, 80%)		[188]
Asimicin and a C_{32} analogue	Total synthesis	A bidirectional outside-in hydroxy mesylate cascade cyclization route and Grubbs cross-metathesis	4-Penten-1-ol and 2-(dec-9-enyl) oxirane	23 Steps, 80%	MOM protection.	[136]
Asimicin and its analogues		Grubbs cross-metathesis	10-Undecenal; 4-penten-1-ol	24 Steps, 80–91%	The truncated asimicin analog of asimicin is some 20 times more active against HCT-116 colon cancer cells than its natural counterpart.	[189]
Asimilobin	First total synthesis	Wittig reaction	*trans*-1,4-Dichloro-2-butene	13 Steps, 79%	The natural product has the opposite absolute configuration on the bis-THF unit of that reported in the literature. 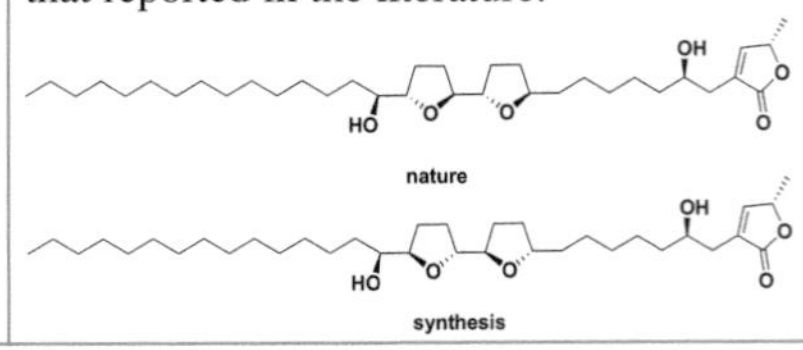	[190]

(−)-Asimilobin and (+)-asimilobin	First total synthesis	Wittig reaction	*trans*-1,5,9-Decatriene and L-glutamic acid	(−)-Asimilobin (12 steps, 79%) and (+)-asimilobin (14 steps, 76%)	By virtue of these synthesis results, the absolute configuration of the bis(THF) unit in naturally occurring (+)-asimilobin should be corrected.	[191]
(10*R*)-Asimin and (10*S*)-asimin	Total synthesis	1. The addition of an enantioenriched γ-OMOM allylic indium reagent 2. The addition of a dialkyl zinc reagent 3. Aldol condensation	6-(Benzyloxy)hex-1-en-3-ol	(10*R*)-Asimin (17 steps, 90%) and (10*S*)-asimin (17 steps, 74%)	1. The natural compound is (10*R*)-asimin. 2. MOM protection.	[192]
Asiminocin, asimicin, asimin, and bullanin		1. Additions of enantiopure allylic indium or tin reagents 2. Sonogashira coupling		Asiminocin (17 steps, 66%), asimicin (17 steps, 57%), asimin (17 steps, 57%), and bullanin (11 steps, 82%)	1. IC_{90} values have not previously been determined for these annonaceous acteogenins. 2. The IC_{90} activities were ca. 10^{-3} μ*M* for asimicin and asimin but only 0.1–1 μ*M* for bullanin and asiminocin. 3. These acetogenins have been reported to exhibit inhibitory activities of ~10^{-12} μg/cm^3 (IC_{50}) against the HT-29 human colon cancer cell line.	[192]
Asimitrin (C-10–C-34 fragment)	First stereoselective synthesis	Stereoselective intramolecular oxymercuration and chelation-controlled Grignard reactions	1,2:5,6-di-*O*-Isopropylidene-*α*-D-glucofuranose	17 Steps, 76%		[193]

(continued)

Table 9 (continued)

Compound name	Total synthesis	Synthesis method	Starting materials	Steps and yields	Other points	Ref.
(+)-(30*S*)-Bullanin	Total synthesis	Sharpless asymmetric dihydroxylation (AD) and S_{E2}' additions	(*E*)-Methyl 8-(benzyloxy)oct-4-enoate and *tert*-butyldimethyl (4-(tributylstannyl) but-3-ynyloxy) silane	25 Steps, 100%	The optical reaction of (30*S*)-bullanin, $[\alpha]_D$ +24°cm^2/g, is in close agreement with the reported value for the mixture, $[\alpha]_D$ +28°cm^2/g.	[194]
(+)-Bullatacin	Total synthesis	Diastereoselective [3+2]-annulation of the highly enantiomerically enriched allylsilane and racemic aldehyde	α-Benzyloxy acetaldehyde; (*E*)-dimethyl(phenyl) (3-(tributylstannyl) prop-1-enyl)silane and decanal	11 Steps, 60%		[195]
10-Hydroxyasimicin	First total synthesis	1. Critical ring-closing metathesis (RCM) step and desymmetrization 2. Hetero-Diels–Alder (HDA) reaction	(*S*,*S*)-Dimethyl-tartrate and butane-1,4-diol	25 Steps, 68%	MOM protection.	[133]
(30*S*)-Hydroxybullatacin, uvarigrandin A and (5*R*)-uvarigrandin A (narumicin I?)	Total synthesis	1. Additions of chiral α-oxygenated allylic stannane and indium reagents 2. Core ring closure reaction 3. Sonogasjhira coupling	(*S*)- or (*R*)-Malic acid	(30*S*)-Hydroxybullatacin (3 steps, 74%), uvarigrandin A (13 steps, 86%) and 5(*R*)-uvarigrandin A (3 steps, 57%)	Spectroscopic properties of synthetic (30*S*)-hydroxybullatacin and uvarigrandin A, as well as their Mosher ester derivatives, were in close agreement to the reported values of the natural substances. The synthetic (5*R*)-uvarigrandin A is possibly identical to narumicin I, but subtle differences in the reported NMR spectra prevented an unambiguous assessment of this point.	[196]

(19*R*,20*S*)-10-Hydroxytrilobacin, (19*S*,20*S*)-4,10-dihydroxysquamocin N, (19*R*,20*R*)-10-hydroxyasimicin and (19*S*,20*R*)- an unnatural acetogenin		Stereodivergent [3 + 2] annulation reaction of tetrahydrofuranyl carboxaldehyde and allylsilane	Methyl-4-pentenoate; methyl adipoyl chloride; 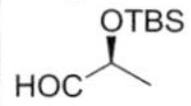 and 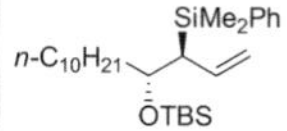	22 Steps, 70–87% (70–87% $\geq$ 98% *ee*)		[197]
Longimicin C	First total synthesis	1. An iterative acetylene–epoxide coupling strategy 2. Sharpless dihydroxylations and intramolecular Williamson etherifications 3. Regioselective epoxide-openings	D-Mannitol	22 Steps, 51%		[198]
Membranacin	Total synthesis	Transition metal-oxo and metal-peroxy-mediated oxidative cyclizations	Ethyl acetoacetate and 1,4-dibromobut-2-ene	17 Steps, 78%		[199]
Mucoxin	First total synthesis	Thiophenyl-directed epoxydiol cyclization, and a one-pot 1,2-*n*-triol cyclization strategy	3-Butynol	32 Steps, 80%	This is particularly advantageous in light of the fact that spectroscopic data for the synthetic and natural materials did not match. →Misassignment of the relative stereochemistry in the C-8–C-17 core.	[140]

(continued)

Table 9 (continued)

Compound name	Total synthesis	Synthesis method	Starting materials	Steps and yields	Other points	Ref.
Rolliniastatin (bis-THF core)		Iodoetherification (modular synthesis)	Tri-*O*-acetyl-D-glucal and *t*-butanol or trifluoroethanol	19 Steps, 68%		[200]
Rolliniastatin 1, rollimembrin, and membranacin		A radical cyclization of β-alkoxyvinyl Sulfoxide-Pummerer rearrangement-allylation protocol	Butane-1,4-diol; D-malic acid and (*R*)-glycidyl tosylate or (*R*)-epichlorohydrin	Rolliniastatin 1 (29 steps, Rollimembrin (29 steps, 66%), and membranacin (25 steps, 74%)		[130]
Squamocin A and squamocin D	Total synthesis	Key reactions are additions of organomagnesium compounds to bi-THF aldehydes	Heptanal; 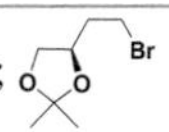 and 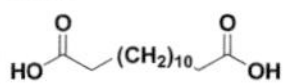	Squamocin A (25 steps, 82%) and squamocin D (25 steps, 79%)		[201]
Squamocin A and squamocin D	Total synthesis	1. Multiple Williamson reaction 2. Addition of organomagnesium compounds to aldehyde functions	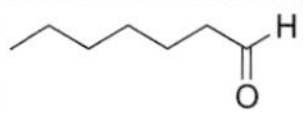 and 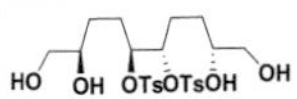	Squamocin A (23 steps, 33%) and squamocin D (23 steps, 39%)		[202]
Squamotacin	First total synthesis	Sharpless asymmetric dihydroxylation (AD) and asymmetric epoxidation (AE) reaction	(+)-Muricatacin	27 Steps, 48%	MOM protection.	[203]

Trilobacin and 36 analogues		Rhenium(VII) oxides mediated mono- or bis-oxidative cyclization, Shi mono- or bis-asymmetric epoxidation, Sharpless asymmetric dihydroxylation, Williamson's type etherification, and Mitsunobu inversion	8,9:12,13-(*E*,*E* and *Z*,*E*)-16-Benzyloxy-5-hydroxy-hexadeca-1,4-olide	43–92% (19,23-bis-*epi*-Trilobacin (78%) and 16,19-bis-trilobacin were (80%)	Using the nonsymmetrical bis-THF lactones, syntheses of two non-natural acetogenins (19,23-bis-*epi*-trilobacin and 16,19-bis-*epi*-trilobacin) were achieved.	[204]
Trilobacin and asimicin (bis-THF core)		A two-directional trategy based on the haloetherification reaction of a bis-5,6-*O*-isopropylidene alkene	Cyclooctadiene	Trilobacin (bis-THF core) (12 steps, 69%) and asimicin (bis-THF core) (14 steps, 70%)		[205]
(+)-Trilobin	Total synthesis	Carbon-carbon bond-forming steps	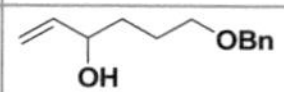	17 Steps, 85%		[206]
(+)-Trilobin	First total synthesis	Sharpless asymmetric dihydroxylation (AD) and asymmetric epoxidation (AE) reaction		22 Steps, 81%	MOM protection.	[207]

(continued)

Table 9 (continued)

Compound name	Total synthesis	Synthesis method	Starting materials	Steps and yields	Other points	Ref.
Uvaricin		1. Palladium-mediated, ligand-controlled double cyclization 2. The C_2-symmetric diene produced was desymmetrized via Sharpless asymmetric dihydroxylation	D-Tartrate		Using chiral DPPBA ligands.	[208]
(+)-Uvaricin	First total synthesis	Sharpless asymmetric dihydroxylation (AD) and Williamson type etherification reaction	Ethyl pentadec-4-enoate and non-8-yn-1-ol	18 Steps, 32%		[209]

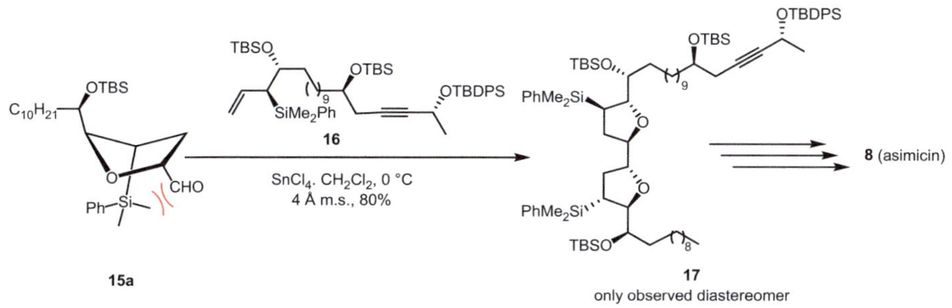

Fig. 16 [3 + 2]-Annulation reactions of chiral allylsilanes and chiral aldehydes to synthesize asimicin (**8**) and its analogues

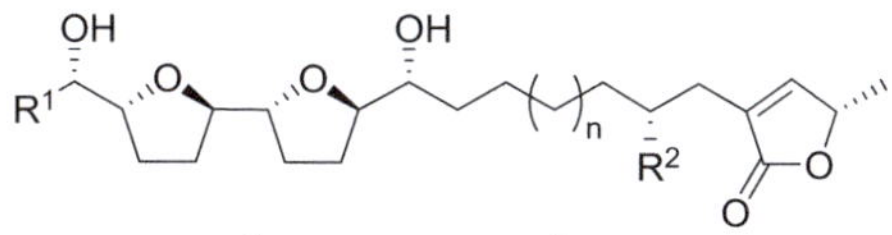

16 $R^1 = CH_3(CH_2)_9$, $R^2 = OH$, n = 7
58a $R^1 = HO(CH_2)_6$, $R^2 = OH$, n = 7
58b $R^1 = HO(CH_2)_{11}$, $R^2 = OH$, n = 7
58c $R^1 = HO(CH_2)_{11}$, $R^2 = H$, n = 6
59 $R^1 = CH_3(CH_2)_7$, $R^2 = OH$, n = 7

three terminal hydroxy analogs **58a-58c** and a truncated analog **59** of asimicin (**8**)

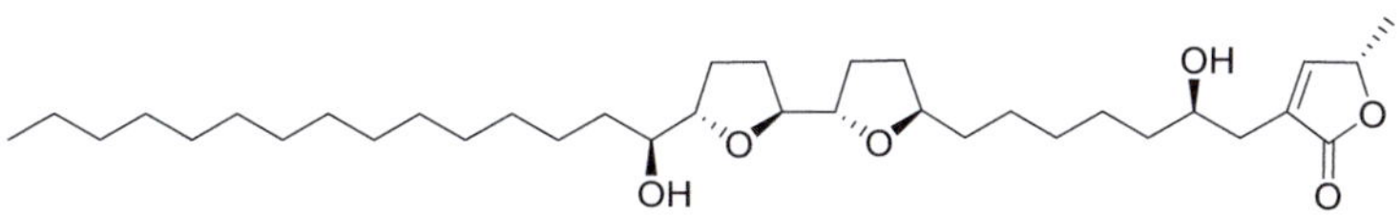

60 (asimilobin)

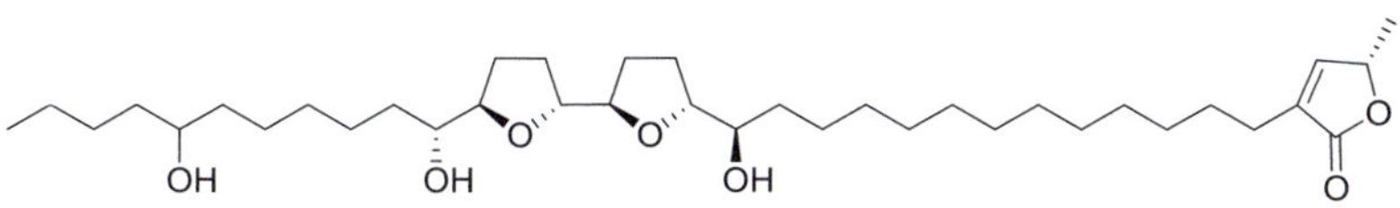

61 (asiminocin)

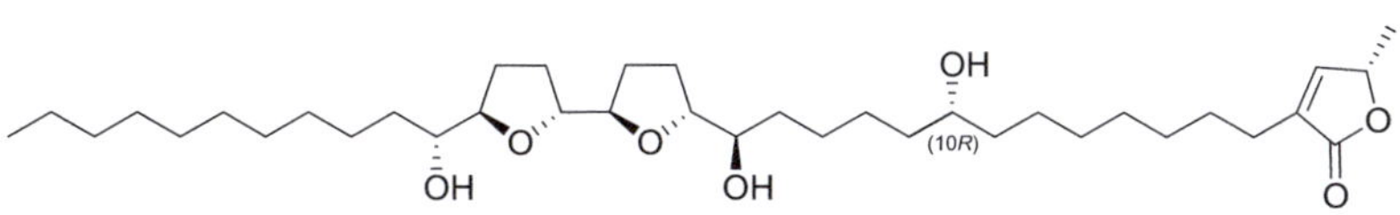

62 (asimin)

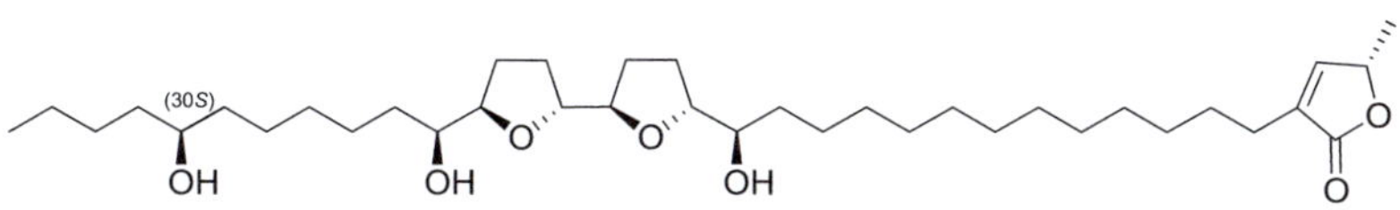

63 (bullanin)

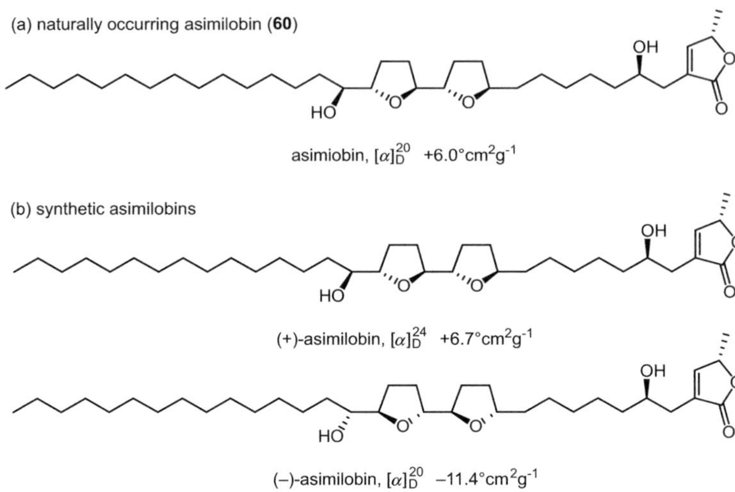

Fig. 17 Structure of asimilobin (**60**)

To investigate the structure of asimilobin (**60**), a bullatacin-type of AGE, Wang et al. published a short and convergent route to synthesize (+)- and (−)-asimilobins via a Wittig reaction [190, 191]. With this approach, the corrected absolute configuration of naturally occurring asimilobin (**60**) was clearly indicated by the optical rotation. Natural asimilobin (**60**), with an positive optical rotation value of $[\alpha]_D$ +6.0°cm^2/g, compares with (+)-asimilobin ($[\alpha]_D$ +6.7°cm^2/g), but not (−)-asimilobin ($[\alpha]_D$ −11.4°cm^2/g) (see Fig. 17).

Marshall et al. developed a modular synthesis for initial synthetsis targets, such as asiminocin (**61**), asimin (**62**), asimicin (**8**), and bullanin (**63**) [210]. This modular synthesis focused on the effects of two pairs of allylic stannanes from parts of these AGEs, namely, their aliphatic termini and the spacer groups. The two parts were attached to the hydrofuran core precursor via allylation and cyclization methods to give bis-THF rings. Then, a Sonogashira coupling method was used to connect the butenolide termini to the bis-THF rings to result in a completed synthesis target of AGEs.

Marshall et al. also reported a study on the total synthesis of asimin (**62**) using an enantioenriched γ-OMOM allylic indium reagent, a dialkyl zinc reagent and aldol condensation, in which 12 chiral centers were constructed precisely via 17 steps along with a good overall yield [192]. Interestingly, the spectroscopic data of synthetic (10*R*)-hydroxyasimin are similiar to those of natural asimin (**62**) (see Fig. 18). In addition, Marshall et al. described a convenient route to synthesize (30*S*)-bullanin (**63**). Its optical rotation of $[\alpha]_D$ +24°cm^2/g is in close agreement with the natural product, $[\alpha]_D$ +28°cm^2/g. With this approach, a Sharpless asymmetric dihydroxylation and SE_2' additions of oxygenated nonracemic allylic

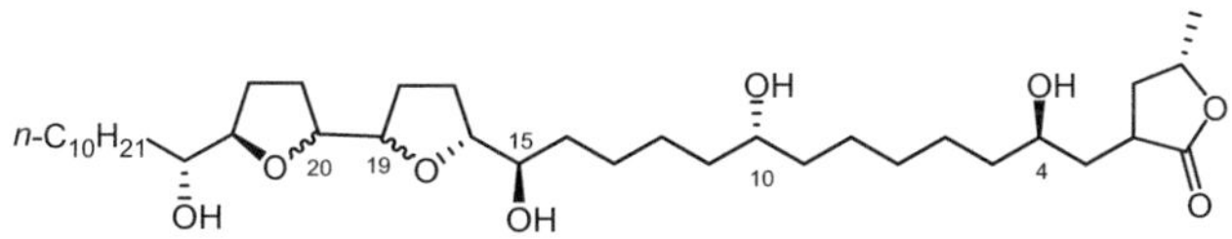

(19*R*,20*S*): 10-hydroxytrilobacin (**68**)
(19*S*,20*S*): 4,10-dihydroxysquamocin N
(19*R*,20*R*): 10-hydroxyasimicin
(19S,20*R*): an unnatural acetogenin

Fig. 18 Structures of stereoisomers of 10-hydroxytrilobacin (**68**)

stannane and indium reagents to γ-oxygenated aldehydes were employed via a 25-step procedure to construct the stereocenters of the THF segment [194]. According to the literature, this approach resulted in a 100% yield of the synthetic sample [194].

Roush et al. completed the total synthesis of bullatacin (**7**) through sequential chelate-controlled [3 + 2]-annulation reactions. The highly enantiomerically enriched allylsilane and racemic aldehyde played a central role in the kinetic resolution obtained [195].

Ley et al. proposed a template approach to the synthesis of 10-hydroxyasimicin (**30**). The bis-THF fragment was prepared via a ring closing metathesis (RCM) reaction from the material, (*S*,*S*)-dimethyltartrate, which was found easy to desymmetrize for further chain extension [133]. The butenolide unit was prepared from butane-1,4-diol under a HDA reaction with a good yield. Then, selective hydrogenation and deprotection produced the target product.

Marshall et al. expedited a previously described four-component modular synthesis on the C-4,C-30-dihydroxylated and C-5-hydroxylated bis-THF units of AGEs [196]. In this work, AGEs such as (30*S*)-hydroxybullatacin (**64**), uvarigrandin A (**65**), and (5*R*)-uvarigrandin A (**66**) were successfully synthesized to confirm their structures. The main features of this method were the addition of chiral α-oxygenated allylic stannane and indium reagents to the acylic core aldehyde precursor, followed by a THF ring closing reaction and Sonogashira coupling attaching the butenolide subunit. Among the products the synthesized (30*S*)-hydroxybullatacin (**64**) and uvarigrandin A (**65**) had physical properties identical to the literature values of the natural compounds [211, 212]. However, the data of synthesized (5*R*)-uvarigrandin A (**66**) were not identical to those of the naturally occurring narumicin I (**67**) as evidenced by the Mosher ester method [31].

A convergent and highly stereoseclective route for the synthesis of 10-hydroxytrilobacin (**68**) and its three diastereomers was published by Roush et al. (see Fig. 18) [197]. In this work, a [3 + 2]-annulation reaction with simple modification from the same precursors was demonstrated as a potential method for the syntheses of these AGEs.

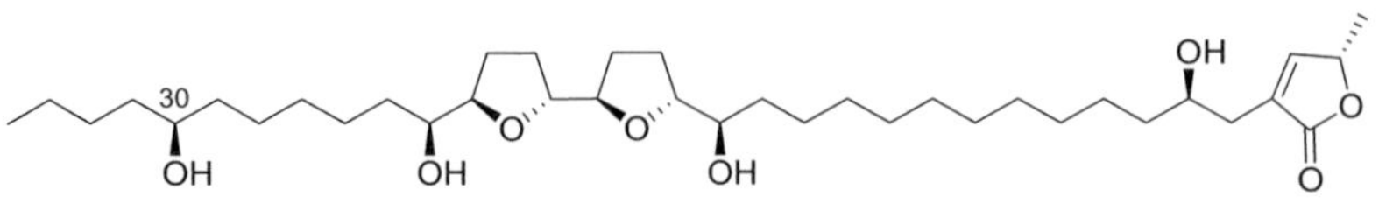

64 ((30*S*)-hydroxybullatacin)

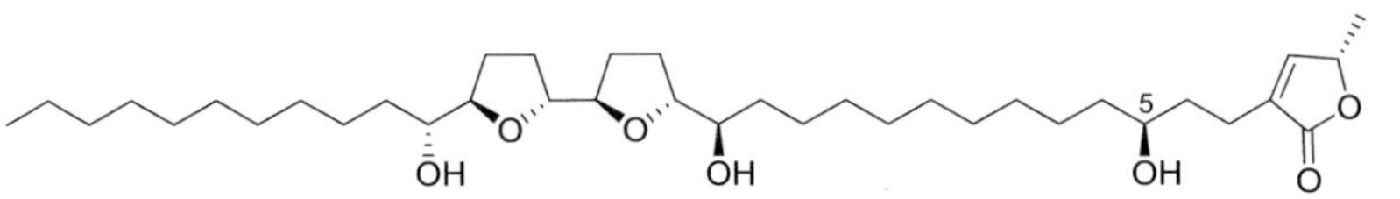

65 (uvarigrandin A)

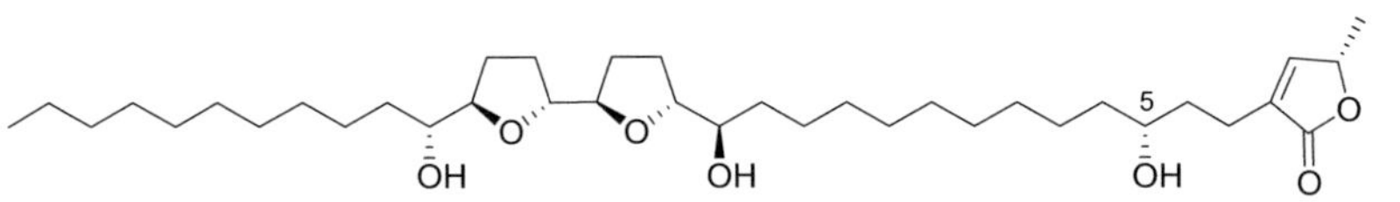

66 ((5*R*)-uvarigrandin A)

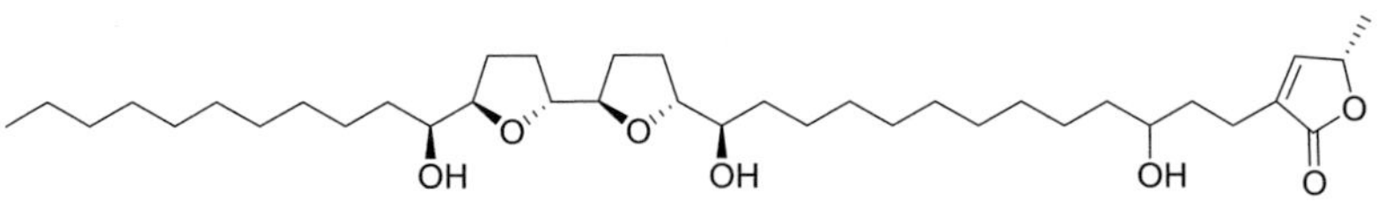

67 ((5*R*or *S*,24*R*)-narumicin I)

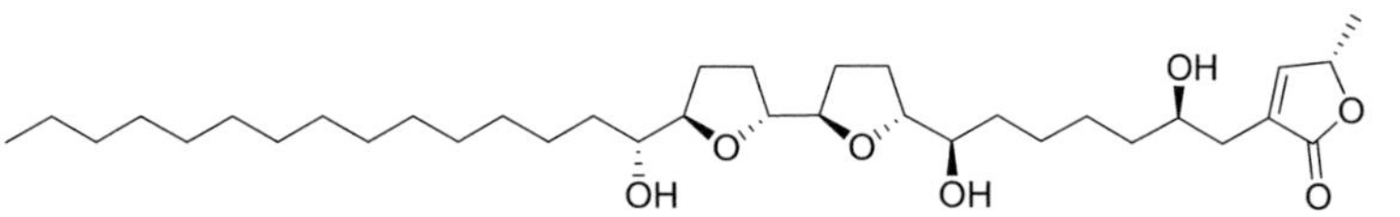

69 (longimicin C)

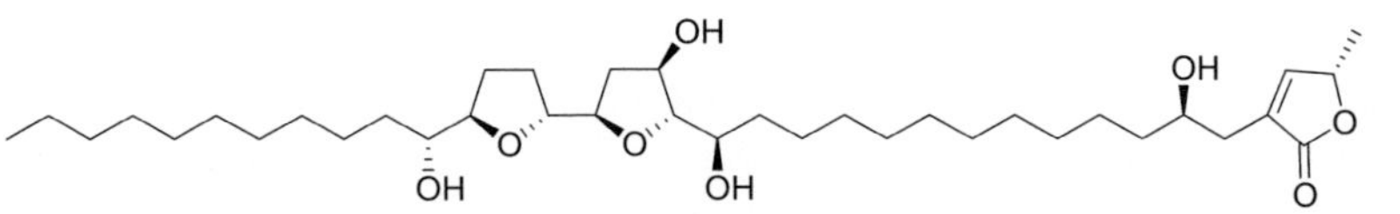

70 (asimitrin)

Yao et al. reported for the first time the total synthesis of longimicin C (**69**) in 2005 [198]. The C_2-symmetrical bis-THF segment of **69** was prepared via a Sharpless dihydroxylation procedure and intramolecular Williamson etherification for the incorporation of the stereocenters. An iterative acetylene–epoxide coupling strategy was used subsequently to assemble all fragments allowing the elaboration of the target compound.

Brown et al. completed the synthesis of membranacin (**31**) using metal-oxo- and metal-peroxy-mediated oxidative cyclizations as the key steps in a 17-step procedure. They achieved a good overall yield. The required starting material, triene, was prepared prior to the introduction of the (2*S*)-2,10-camphorsultam auxiliary [199].

Mucoxin (**33**) was obtained as the first AGE with a hydroxy group substituted on the bis-THF rings [139]. In 2006, Borhan et al. studied the synthesis of the proposed mucoxin (**33**) structure using a neighboring-group-directed regioselective cyclization approach, in which a methylene-interrupted epoxydiol and the one-pot 1,2,-*n*-triol cyclization were employed for constructing this AGE [140]. When the synthetic product was compared to the naturally occurring compound by ^{1}H NMR spectroscopy, differences (>0.1 ppm) in the chemical shifts of the bis-THF (C-8–C-17) region of the molecule were evident. To confirm the exact configuration of mucoxin (**33**), exciton coupled circular dichroism (ECCD) spectroscopy, and a 1D-NOESY NMR measurement were employed to reveal that the stereochemical discrepancies are due to a stereochemical misassignment of the natural mucoxin (**33**) (see Fig. 19). Mohapatra et al. synthesized asimitrin (**70**) using the commercially available carbohydrate, 1,2:5,6-di-*O*-isopropylidene-*α*-D-glucofuranose, as the starting material. This was further elaborated with respect to the adjacent *trans*-bis-THF subunit via stereoselective intramolecular oxymercuration and chelation-controlled Grignard reactions [193]. The strategy was based on a retrosynthesis reaction, in which the butenolide moiety was coupled with the bis-THF fragment.

Mootoo et al. reported the synthesis of the bis-THF core of rolliniastatin (**28**) from pyranoside precursors (see Fig. 20) [200]. The C-6 allylated 2,3-dideoxypyranoside precursors were prepared via Ferrier reaction of tri-*O*-acetyl-D-glucal and *t*-butanol or

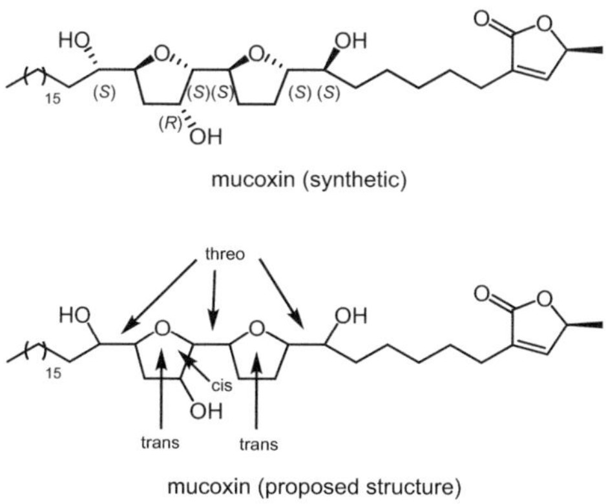

Fig. 19 Structures of synthetic mucoxin and proposed mucoxin (**33**)

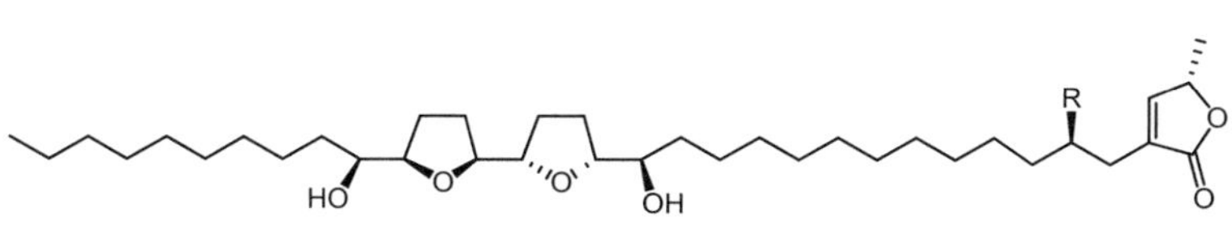

28 ((+)-rolliniastatin, R = OH)
31 ((+)-membranacin, R = H)

Fig. 20 Structures of AGEs with the bis-THF core as that of rolliniastatin (**28**)

trifluoroethanol. Iodoetherification reaction of the precursors was employed further to obtain a highly functionalized and selective *cis*-2,5-disubstituted THF, allowing a 68% yield of rolliniastatin (**28**) to be obtained.

Lee et al. synthesized under stereocontrol rolliniastatin (**28**), rollimembrin (**29**), and membranacin (**31**) [130]. A radical cyclization of β-alkoxyvinyl sulfoxide-Pummerer rearrangement-allylation methodology was developed to demonstrate the synthesis of the compound with a *threo,cis,threo,cis,erythro*-bis-oxolane moiety flanked by dihydroxy groups. The two major segments were then coupled by allylation-olefin cross metathesis. This approach enables an efficient and selective production of the bis-THF type of AGEs.

5.4 Non-adjacent Bis-THF Annonaceous Acetogenins

Methods of chemical synthesis for non-adjacent bis-THF structures are not yet well developed (Table 10). Since 1998, six naturally occurring structures have been investigated by total synthesis in detail, including bullatanocin (squamostatin C) (**71**), (+)-4-deoxygigantecin (**72**), (+)-14-deoxy-9-oxygigantecin (**73**), (+)-gigantecin (**74**), squamostatin-D (**75**), and (+)-sylvaticin (**76**) (see Fig. 21).

Bullatanocin (squamostatin C) (**71**) is a typical example of a non-adjacent 2,5-*trans*-disubstituted THF. Mootoo et al. described a preparation of the bis-THF element of this AGE using the construction of THF-allylic alcohol moieties (see Fig. 22), which underwent iodoetherification of two 1,2-*O*-isopropylidene-5-alkene precursors that emerged from two relatively simple mono-THF structures [213, 214]. This approach showed enantiodivergence and resulted in a good yield.

The potent antitumor agent AGE, (+)-gigantecin (**74**), has remained a big challenge to investigators because of the unclear configuration of the THF moieties (see Fig. 23). Crimmins et al. chose an enantioselective methodology utilizing a modified asymmetric aldol reaction with chlorotitanium enolates of oxazolidinone glycolates to synthesize three key subunits of this molecule, namely, the butenolide, the C-9–C-16 fragment, and the C-17–C-34 fragment. These were synthesized individually using different starting materials (see Fig. 24a) [216]. The fragments were assembled to give (+)-gigantecin (**74**) as the final product achieving a 71% yield. Furthermore, Hoye et al. reported a more convenient one-pot reaction to synthesize gigantecin (**74**) by coupling three component olefin metatheses via a 13-step protocol, resulting in an 87% yield (see Fig. 24b) [144]. This method was also employed reversely to construct 14-deoxy-9-oxygigantecin (**73**), a constitutional isomer of gigantecin (**74**).

(+)-4-Deoxygigantecin (**72**) possesses a similar structure to gigantecin (**74**) (see Fig. 23). Makabe et al. used a convergent route to determine the absolute configuration of (+)-4-deoxygigantecin (**72**) (see Fig. 24c) [215], in which the synthesis of (+)-4-deoxygigantecin (**72**) from (−)-muricatacin (**22**) and (−)-(*S*)-ethyl lactate had been reported previously [220]. The optical rotation value of the synthetic

Table 10 Synthesis topics on non-adjacent bis-THF AGEs from 1998 to 2011

Compound name	Total synthesis	Synthesis method	Starting materials	Steps and yields	Other points	Ref.
Bullatanocin (squamostatin C) [213]		Olefin cross-metathesis as the segment coupling (The plan centers on the olefin cross-metathesis of THF allylic alcohol two derivatives as the key segment coupling step and the assembly of two derivatives through the iodoetherification of 1,2-*O*-isopropylidene-5-alkene precursors)	Ethyl (*E*)-4,6-heptadienoate	24 Steps, 95%		[213]
Bullatanocin (squamostatin C)	First total synthesis	Olefin cross-metathesis and Wittig olefination as the segment-coupling reactions.	Ethyl (*E*)-4,6-heptadienoate	13 Steps, 71%	The synthesis confirms the structure of the natural product.	[214]
(+)-4-Deoxy-gigantecin	Total synthesis	Retrosynthetic synthesis with Pd(0)-catalyzed cross coupling reaction	(−)-Muricatacin and (*S*)-(−)-ethyl lactate	30 Steps, 95%	1. The synthesis confirms the structure of the natural product. 2. MOM protection.	[215]
(+)-Gigantecin	First total synthesis	The synthesis exploits a modified asymmetric aldol protocol using chlorotitanium enolates of oxazolidinone glycolates.	Commercially available benzyl glycidyl ether	19 Steps, 71%		[216]
(+)-Gigantecin or (+)-14-deoxy-9-oxygigantecin	Total synthesis	1. A three-component ring-closing/cross-metathesis sequence that differs only in the ordering of the RCM vs. CM events 2. Another notable aspect	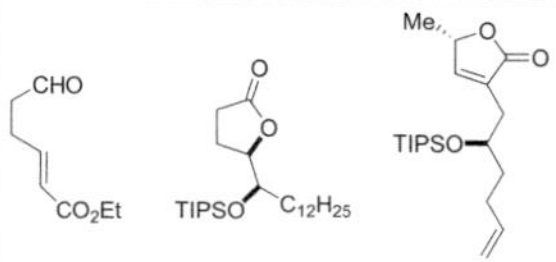	(+)-Giggantecin (13 steps,87%) or (+)-14-deoxy-9-oxygigantecin (14 steps, 48%)		[144]

(continued)

Table 10 (continued)

Compound name	Total synthesis	Synthesis method	Starting materials	Steps and yields	Other points	Ref.
		is the use of in situ epoxide-closing and -opening of iodohydrins with dimethylsulfonium methylide to provide inverted allylic alcohols				
Squamostatin-D	Total synthesis	1. Sharpless asymmetric dihydroxylation (C-19/C-20) 2. BF_3-Promoted addition of a γ-alkoxy allylic stannane (C-23/C-24) 3. Addition of a γ-alkoxy allylic indium chloride C-15/C-16 4. Addition of an organozinc reagent catalyzed by a chiral titanium triflic amide C-12	(*S*)-Lactate; ethyl undec-10-enoate and 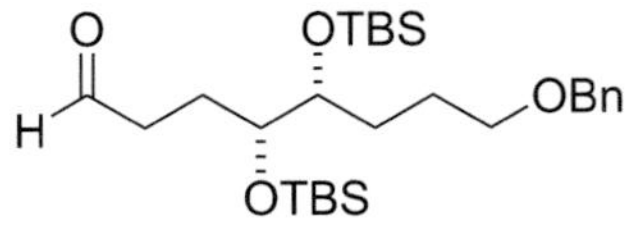 (4*R*,5*R*)-8-(benzyloxy)-4,5-bis(*tert*-butyldimethylsilyloxy)octanal	14 Steps, 97% (>90% *ee*)		[217]
(+)-*cis*-Sylvaticin and (+)-sylvaticin	First total synthesis	1. Double oxidative cyclization of a protected tetraol onto a diene unit that forms two rings in one reaction 2. The *trans*-THF ring of sylvaticin was prepared by utilizing a one-pot hydride shift/intramolecular oxo-carbenium ion reduction protocol	(*E*,*E*)-Tetradecatetraene and (*E*,*E*,*E*)-cyclododecatriene	(+)-*cis*-Sylvaticin (13 steps, 87%) and (+)-sylvaticin (19 steps, 76%)	A similar sequence on the C-4,36 bis-epimer of sylvaticin (for which the NMR data, but not specific rotation data, matched the literature) gave a synthetic sample that also had a poor match of its di-(*R*) and di-(*S*)-Mosher esters.	[218]

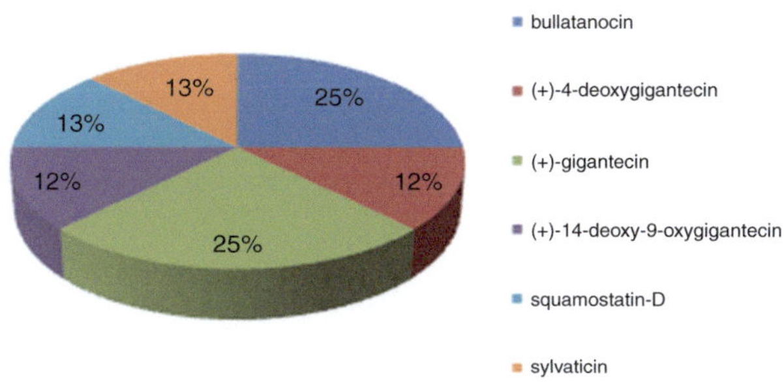

Fig. 21 Synthesis topics on bis-THF AGEs (nonadjacent type) since 1998

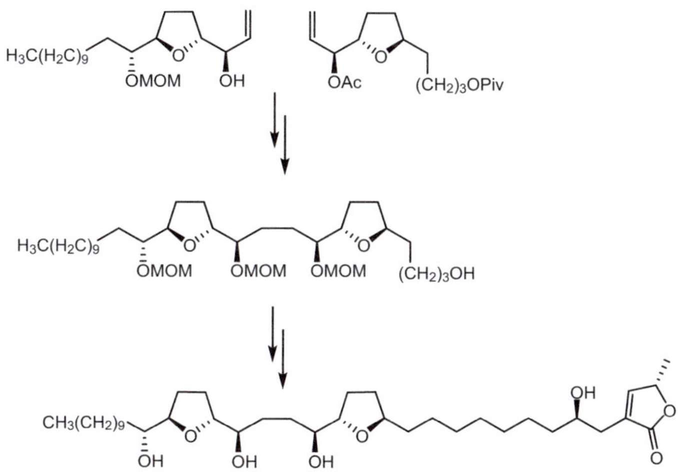

71 (bullatanocin = squamostatin C)

Fig. 22 Synthesis of the non-adjacent bis-THF core of bullatanocin (**71**) using THF-allylic alcohol moieties

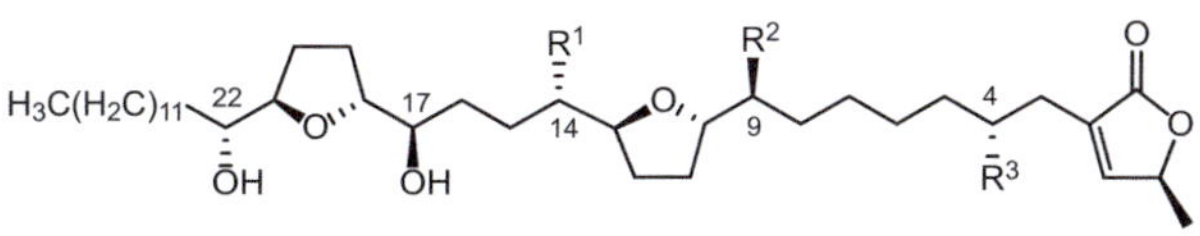

gigantecin: R^1 = OH, R^2 = H, R^3 = OH
14-deoxygigantecin: R^1 = H, R^2 = OH, R^3 = OH
4-deoxygigantecin: R^1 = OH, R^2 = H, R^3 = H

Fig. 23 Structures of gigantecin (**74**), 14-deoxy-9-oxygigantecin (**73**), and 4-deoxygigantecin (**72**)

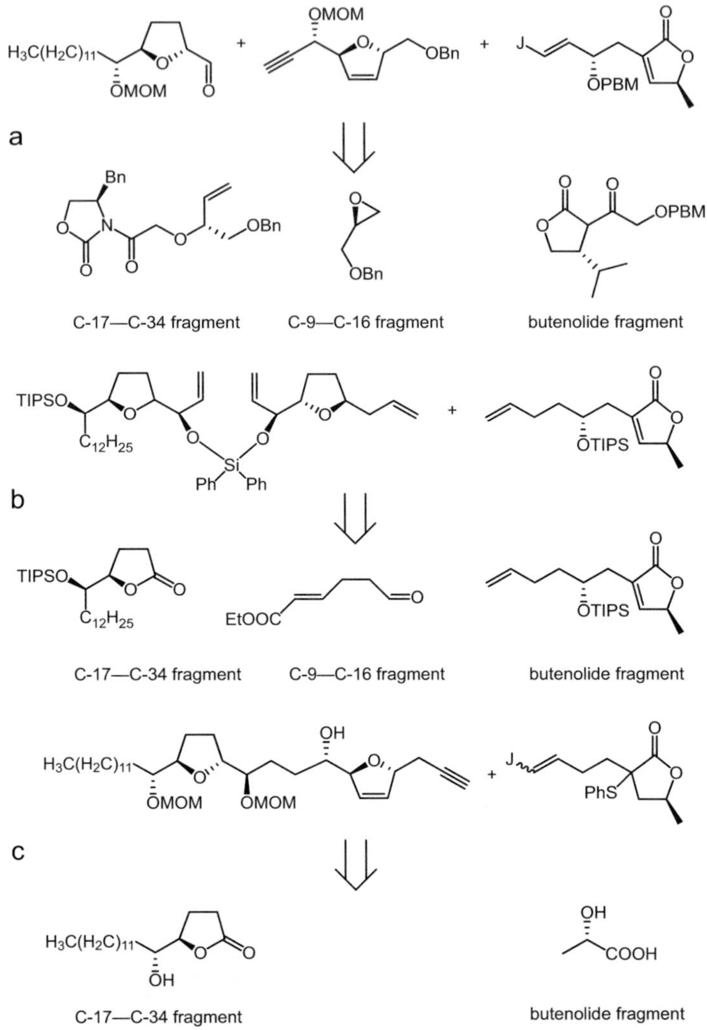

Fig. 24 Synthesis strategies of (**a**) (+)-gigantecin (**74**) by Crimmins and She [216], (**b**) (+)-14-deoxy-9-oxygigantecin (**74**) by Hoye et al. [144], and (**c**) (+)-4-deoxygigantecin (**73**) by Makabe et al. [215]

compound ($[\alpha]_D^{23}$ +16.0°cm^2/g) matched with that of the natural product ($[\alpha]_D$ +15.5°cm^2/g).

Marshall et al. proposed a synthesis strategy for squamostatin D (**75**) by constructing chiral centers incorporating several groups and steps. The stategy included a C-36 methyl group from (*S*)-lactic acid, a γ-alkoxy allylic stannane (C-23/C-24) by a BF_3-promoted addition, a Sharpless asymmetric dihydroxylation (C-19/C-20), a γ-alkoxy allylic indium chloride (C-15/C-16) addition, and an addition of an organozinc reagent catalyzed by a chiral titanium triflic amide (C-12) (see Fig. 25) [217].

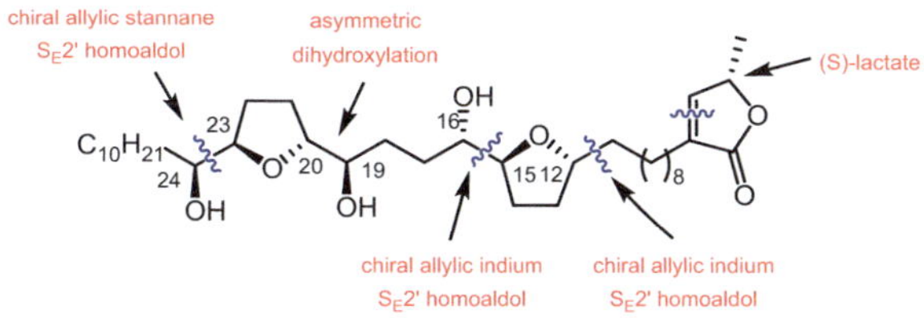

75 (squamostatin D)

Fig. 25 Retrosynthesis analysis of squamostatin D (**75**)

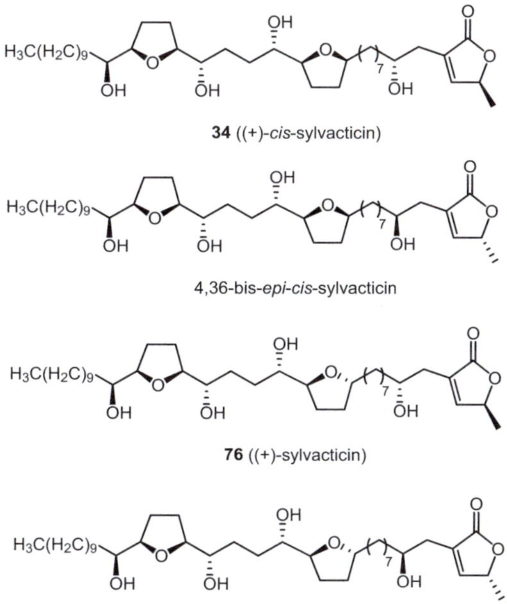

34 ((+)-*cis*-sylvacticin)

4,36-bis-*epi*-*cis*-sylvacticin

76 ((+)-sylvacticin)

4,36-bis-*epi*-sylvacticin

Fig. 26 Structures of sylvacticin (**76**) and its epimers

Both *cis*-sylvaticin (**34**) and sylvaticin (**76**) (Fig. 26) show potent antitumor activity containing the non-adjacent THF rings. Sylvaticin (**76**) bears one *cis*- and one *trans*-THF ring, while *cis*-sylvaticin (**34**) possesses two *cis*-THF moieties. Donohoe et al. used an oxidative cyclization strategy and a cross-metathesis reaction as efficient and concise routes for the synthesis of these AGEs [218]. The starting material, (*E*,*E*)-tetradecatetraene, was utilized to protect the tetraol group for constructing the two THF rings with >95% stereoselectivity. In addition, the *trans*-THF ring of sylvaticin (**76**) was prepared via a one-pot hydride shift/intramolecular oxo-carbenium ion reduction methodology using *cis*-THF as precursor [221]. This work provides a shortcut for producing both natural products, in which *cis*-sylvaticin (**34**) was synthesized in only 13 steps with a commercially available diene product while 19 steps were needed for sylvaticin (**76**). The optical rotation values and NMR data of the synthesized products and their Mosher esters were similar to those reported in literature [141], which confirmed the configurations of these two compounds. In addition, the C-4,C-36 bis-epimers of *cis*-sylvaticin (**34**) and sylvaticin (**76**) showed a poor consistency of their optical rotation values with those in the literature.

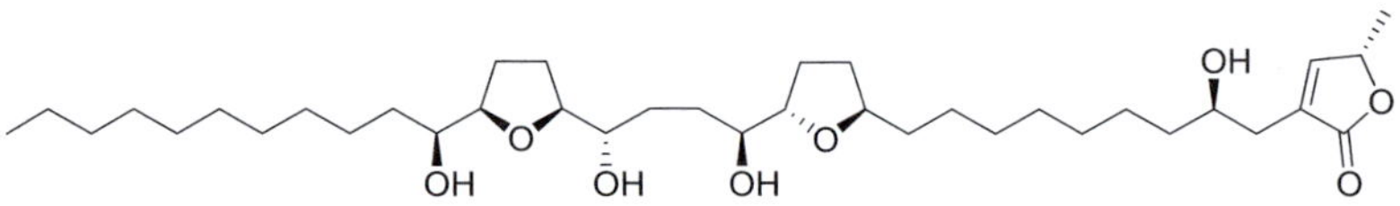

76 (sylvaticin)

5.5 *Other AGEs*

Except for the above mentioned type AGEs, we summarize procedures for the synthesis of five classes of other AGEs, like the adjacent THF–THP, nonadjacent THF–THP, THP, tri-THF, and bis-lactone types of AGEs, performed during 1998–2011. In the subgroups, synthesis of the THF–THP type AGEs have been reported the most (see Fig. 27). Related studies for other AGEs are listed in Table 11.

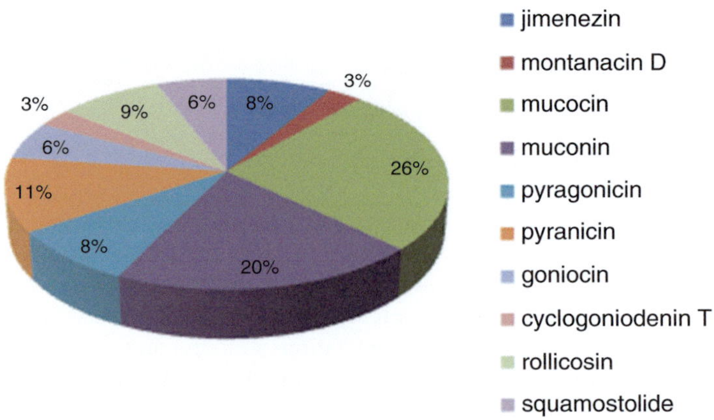

Fig. 27 Analysis of other AGEs syntheses from 1998 to 2011

Table 11 Synthesis topics on other AGEs from 1998 to 2011

Compound name	Total synthesis	Synthesis method	Starting materials	Steps and yields	Other points	Ref.
THF–THP (adjacent type)						
Jimenezin	Total synthesis	1. A highly stereoselective intramolecular allylboration 2. An intramolecular Williamson reaction	(*S*)-Glycidol	24 Steps, 99%		[145]
Jimenezin (19α-H and 19β-H)	First total synthesis	1. A stereoselective condensation between the pyranyl aldehyde and the acetylene derivative 2. A palladium-catalyzed coupling reaction	D-Galactose, L-arabinose, and L-rhamnose	Jimenezin [19α-H (12 steps, 56%) and 19β-H (13 steps, 68%)]	Natural compound is (19α-H)-jimenezin.	[222, 223]
(+)-Muconin	Total synthesis	A novel α-C-H hydroxyalkylation and α′-C-H oxidation of tetrahydrofuran	(−)-Muricatacin	17 Steps, 96%		[229]
Muconin	Total synthesis	The key reactions include successive ether-ring formation reaction under acidic and basic conditions or one-pot double cyclization promoted by TBAF and stereoselective reduction of acyclic ketones adjacent to the 2,6-*cis*-THP with $Zn(BH_4)_2$		25 Steps, 78%		[224]

(continued)

Table 11 (continued)

Compound name	Total synthesis	Synthesis method	Starting materials	Steps and yields	Other points	Ref.
Muconin	Total synthesis	1. Ether-ring formation reaction under acidic and basic conditions 2. Stereoselective reduction of acyclic ketone	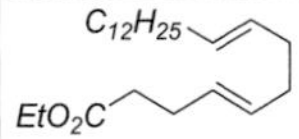	25 Steps, 65%		[225]
(+)-Muconin	Total synthesis	A Pd(0)-mediated cross diyne coupling reaction	D-Glutamic acid	26 Steps, 82%	MOM protection.	[226]
(+)-Muconin	Total synthesis	Dess-Martin oxidation/L-selectride reduction	D-Glutamic acid	25 Steps, 82%		[227]
Muconin	Total synthesis	Hydrolytic kinetic resolution (HKR) of terminal epoxides catalyzed by cobalt complex	Tetradecene oxide, epichlorohydrin, and propylene oxide	19 Steps, 70%		[228]
(+)-Muconin		By means of the KSAE (Katsuki-Sharpless asymmetric epoxidation) and SAD (Sharpless asymmetric dihydroxylation) reaction	5-Hexen-1-ol	15 Steps, 70%		[230]
THF–THP (nonadjacent type)						
Montanacin D and (4*S*,8*R*)-montanacin D 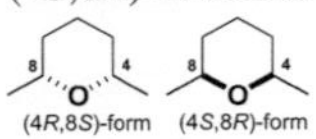	First total synthesis	Cross-metathesis: intermolecular metathesis of an α,β-unsaturated ketone carrying a tetrahydropyranyl lactone with a tetrahydrofuran derivative	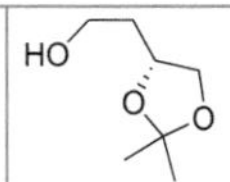	22 Steps, 83–86%	Natural compound montanacin D is the (4*R*,8*S*)-isomer.	[231]

Mucocin		1. Cross-metathesis 2. Julia–Kocienski olefination		10 Steps, 51%		[240]
(−)-Mucocin	Total synthesis	1. The temporary silicon-tethered (TST) ring-closing metathesis (RCM) cross-coupling reaction 2. The enantioselective alkyne/aldehyde addition 3. Bismuth tribromide-mediated reductive etherification		12 Steps, 95%	First application of the temporary silicon-tethered (TST) ring-closing metathesis (RCM) cross-coupling reaction for an acetogenin.	[239]
Mucocin	Total synthesis	1. Chiron approach 2. Palladium-catalyzed cross-coupling reaction	D-Galactose, 2,5-anhydro-D-mannitol, and L-rhamnose	30 Steps, 77%		[233]
(−)-Mucocin	Total synthesis	1. 6-*endo*-Epoxide cyclization 2. Wittig-type reaction	(*E*)-Dihydromuconic acid	(−)-Mucocin (53%) and 16-*epi*-mucocin (75%)	Naturural compound is (−)-mucocin.	[234]
Mucocin	Total synthesis	1. Palladium-catalyzed cross coupling reaction 2. Radical cyclization	L-Rhamnose	19 Steps, 76%	MOM protection.	[235]
Mucocin	Total synthesis	1. Naked carbon skeleton strategy 2. Double AE reaction and double AD reaction	Cyclododecatriene	20 Steps, 64%	Simultaneous two-ring closure reactions provided both the THP and THF rings in a single step.	[232]
Mucocin	Total synthesis	1. Swern reaction 2. L-Selectride reduction	D-Galactose and L-rhamnose	31 Steps, 76%	MOM protection.	[238]
(−)-Mucocin	Total synthesis	1. Highly convergent synthetic strategy 2. Metalloorganic coupling reaction 3. Sharpless epoxidation 4. Dess–Martin oxidation	β-Ketoester and 	32 Steps, 91%	(−)-Mucocin was found to be identical to the naturally occurring product in respect to the spectroscopic data.	[237]

(continued)

Table 11 (continued)

Compound name	Total synthesis	Synthesis method	Starting materials	Steps and yields	Other points	Ref.
THP						
Pyragonicin		Olefin cross-metathesis	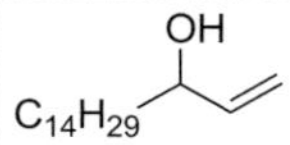	19 Steps, 73%	A novel MOM-migrating reaction found in a cyclization reaction is also discussed.	[241]
Pyragonicin		1. Asymmetric Horner-Wadsworth-Emmons (HWE) methodology 2. A key feature of the synthesis is a mild, stereoselective coupling reaction using Carreira's asymmetric acetylide addition with a substrate bearing an adjacent unprotected hydroxy group and a base-sensitive butenolide moiety	Cyclohexadiene	9 Steps, 34%	Zinc-mediated stereoselective coupling.	[150]
Pyranicin		1. Achmatowicz oxidation—Kishi reduction 2. Fu's alkyl-alkyl Suzuki coupling		13 Steps, 89%	Pyranicin shows activity against the PACA-2 (pancreatic cancer) cell line (ED_{50} 1.3 ng/cm^3).	[244]
Pyranicin and pyragonicin		1. An asymmetric HWE desymmetrization 2. An uncommon, yet highly efficient, protective group [(TMS)$(CH_2)_2$] 3. Carreira's asymmetric acetylide additions to construct 1,4- and 1,6-diols	Cyclohexadiene	Pyranicin (31 steps, 72%) and pyragonicin (27 steps, 34%)		[243]

Pyranicin	Total synthesis	1. Asymmetric Horner-Wadsworth-Emmons (HWE) reactions followed by a Pd-catalyzed allylic substitution 2. The C-10/C-15 1,6-diol motif was installed using Carreira's asymmetric acetylide addition methodology	Cyclohexadiene	32 Steps, 72%		[149]
Pyranicin	First total synthesis	1. SmI_2-induced reductive cyclization 2. Mitsunobu lactonization 3. Wittig reaction	2,3-*O*-Isopropylidene-D-threitol allylmagnesium bromide and chiral acetylene 	30 Steps, 95%		[242]
Tri-THF						
Goniocin	Total synthesis	Rhenium oxo complexes involves tandem cyclization	4,8,12-Trienol substrate		A mixture of trifluoroacetyl perrhenate and trifluoroacetic anhydride (TFAA) has proven to be an effective reagent for promoting triple-oxidative cyclizations.	[245]

(continued)

Table 11 (continued)

Compound name	Total synthesis	Synthesis method	Starting materials	Steps and yields	Other points	Ref.
Goniocin and cyclogoniodenin T	First asymmetric total synthesis	Sharpless asymmetric dihydroxylation (AD)8 and epoxidation (AE)9 reactions as well as the Williamson etherification reaction	Polyepoxide	Goniocin (17 steps, 58%) and cyclogoniodenin T (17 steps, 17%)	The synthetic materials, goniocin and cyclogoniodenin T, were found to be identical to the naturally occurring compounds, thereby confirming their proposed absolute configurations.	[247]
Two butenolides						
Rollicosin	First total synthesis	1. A highly regio- and stereoselective tandem RCM/CM reaction for construction of the east-wing lactone and incorporation of alkyl spacer 2. Establishment of the C-4 stereocenter and addition of the west-wing lactone were achieved by Sharpless asymmetric dihydroxylation and enolate alkylation	Hexa-1,5-diene-3,4-diol	12 Steps, 86%		[152]
(4*R*,15*R*,16*R*,21*S*)- and (4*R*,15*S*,16*S*,21*S*)-Rollicosin	Total synthesis	Convergent synthesis	4-Pentyn-1-ol; 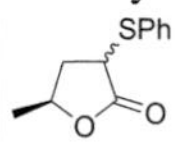 and 5-hexen-1-ol	(4*R*,15*R*,16*R*,21*S*)-Rollicosin (23 steps, 85%)	Both compounds showed inhibitory activity against the bovine heart mitochondrial complex I.	[248]

Squamostolide	Total synthesis	Tandem ring-closing metathesis (RCM)/cross metathesis (CM) step in which lactone formation and fragment coupling	Mannitol	9 Steps, 56%		[219]
Squamostolide, (4*R*,15*R*,16*R*,21*S*)- and (4*R*,15*S*,16*S*,21*S*)-rollicosin	Total synthesis	A convergent stereoselective synthesis with Pd-catalyzed cross-coupling reaction	4-Pentyn-1-ol; 5-hexen-1-ol or 1,6-hexanediol; 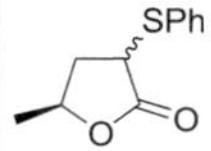	(4*R*,15*R*,16*R*,21*S*)-Rollicosin (23 steps, 66%); (4*R*,15*S*,16*S*,21*S*)-rollicosin (23 steps, 78%) and squamostolide (15 steps, 62%)	These compounds showed weak activity against bovine heart mitochondrial NADH-ubiquinone oxidoreductase.	[153]
Solamin, reticulatacin, asimicin, bullatacin, trilobin, trilobacin, squamotacin, rolliniastatin, uvaricin, rollidecins C and D, mucocin, goniocin, and cyclogoniodenin T, as well as ensembles of non-natural analogues		Sharpless asymmetric dihydroxylation (AD) and the asymmetric epoxidation (AE) reactions	Cyclohexadiene	20 Steps	Oxidative polycyclization reaction with rhenium (VII) oxides.	[246]

5.5.1 Adjacent Type THF–THP AGEs

Takahashi et al. successfully conducted the first total synthesis of jimenezin (**36**) and its 19-epimer (**36a**) through a convergent route by Pd-catalyzed cross-coupling reaction of a THP–THF segment and a vinyl iodide butenolide, of which both were synthesized from carbohydrates, L-rhamnose and D-galactose [222, 223]. However, the physical data of H-19α of jimenezin (**36a**) were different from the natural jimenezin (**36**) so that the latter compound was corrected with a H-19β configuration. To resolve the stereochemistry of the THP moiety of jimenezin (**36**), Bandur et al. synthesized (−)-jimenezin via a stereocontrolled process, such as the intramolecular allylboration for building the THP ring and an intramolecular Williamson reaction for closing the THF ring [146]. The synthesized product, (−)-jimenezin, was found to be identical with respect to its spectroscopic data with the naturally occurring jimenezin (**36**) (Fig. 28).

Another example is muconin (**77**) due to its broad bioactivities and unique structure. Schaus et al. synthesized **77** by an advanced methodology from chiral building blocks. It was attempted to synthesize the THF–THP fragment using the hydrolytic kinetic resolution (HKR) of tetradecene oxide and to synthesize the butenolide segment via HKR of racemic epoxide [224]. Yang and Kitahara et al. succeeded in the total synthesis of **77** via a convergent route, in which two key building blocks I and II were derived from D-glutamic acid (see Fig. 29) [225, 226]. Also, Takahashi et al. proposed a new synthetic strategy of **77** via a successive ether-ring formation reaction under acidic and basic conditions or a

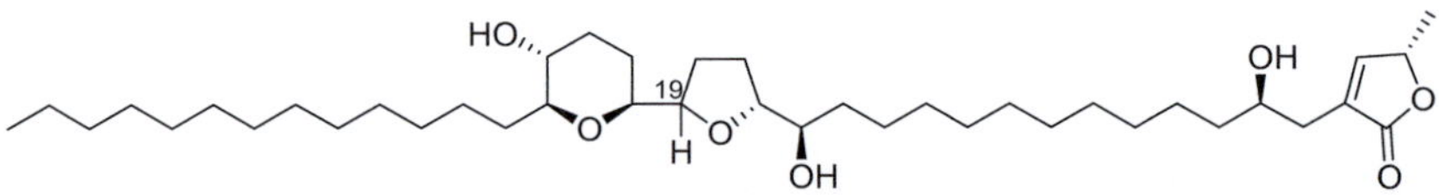

36 19-Hβ
36a 19-Hα

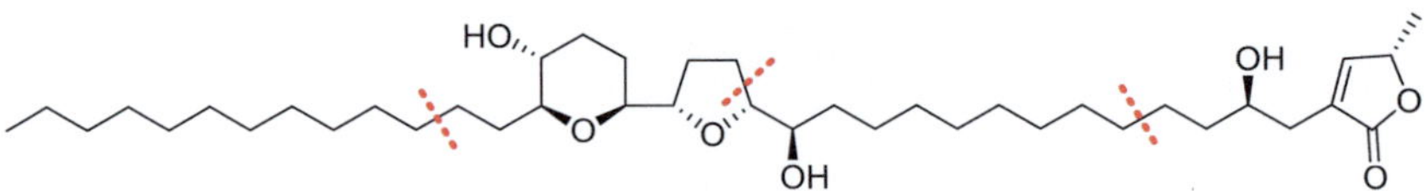

36 (jimenezin)

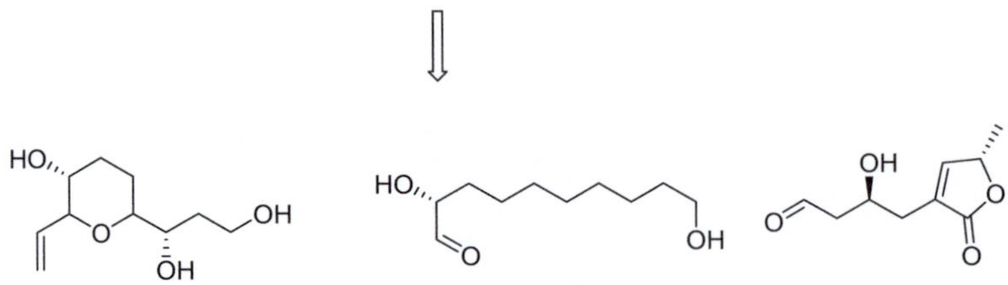

Fig. 28 Jimenezin (**36**) and its 19-Hα isomer and retrosynthesis analysis of jimenezin (**36**)

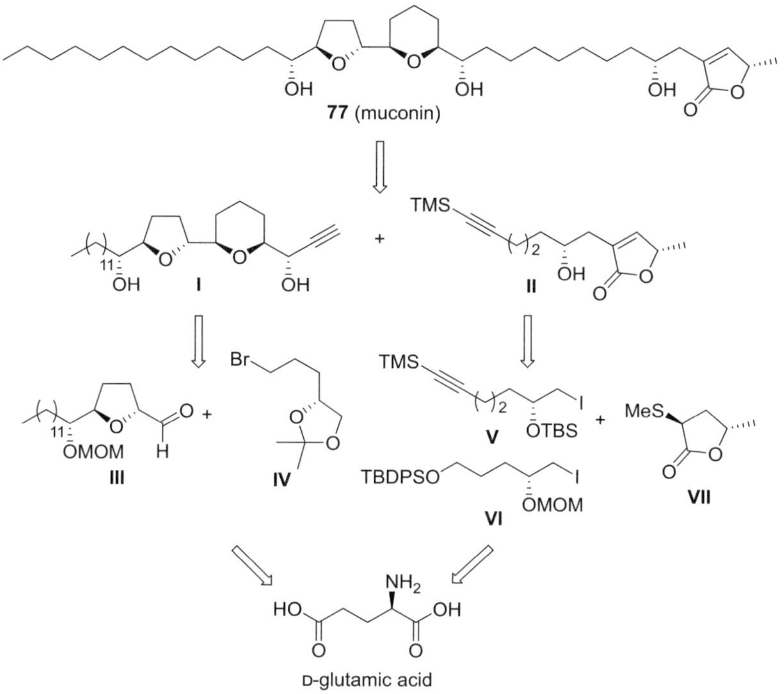

Fig. 29 Retrosynthesis analysis of muconin (**77**)

one-pot double cyclization utilizing the TBAF reagent and diasteroselective reduction of acyclic ketones with $Zn(BH_4)_2$ [227, 228]. Yoshimitsu et al. designed a synthesis strategy for **77** from (−)-muricatacin (**22**), in which the novel α-C-H hydroxyalkylation and α′-C-H oxidation of tetrahydrofuran were developed and achieved as key processes [229].

Crisóstomo et al. reported the synthesis of muconin (**77**) based on Sonogashira coupling between a terminal alkyne with the THF–THP fragment and iodine in the γ-lactone fragment [230]. The precursor of the THF–THP fragment was synthesized from 5-hexen-1-ol as the starting material, and the stereroselective *exo*-cyclization of the THF ring was mediated by means of the Katsuki-Sharpless asymmetric epoxidation and Sharpless asymmetric dihydroxylation. The unique characteristic of the method was established based on the consecutive enantioselective reaction to ensure the high enantiomeric purity of the products.

5.5.2 Non-adjacent THF–THP Type

In 1999 Wang et al. published the first non-adjacent THF–THP AGE, montanacin D (**78**), but the absolute configuration of the THP ring of the naturally occuring AGE was undefined. Takahashi et al. reported the total synthesis of montanacin D (**78**) and its (4*S*,8*R*) isomer for the first time (Fig. 30) [231]. They designed a cross-metathesis strategy for the synthesis of an α,β-unsaturated ketone bearing a THP lactone unit with a THF derivative (Fig. 31). Thus, the synthesized product suggested that the naturally occurring montanacin D (**78**) should be assigned from its spectroscopic data to be of the (4*R*,8*S*) configuration (Fig. 30).

Neogi et al. demonstrated the total synthesis of mucocin (**37**) from cyclododecatriene and verified the proposed structure [232]. Through the "naked" carbon

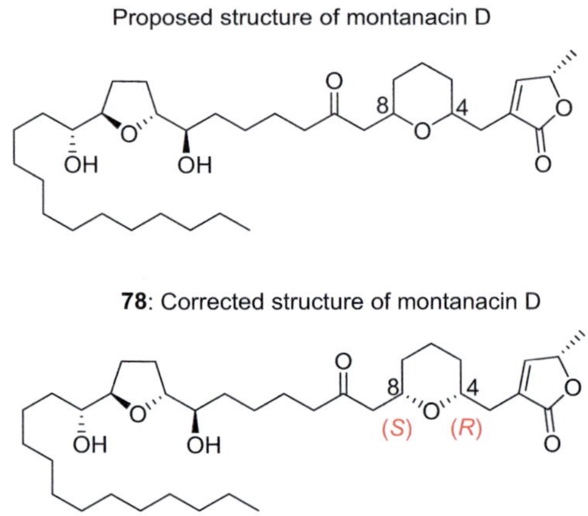

Fig. 30 Structure of montanacin D (**78**)

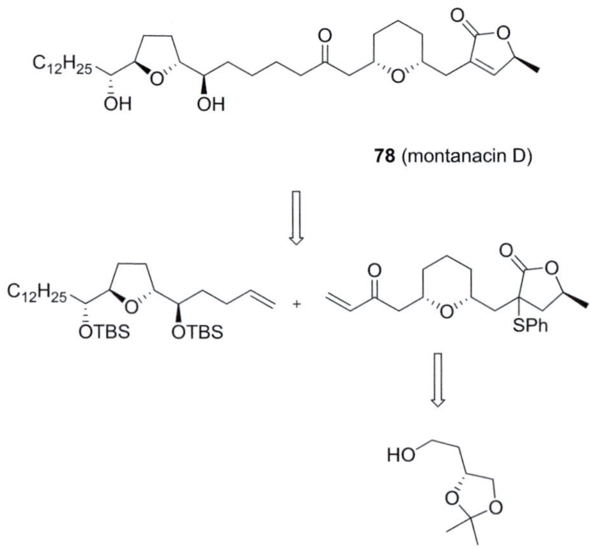

Fig. 31 Retrosynthesis analysis of montanacin D (**78**)

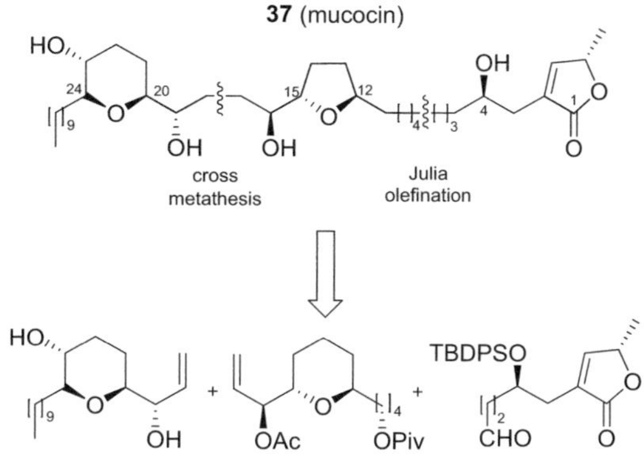

Fig. 32 Retrosynthesis analysis of mucocin (**37**) by Zhu and Mootoo et al.

skeleton strategy, the seven stereocenters on the center fragment were prepared from double AE and double AD reactions. The THF and THP ring were closed and produced in one step as the key feature. In the continued conformational studies of mucocin (**37**), Takahashi and Nakata et al. proposed a series of synthesis methodologies for mucocin (**37**) [233–236]. Takahachi et al. synthesized mucocin (**37**) by using carbohydrates as the key starting point, in which the THP and THF moieties were synthesized from β-selective C-glycosidation and a chelation-controlled addition of an ethynyl group, respectively, and the γ-lactone was prepared via radical cyclization from commercial rhamnose.

Baurle et al. synthesized mucocin (**37**) by the addition of a THP organometallic compound to a THF aldehyde by an appropriate metalloorganic coupling reaction [237]. Four years later, Evans et al. synthesized **37** by means of a temporary silicon-tethered ring-closing metathesis homo-coupling reaction and an enantioselective alkyne/aldehyde addition [239]. Zhu and Mootoo et al. developed a modular synthesis of **37** by olefinic coupling reaction [240]. The cross metathesis reaction on the terminal alkene moiety of THP and THF segments was employed to assemble the non-adjacently-linked five and six membered rings. Then, the nonadjacent THP and THF rings and the butenolide unit were connected by Julia–Kocienski olefination, which was similar to the Wittig-reduction method for bullantanocin synthesis (Fig. 32). The synthesized product showed the identical spectroscopic data with natural mucocin (**37**).

5.5.3 THP Type

Strand and Rein et al. synthesized pyragonicin (**39**) by stereoselective coupling and hydrogenation of the key asymmetric Horner-Wadsworth-Emmons (HWE) approach (Fig. 33) to confirm the proposed structure and for further bioactivity

39 (pyragonicin)

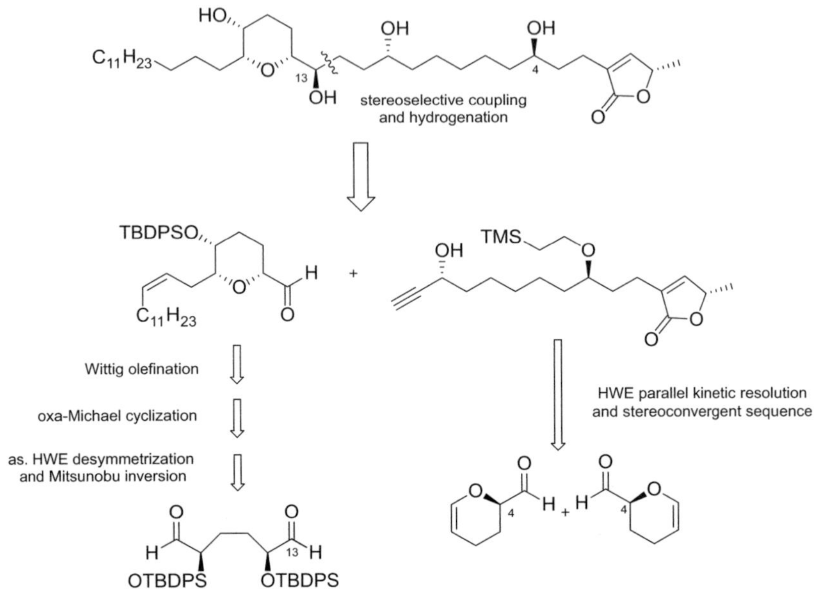

Fig. 33 Retrosynthesis analysis of pyragonicin (**39**) by Strand and Rein et al.

studies [150]. The major characteristic of the coupling reaction was to use an adjacent unprotected hydroxy group and a base-sensitive butenolide moiety to undergo Carreira's asymmetric acetylide addition, which generally can be used for a fragment with a 1,4-diol subunit. The spectroscopic data of synthesized product were the same as for the proposed pyragonicin (**39**).

In 2006, Takahashi et al. synthesized pyragonicin (**39**) via an olefin cross-metathesis between THP fragment and terminal γ-lactone unit in the presence of a Grubbs second-generation catalyst (Fig. 34) [241]. The THP fragment was obtained from asymmetric dihydroxylation and 6-exo-cyclization, and the γ-lactone unit was prepared by an alkylation of γ-lactone with iodide. A twofold yield increase in comparison with that of the previous study was obtained [150].

Pyranicin (**38**) is the first AGE with a THP moiety and an *axial* hydroxy group on the THP ring. Takahashi et al. showed the first total synthesis of **38** in a stereocontrolled manner [242]. A retrosynthesis analysis suggested **38** may be synthesized by a Wittig reaction of the phosphonium salt of a 16,20-*syn*-19,20-*cis*-THP ring, which could be cyclized via a SmI_2-induced reduction of a β-alkoxyacrylate derivative, and an aldehyde of the lactone unit. Strand and Rein et al. synthesized **38** from cyclohexadiene with an overall yield of 6.3% [149] by a similar principle to one Takahashi et al. proposed [150]. Strand and co-workers

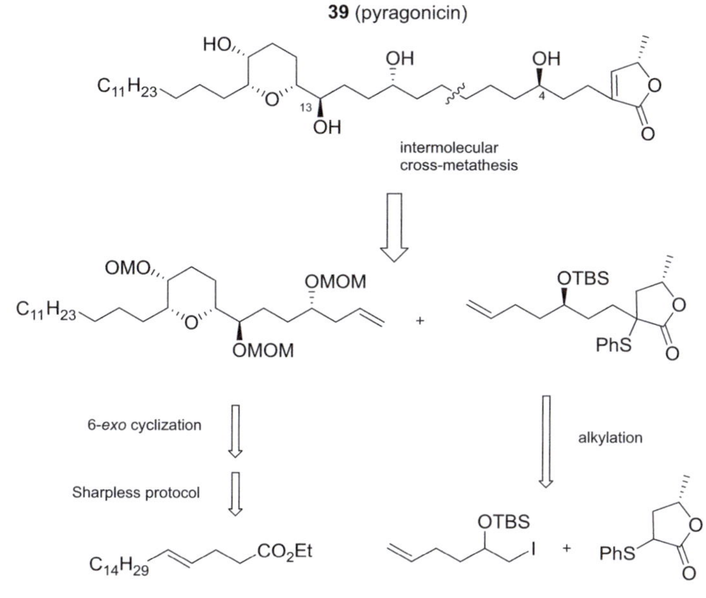

Fig. 34 Takahashi's retrosynthesis analysis of pyragonicin (**39**)

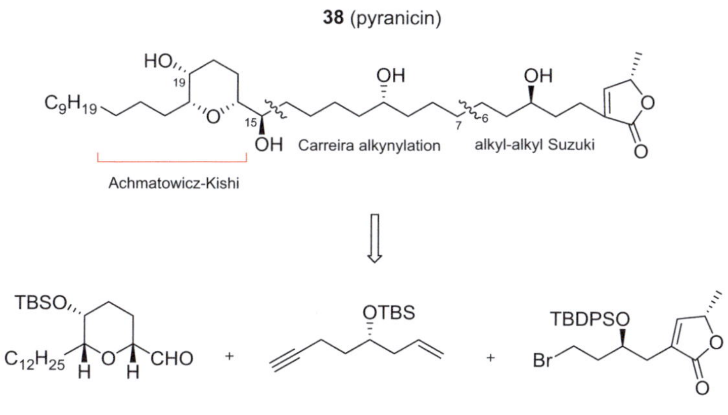

Fig. 35 Pyranicin (**38**) and its overall synthesis strategy

further summarized the divergence en route to pyragonicin (**39**) and pyranicin (**38**) in detail [243]. Soon afterwards, Griggs and Phillips proposed a convenient 13-step and efficient synthesis of pyranicin (**38**) (Fig. 35), in which the pyran subunit was prepared by an Achmatowicz oxidation-Kishi reduction and the other two subunits were made by Fu's alkyl-alkyl Suzuki coupling reaction and Carreira alkynylation to give the desired compound [244]. The physical data of the synthesized compound were in excellent agreement with those of naturally occurring pyranicin (**38**).

5.5.4 Tri-THF Type

Goniocin (**79**) and cyclogoniodenin T (**80**) are a new subgroup of the AGEs that possesses three adjacent THF rings. Initially, Sinha et al. planned to synthesize goniocin (**79**) using rhenium oxide in tandem oxidative polycyclizations [245, 246]. A mixture of trifluoroacetylperrhenate and trifluoroacetic anhydride (TFAA) was proven to increase the efficiency of triple-oxidative cyclizations (Fig. 36). However, they obtained 17,18-bis-*epi*-goniocin (**79a**) rather than **79**. Later on, Sinha et al. proposed an asymmetric total synthesis of goniocin (**79**) and cyclogoniodenin T (**80**) from a trienol and its *ent*-form via a tandem epoxide opening cascade of Sharpless asymmetric dihydroxylation (AD) and epoxidation (AE) reaction as well as Williamson etherification [247]. Both synthetic materials were found to be

Fig. 36 Sinha's strategy to synthesize the stereoisomer **79a** of goniocin (**79**)

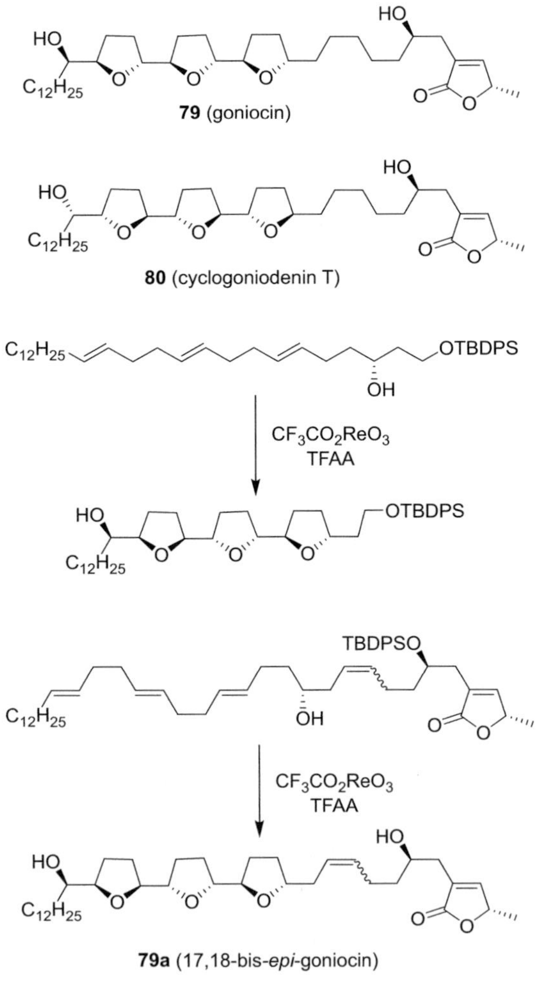

identical with the naturally occurring compounds **79** and **80**, thereby confirming the proposed absolute configurations. The only disadvantage of this work was the low yield obtained.

5.5.5 Bis-lactone Type

Rollicosin (**23**) and squamostolide (**24**) are of the characteristic subclass of AGEs that are composed of two terminal γ-lactone rings linked to an alkyl chain. They can be regarded as derivatives of classical THF-containing AGEs stemming from the oxidative cleavage in biosynthesis. Quinn et al. synthesized rollicosin (**23**) using the tandem ring-closing/crossmetathesis (RCM/CM) strategy [152], in which the allyl butenolide segment was obtained by a site-selective initiation of the catalyst as the key step. Sequentially, the C-4 stereocenter of **23** was introduced by an AD-mix-β reaction. Coupling the triflate with the enolate, oxidation, and deprotection, **23** was obtained in 9% yield over 12 steps from the C_2-symmetric dienediol. To obtain **23**, Makabe et al. reported a synthesis of **23** and its (4*R*,15*S*,16*S*,21*S*)-configured isomer (Fig. 37) [248]. The retrosynthesis analysis indicated that rollicosin (**23**) could be dissected into two building blocks, a hydroxy lactone with a terminal alkyne fragment and an iodine substituted α,β-unsaturated lactone. Both building blocks could be prepared from the corresponding primary alcohols. By palladium-catalyzed coupling of the two building blocks the target products were obtained. The ^{1}H and ^{13}C NMR spectra of the (4*R*,15*R*,16*R*,21*S*) product showed good agreement with those of the naturally occurring rollicosin (**23**). Interestingly, both synthetic compounds displayed the same activity ($IC_{50} = 0.6\ \mu M$) with the bovine heart mitochondrial complex I. The same approach was also applied to the rollicosin analogue, squamostolide (**24**), resulting in a good yield [153]. In addition,

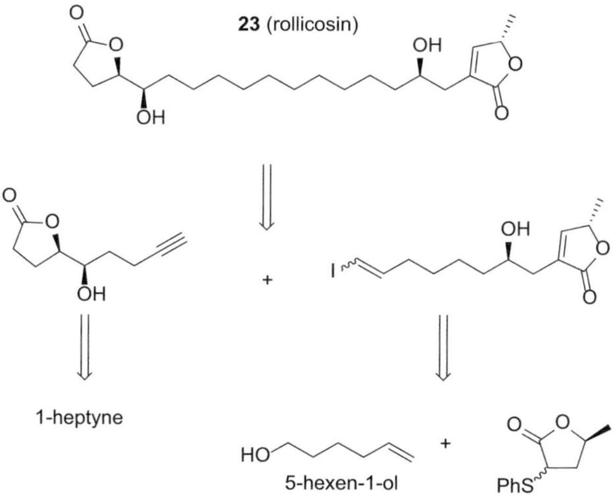

Fig. 37 Retrosynthesis of rollicosin (**23**) [91, 152]

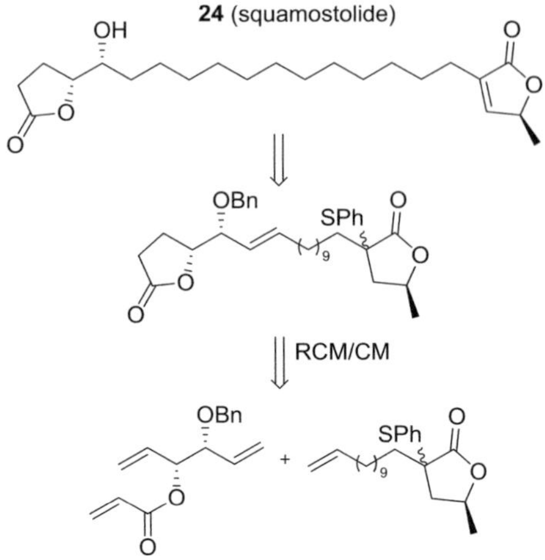

Fig. 38 Retrosynthesis of squamostolide (**24**) [121, 153]

Wu and Quinn et al. proposed a total synthesis of **24** [219]. Quinn et al. synthesized **24** from D-mannitol as the starting material over nine steps, in which the catalyst ($L_nRu{=}$) plays an important role in the inherent selectivity of the five-membered ring formation, which is the key step for the highly selective tandem ring-closing/cross metathesis reaction used (Fig. 38).

6 Biological Activity and Mechanism of Action of Annonaceous Acetogenins

In this section, the biological function and mechanism of action of AGEs are both discussed in addition to the clarification of the linkage between the mitochondrial respiratory chain and cellular apoptosis [249]. Apoptosis, also known as programmed cell death, is a normal physiological process that selectively and desirably destroys cells and tissues without triggering any inflammatory response, as opposed to necrotic cell death. Instead of focusing on the well-established inhibition of bullatacin (**7**) on mitochondrial complex I, this compound was found capable of inducing apoptotic cell death. This is based on the morphological changes observed with bullatacin (**7**)-treated Hep 2.2.15 cells, as determined by double staining using fluorescein-isothiocyanate (FITC)-labeled annexin V and propidium iodide (PI) [250]. These findings have opened a new means of exploring the mechanism of action of AGEs on apoptosis.

Although AGEs have exhibited diverse biological activities, including antibacterial [251], insecticidal [252], cytotoxic [1], and immunosuppressive effects [253], investigators have developed a particular interest on their potential anticancer effects and on the underlying mechanisms of activity. However, both the cytotoxic and pesticidal activities are related to ATP generation and NADH-oxidation in the mitochondria. Therefore, several studies have focused on the interaction of AGEs with mitochondrial complex I.

6.1 Pesticidal Activities

Annonaceous plants have potential use as pesticides, and McLaughlin et al. were the first group to report this application. They determined the pesticidal potencies of extracts from the paw paw tree (*Asimina triloba*) by the brine shrimp lethality test (*Artemia salina* L. larvae). Also, similar pesticidal activities were found against the striped cucumber beetle (*Acalymma vittatum* F.), the Mexican bean beetle (*Epilachna varivestis* Mulsant), mosquito larvae (*Aedes aegypti* L.), blowfly larvae (*Calliphora vicina* Meigen), the melon aphid (*Aphis gossyphii* Glover), the two-spotted-spider mite (*Tetranychus urticae* Koch) and the free-living nematode (*Caenorhabditis elegans* Maupas). During preliminary screening work, asimicin (**8**) was isolated and its pesticidal action was evaluated [135]. Utilizing a bioactivity-guided isolation and fractionation method, bullatacin (**7**) was purified and its pesticidal effects were observed at a fairly low concentration (1 ppm), whereas bullatacinone (**81**) lacked any discernible pesticidal activity at the concentrations tested [254]. McLaughlin et al. conducted a controlled study to investigate the pesticidal potencies of extracts from various parts of plant of the paw paw tree (*Asimina triloba*) using the brine shrimp test [255].

McLaughlin et al. further evaluated the pesticidal properties of 44 AGEs using a yellow fever mosquito larvae (YFM) assay [256]. The results clearly demonstrated that many AGEs possess pesticidal properties. In addition, the results focused on adjacent bis-THF AGEs with three hydroxy groups. For example, bullatacin (**7**) and trilobin (**82**) were the most potent AGEs in this aspect. They further used AGEs of the mono-THF, adjacent bis-THF, and nonadjacent bis-THF types as insecticidal baits for testing the potent toxicity of these compounds against insecticide-susceptible and -resistant German cockroaches. The resultant pesticidal activities were compared with conventional synthetic insecticides [257]. Ohsawa et al. evaluated the insecticidal activities of AGEs from the seeds of the pond apple, *A. glabra* with a micro-sprayer on the cabbage leaf or with a filter paper [258]. Guadano et al. also found that annonacin (**42**) showed antifeedant effects on *L. decemlineata*, whereas squamocin was toxic to *L. decemlineata* and *M. persicae*. They also proved that both AGEs were not mutagenic but indeed toxic to the insects in the absence of a metabolic energy-activating system [259].

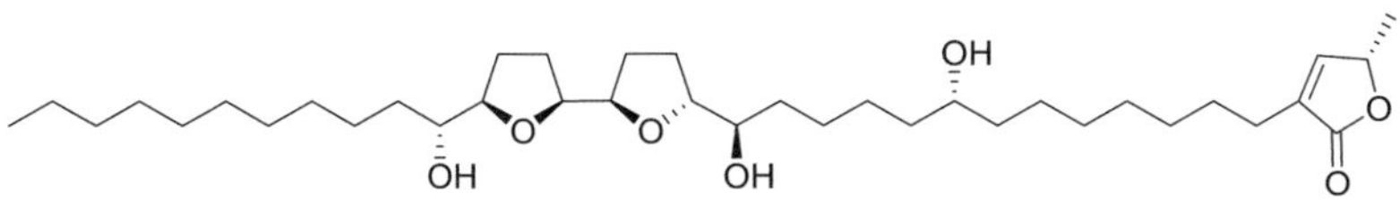

82 (trilobin)

Londershausen et al. also determined how extracts of the ground seeds of *A. squamosa* revealed interesting insecticidal properties, in which AGEs were determined to be the active components through an activity-monitored fractionation procedure. The measurement of ATP-levels (at the LT_{50} value) in *Plutella xylostella* under treatment with squamocin (**6**) and antimycin A were 1.45 and 1.35 μmol/g weight [4]. Further studies demonstrated that squamocin (**6**) showed an inhibitory effect on NADH-cytochrome *c*-reductase and complex I of insect mitochondria with IC_{50} values of 4–8 μmol/g protein and 0.8 μ*M*, respectively. Similar results were observed for the inhibition of squamocin (**6**) on complex I in bovine heart muscle (IC_{50} <0.1 μ*M*) or *Neurospora crassa* cells (IC_{50}: 0.3 μ*M*). However, no effects on other coupling sites of mitochondrial complexes were observed [4]. These experimental results were established by Lewis et al. in 1993 [260]. Friedrich et al. and McLaughlin et al. simultaneously reported that the insecticidal action is attributed to the inhibition of AGEs on mitochondria complex I [261, 262]. Friedrich et al. found the inhibition of mitochondrial and bacterial NADH:ubiquinone oxidoreductase (complex I) by AGEs did not rely on pure competition [261]. They demonstrated how AGEs also affect the electron-transfer step from the high-potential iron-sulphur cluster to ubiquinone by directly acting on the ubiquinone-catalytic site of complex I [261].

Recently, a Brazilian group has found that two AGEs, annonacinone (**83**) and corossolone (**84**) from *A. muricata*, showed IC_{50} values ranging from 25.9 to 37.6 μg/cm^3 against the promastigote form of *Leishmania chagasi*, whereas the IC_{50} values ranged from 13.5 to 28.7 μg/cm^3 against the amastigote form of this protozoan [263].

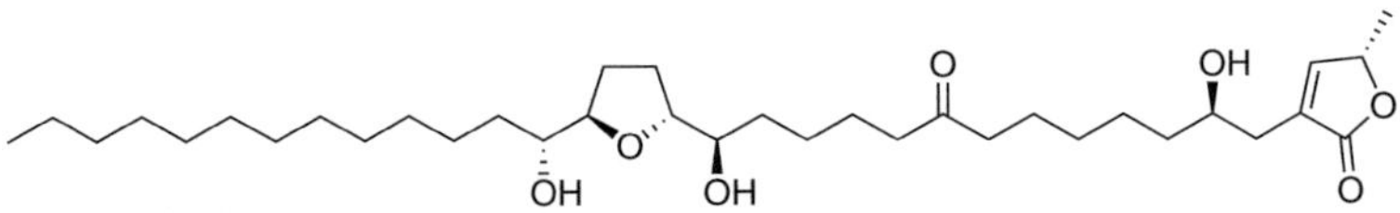

83 (annonacinone)

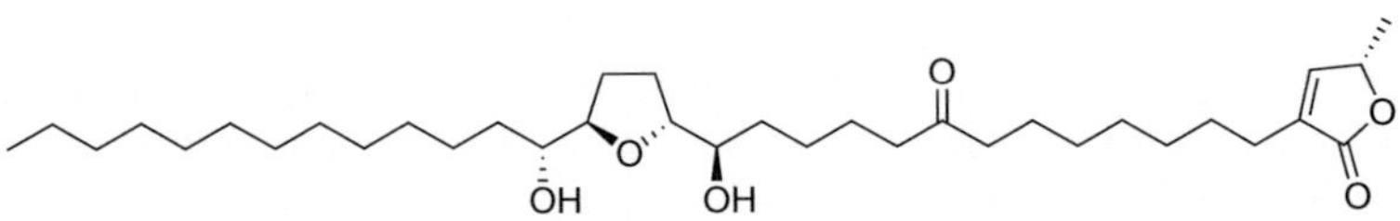

84 (corossolone)

6.2 Cytotoxic, Cancer-Related, and Ionophore Activities (Anticancer Activity)

Jolad et al. first reported the significant in vivo cytotoxic activity of uvaricin (**1**) using the murine P-388 lymphocytic leukemic test system [1]. In 1993, McLaughlin et al. reported the usage of normal mice bearing L1210 murine leukemia and athymic mice bearing A2780 conventional ovarian cancer xenografts as models to study the cytotoxic action of bullatacin (**7**) and its analogues. These compounds also demonstrated potential insecticidal activity in insect-derived Sf9 cells. It was postulated that the toxicity of AGEs in both cases is due to the strong inhibitory effect on the mitochondrial electron transport with specific action at complex I [264]. Degli Esposti et al. first used mammalian mitochondria to study the action of AGEs on NADH:ubiquinone reductase (complex I). They reported that bullatacin (**7**) inhibited the proton pumping function of complex I with similar efficiency under steady-state and non-steady-state conditions, while comparing to the action of rotenone and piericidin [265]. Their ability to inhibit mitochondrial complex I, the main gate for cellular energy production, has helped promote AGEs as candidates for the development of a new class of antitumor drugs with a different mechanism of action than conventional cancer chemotherapeutic agents.

Besides blocking NADH:ubiquinone oxidoreductase (complex I) in the electron transport system, AGEs are also considered as powerful inhibitors of NADH oxidases peculiar to the plasma membranes of cancer cells. Both mechanisms of action led to the inhibition of ATP production and may also be accountable for the observation on the effectiveness of AGEs killing multiple-drug resistant (MDR) tumors than their non-resistant counterparts. This effect was due to the requirement of ATP for the MDR pumps on cell membranes. In addition, Oberlies et al. observed that AGEs can inhibit selectively cell growth in in vitro cell inhibition assays against three murine (P388, PO3, and M17/Adr) and two human (H8 and H125) cancer cell lines [266]. Interestingly, the work of McLaughlin et al. showed that certain members of this class of natural products exhibit inhibitory activity against some drug-resistant cancers. Currently, MDR cancers are hard to cure due to the mechanism developed by the cancer cells in overcoming the anticancer agents being administered therapeutically. Owing to the biochemical differences between MDR and parental cancer cells, the ATP-dependent P-glycoprotein mediated pumps (P-gp) found in MDR cancer cells require a higher demand for ATP. Also, Oberlies et al. tested the effect of bullatacin (**7**) on two cell lines, MDR human mammary adenocarcinoma (MCF-7/Adr) cells and parental, non-resistant wild-type (MCF-7/wt) cells [267]. Thus, ATP depletion could be another mode of action that offers an advantage for AGEs to be developed into novel chemotherapeutic agents for MDR tumors.

McLaughlin et al. also proposed a model for explaining the action of AGEs [268]. They suspected that the lactone ring alone could directly interact with the binding to complex I, while the THF rings with flanking OH groups function as hydrophilic anchors at the membrane surface that allow lateral diffusion (or random distribution) of the lactone ring in the interior membrane. To verify this model, Kuwabara et al. synthesized a series of analogues with two terminal γ-lactone rings [269]. However, the bioassay results showed that these analogues did not have the same degree of effectiveness as the AGEs. For the AGEs abundant in plants from Taiwan, our group has collaborated with biochemists and pharmacologists countrywide to clarify their mechanism of actions. Yuan et al. found that annonacin (**42**) could arrest T24 bladder cancer cells at the G1 phase and cause cytotoxicity in a Bax- and caspase-3-related pathway [270]. In addition, squamocin (**6**) was also observed to arrest the same cancer cells at the G1 phase and resulted in a selective cytotoxicity in S-phase-enriched T24 cells via the same pathway of cleaving the functional protein of PARP thus inducing cellular apoptosis [271]. Squamocin (**6**) was also found to inhibit the proliferation of K562 cells via G2/M arrest in association with the induction of p21 and p27 and the reduction of Cdk1 and Cdc25C kinase activities [271].

6.3 Neurotoxic Activities

In 1999, Caparros-Lefebvre et al. found an unexpectedly high percentage of atypical parkinsonism in Guadeloupe, French West Indies, by a 1-year epidemiological study from August 1995 to August 1996. The study reported that progressive supranuclear palsy (PSP) and atypical parkinsonism were apparent in those patients who had drunk herbal tea or eaten the fruits from annonaceous plants (custard apple or paw paw family) in Guadeloupe [272]. Originally, a relationship was speculated between dietary neurotoxins from the tropical herbal teas and fruits of *Annona muricata*, *A. squamosa*, and *A. reticulata* and atypical parkinsonism of patients as being due to chronic poisoning by benzyl-tetrahydro-isoquinolines thought to be potent dopaminergic neurotoxins [272–274]. In addition, in 2002, Lannuzel et al. found that both a crude extract and pure compounds, such as the alkaloids, coreximine, and reticuline, from *A. muricata* root bark could affect physiological functions and the survival of mesencephalic dopaminergic neurons. This finding matched closely with Caparros-Lefebvre's hypothesis [275]. On the other hand, the major AGE, annonacin (**42**), in *A. muricata*, a potent inhibitor of mitochondrial respiratory chain at the level of complex I was investigated for potential neurotoxic activity in vitro. Annonacin (**42**) was highly toxic to the dopaminergic and other mesencephalic neurons, and **42**-treated neurons, by impairment of energy production [276]. Subsequently, in 2004, Champy et al. found that neurodegeneration in the rat brain was induced under a chronic systemic exposure of annonacin (**42**) intravenously for 28 days [277]. The aforementioned in vitro and in vivo studies provided evidence that neurotoxicity between the consumption of annonaceous products and the occurrence of atypical Parkinson's disease. Similarly, atypical Parkinson's disease also occurred in people who consumed traditionally annonaceous fruits in Guam and New Caledonia [278]. Moreover, a temperate annonaceous plant in Eastern United States named paw paw fruit (*A. triloba*) also contains a high percentage of annonacin (**42**). Its ethyl acetate extract possesses about 10% annonacin (**42**), which induced 50% death of cortical neurons in an in vitro experiment [279]. Did this side effect come from the alkaloids and/or acetogenins of annonaceous plants? If true, is the edible pulp of various ripe annonaceous fruits suitable for dietary food? The edible pulp of these plants are important and delicious fruit resources of many countries in the world. Meanwhile, the seeds of these plants are well-known for their toxicity and abundance of acetogenins. The content of annonaceous acetogenins and dopaminergic alkaloids in the pulp of the fruits should be investigated in future studies to clarify their potential for chronic neurotoxicity. As of now, no reports on the chemical composition of the edible pulp of annonaceous fruits are available.

6.4 *Other Activities*

6.4.1 Anti-inflammatory Effects

In a search for phytochemicals with anti-inflammatory activity, 15 representative AGEs were evaluated for their COX-2 inhibitory activity. Among them, isodesacetyluvaricin (**9**), from the Formosan tropical fruit tree, *Annona glabra*, exhibited the most potent inhibitory activity. Reverse transcription PCR in cultures of A431 human epidermoid carcinoma cells and luciferase assays on lipopolysaccharide-stimulated expression of COX-2 in RAW 264.7 mouse leukemic monocyte-macrophages revealed that the treatment of isodesacetyluvaricin (**9**) reduced the activities of two COX-2 promoter-transcription factors: cAMP response element-binding factor and nuclear factor of activated T-cells. In these tests, isodesacetyluvaricin (**9**) did not affect cell proliferation, as measured by a colorimetric assay, or intracellular store-operated calcium influx, as determined by fluorescence imaging. Thus, isodesacetyluvaricin (**9**) may serve as a lead compound for targeting inflammatory diseases as well as angiogenesis and cancer metastasis [280].

6.4.2 Promotion of Biofilm Formation in Microbes

An Argentinian group found that *Pseudomonas aeruginosa* PA100 and *P. plecoglossicida* J26 increased their biofilm formation when squamocin (**6**) was added to the culture medium [251]. Although the evaluated AGEs possessed similar structures as the autoinducer, AHL, bioassays using *C. violaceum* showed that squamocin (**6**) is not an autoinducer agonist. It was proposed that this compound is indirectly involved in a quorum sensing mechanism by inducing a stress-related increase in AI production for a given incubation time. Therefore, the exacerbation of biofilm formation was found to be due to increased production of AI-1 [281]. Squamocin (**6**) and laherradurin (**85**) stimulated *P. plecoglossicida* J26 biofilm formation, which led to an increase in consumption of naphthalene in the absence of planktonic cells. The authors proposed that the AGEs, squamocin (**6**) and laherradurin (**85**), can be used as biofilm formation promoters to allow more efficient, safe, and durable naphthalene bioremediation processes [282].

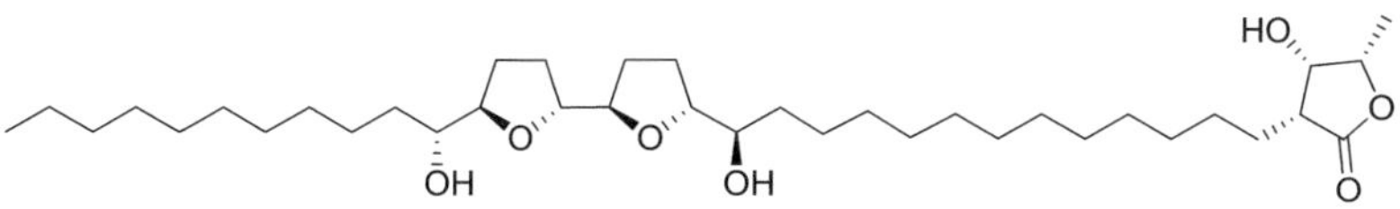

85 (laherradurin)

6.4.3 Interaction of AGEs with Membranes

With regard to the interaction of AGEs with membranes, McLaughlin's group used NMR spectroscopy to measure the space between AGEs and a bilayer membrane. ^{1}H difference NOE NMR spectra indicated that the lactone rings of asimicin (**8**) and parviflorin (**86**), of which the latter has two fewer carbons in its alkyl chain, were located below the glycerol backbone in the membrane [268]. An Argentinian group designed an FTIR experiment and provided molecular dynamics simulations of the interactions of AGEs with artificial lipid bilayers. According to their results, AGEs can interact to different extents with the phosphate and carbonyl groups of membranes in the liquid crystalline state. The THF rings, through their flanking hydroxy groups, form the hydrogen bond interactions that act as hydrophilic anchors in the lipid membrane [283, 284]. Our group used dipalmitoylphosphatidylcholine (DPPC) as a monolayer membrane to measure the change of surface pressure of membrane after treatment with squamocin (**6**). These preliminary data showed that AGEs can disrupt the integrity of such membranes (see Fig. 39). These studies should be helpful in clarifying the mechanisms of action of AGEs in the membrane environment.

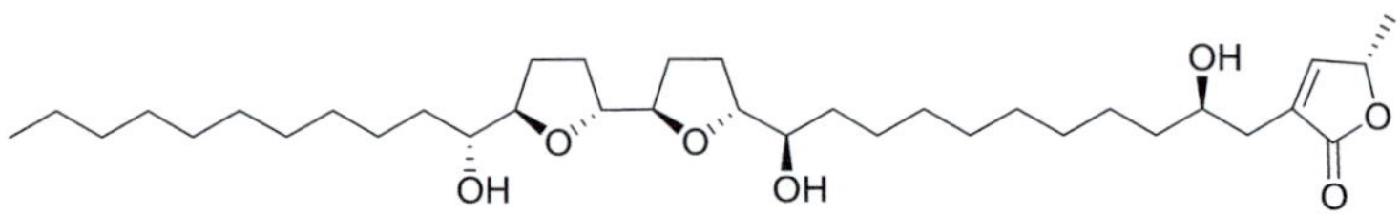

86 (parviflorin)

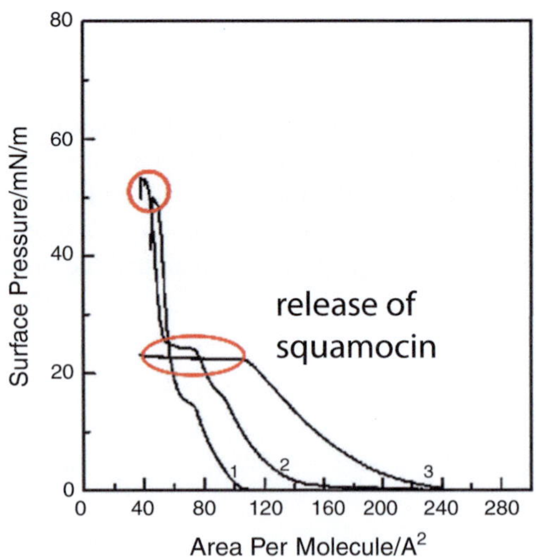

Fig. 39 Changes of membrane surface pressure after treatment with squamocin (**6**). (1) Pure DPPC, (2) squamocin (**6**) (C001)/DPPC = 1/4, 3) squamocin (**6**) (unpublished data)

6.4.4 AGEs as Cation Ionophores

Although many research investigators have examined the mechanisms of actions of the AGEs, such as the inhibitory action of mitochondria complex I (NADH: ubiquinone oxidoreductase) [285], induction of programmed cell death by the expression of pro-apoptotic Bax, Bad, and caspase-3 [270], and the structure-activity relationships of either natural, semi-synthetic or synthetic compounds, the diverse bioactivities of the various types of AGEs still remain to be elucidated in more detail. Several researchers have investigated the physicochemical features of various AGEs, and have provided direct evidence for new structure-activity relationships for these compounds.

Sasaki et al. first reported the ionophore activity of AGEs, and NMR studies revealed that the structurally related analogues of AGEs form supramolecular complexes with metal cations [286]. These studies indicated that hydroxylated bis-THF derivatives, the structural components of the more potent cytotoxic AGEs, may form supramolecular complexes with metal cations. In particular, some formed 2:1 ligand:metal complexes with calcium cations with high selectivity [286]. Earlier, in 1995, however, Araya et al. evaluated the ion-transport and ion-binding activities of AGEs using a W-08 apparatus and did not find any particular activity [287], Sasaki et al. later indicated that the two AGEs bullatacin (**7**) and asimicin (**8**) and their structurally related analogues assembled with bivalent cations, such as Ca^{2+} and Mg^{2+} [288, 289]. Peyrat et al. evaluated the ^{13}C NMR longitudinal relaxation times (T_1) for both annonacin (**42**) and squamocin (**6**) in the absence and presence of Ca^{2+} ions to assess the structural changes that accompany complexations. They considered that the potent cytotoxic activities shown by the THF-γ-lactone derivatives could be explained by their ionophoric ability. Their results also showed differences in the stoichiometry of the complexes for mono-THF AGEs and bis-THF AGEs with Ca^{2+} ions [290].

In biological studies of living cells, a point often considered was how AGEs could play a role in the bioavailability of the cations in cell membranes due to their amphiphilic nature. While culturing smooth muscle cells of the human coronary artery, our group observed that squamocin (**6**) could induce a transient but strong increase in the large-conductance Ca^{2+}-activated K^+ channels [291]. In a whole-cell configuration, this AGE (0.3–100 μ*M*) induced a Ca^{2+}-activated K^+ current ($I_{K(Ca)}$) in a concentration-dependent manner with an EC_{50} value of 4 μ*M*. When cells were exposed to a Ca^{2+}-free solution, squamocin (**6**) (3 μ*M*) induced a transient increase in $I_{K(Ca)}$. In the continued presence of squamocin (**6**), an additional increase in extracellular Ca^{2+} (1 m*M*) caused a significant increase in $I_{K(Ca)}$. In the cell-attached configuration of single-channel recordings, squamocin (**6**) applied to the bath increased the activity of large-conductance Ca^{2+}-activated K^+ (BK_{Ca}) channels without altering the single-channel conductance. These findings provide evidence that squamocin (**6**) can activate $I_{K(Ca)}$ in coronary arterial smooth muscle cells. The initial transient activation of $I_{K(Ca)}$ may reflect squamocin (**6**)-induced Ca^{2+} release from intracellular Ca^{2+} stores, whereas the sustained activation of $I_{K(Ca)}$ may arise from the squamocin (**6**)-induced Ca^{2+} influx across the cell membrane. The stimulatory effects of squamocin (**6**) on these channels would affect the functional activity of vascular smooth muscle cells [287].

We speculated that AGEs may use their hydrophilic centers (THF rings with flanking hydroxy groups) to bind with cations like Ca^{2+} and surround the ion core by the peripheral hydrophobic regions (long chains). This arrangement allows the AGE molecules to dissolve effectively in the membrane and diffuse transversely into cells as ionophores. The interaction was clarified between mono-THF AGEs and Ca^{2+} by isothermal titration calorimetry, which is a powerful and sensitive technique for measuring the heat of interaction of reacting species in dilute solution. Interestingly, it was found that the mono-THF AGEs, annonacin (**42**) and uvariamicin-I (**87**), interacted with Ca^{2+} by an exothermic process, indicating the formation of AGE-calcium complexes [37]. Furthermore, our group used various types of AGEs, including three mono-THF AGEs, two adjacent bis-THF AGEs, two non-adjacent bis-THF AGEs, and one linear AGE to clarify the relationship between the Ca^{2+}-chelating ability and their cytotoxicity. From the results, NMR spectroscopy and isothermal titration calorimetry showed that calcium ions are chelated by the hydroxylated THF ring of acetogenins, which results from formation of complexes that aid the Ca^{2+} cations in penetrating cell membranes and in elevating the intracellular calcium level (see Fig. 40). This disruption of intracellular calcium homeostasis induces mitochondrial depolarization and mediates cell toxicity [292].

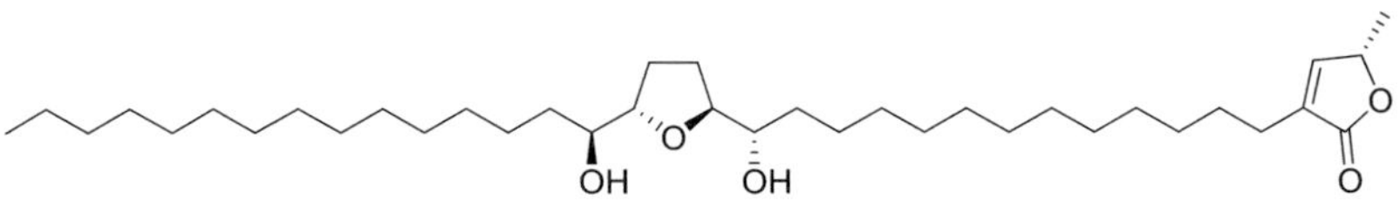

87 (uvariamicin-I)

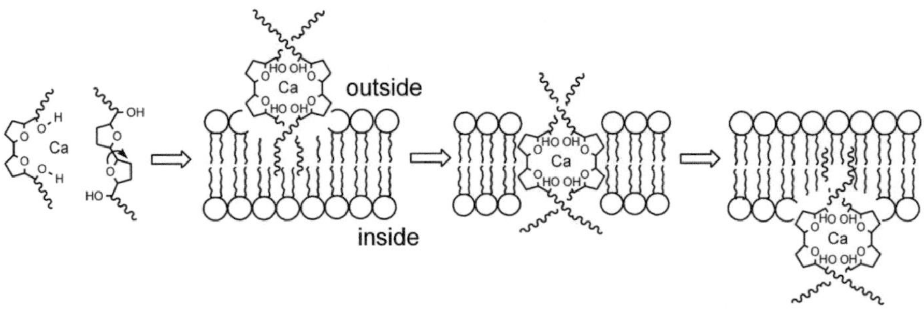

Fig. 40 Calcium-chelation model for an adjacent bis-THF AGE/Ca^{2+} complex that enhances cell-membrane penetration. The lipid bilayer can be that of the cell membrane or the mitochondrial membrane

7 Medicinal Chemistry of Annonaceous Acetogenins (Annonaceous Acetogenin Mimics)

7.1 Structure-Activity Relationship

The structure-activity relationship studies of AGEs have proven to be interesting to medicinal and natural products chemists alike. Miyoshi et al. noted that the alkyl spacer between the γ-lactone and hydroxylated THF ring moieties play an important role for AGEs to elicit potent inhibitory activities on the NADH oxidase [293]. They summarized several structure-activity relationship observations of AGEs as follows: (1) the adjacent bis-THF ring moiety is not an essential structural factor for inhibition, and the mono-THF-containing compounds can maintain potent activities; (2) this stereochemical configuration of the THF ring moiety is also not essential for potent activity irrespective of the number (one or two) of THF rings; (3) The THF rings of the AGEs have strong interactions with the interface of lipid bilayers irrespective of the configuration in the THF region; (4) the length of the alkyl side chain is very important for the elicitation of potent activity [294]. Takada et al. also tested the NADH oxidase activity of two naturally occurring AGEs, bullatacin (**7**) and diepomuricanin (**88**), and several synthesized analogues in a comparison with piericidin A [295]. They concluded that both ring moieties, the γ-lactone ring and the tetrahydrofuran ring, acted in a cooperative manner on the enzyme and that the optimal length of the alkyl spacer is 13 carbon atoms. These results supported the above hypothesis that Miyoshi et al. offered.

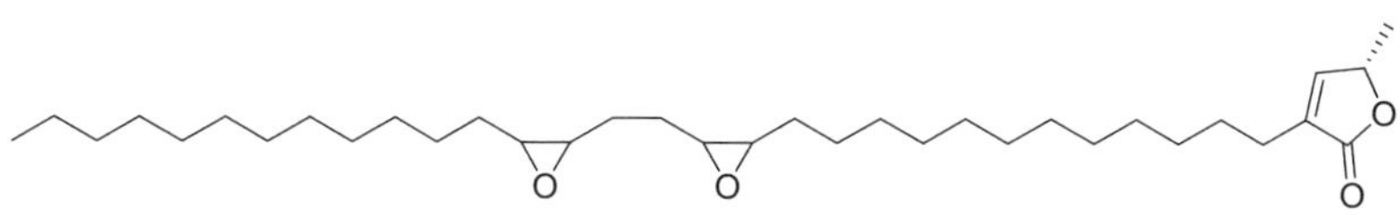

88 (diepomuricanin)

To consider solely the role of the THF ring moieties, Murai et al. synthesized Δlac-AGE (**89**) (an acetogenin derivative without the α,β-unsaturated γ-lactone ring). This was also shown to be a novel type of inhibitor that acts at the terminal electron transfer step of mitochondrial NADH:ubiquinone oxidoreductase (complex I). They also prepared a photolabile Δlac-AGE (**90**) connected to a biotin probe to trace the labeled peptide without the use of a radioisotope. This photolabile Δlac-AGE elicited potent inhibition of bovine heart mitochondrial complex I at nanomolar levels (see Fig. 41) [296]. Ichimaru et al. further synthesized a series of Δlac-AGEs, in which the stereochemistry around the hydroxylated tetrahydrofuran ring moiety was systematically modified, and examined their inhibitory effects on complex I. The results revealed that the bis-THF ring analogues are much more potent than are the mono-THF ring analogues and that the stereochemistry around the bis-THF ring moiety plays a significant role in the inhibitory effects on

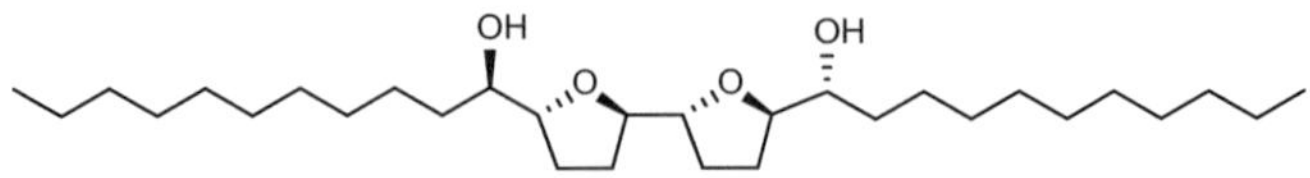

89 (${}^{\Delta}$lac-AGE)

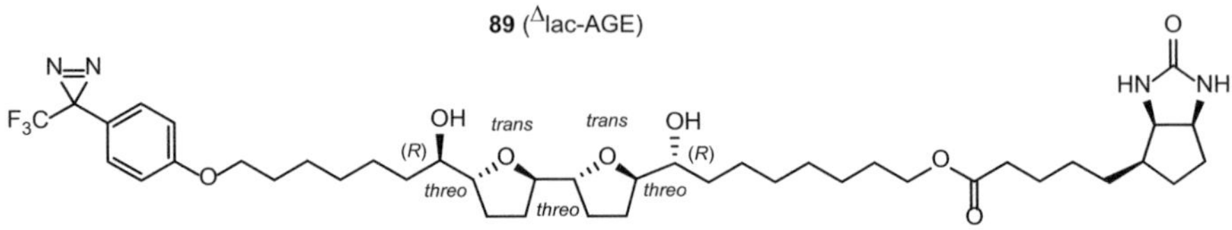

90 (photolabile ${}^{\Delta}$lac-AGE)

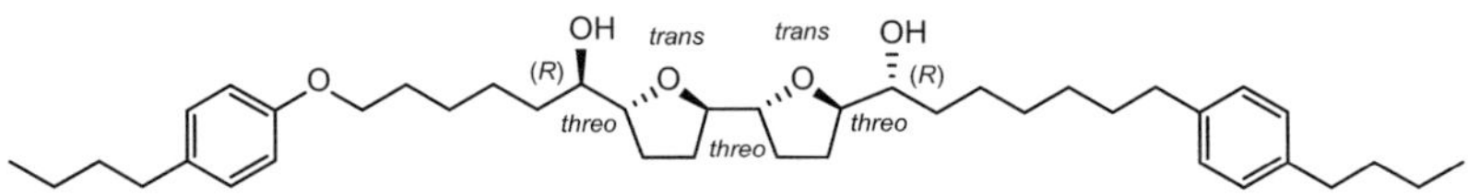

91 bis-6-(4-butylphenyloxy)-1-hydroxyhexyl-${}^{\Delta}$lac-AGE)

Fig. 41 Structures of Δlac-AGE (**89**) and modified analogues **90** and **91**

complex 1 [297]. The synthetic bis-6-(4-butylphenyloxy)-1-hydroxyhexyl Δlac-AGE (**91**) showed a similar IC_{50} value to that observed for bullatacin (**7**) in terms of the reduction of NADH oxidase activity (0.60–0.65 mmol NADH/min/mg of protein) in submitochondrial particles (see Fig. 41). Intriguingly, Ichimaru et al. demonstrated that the inhibitory site of complex I on which the Δlac-AGEs act might be different from that of the natural AGEs [296, 297].

In addition to the total syntheses of various AGEs, some specialized analogues were designed to improve the bioactivities through, for example, modifications of the γ-lactone ring, the THF ring, and the hydroxy group moieties on the aliphatic chain.

7.2 *Modifications of the γ-Lactone Ring Moiety*

Hoppen et al. designed and prepared quinone-mucocin (**92**) and quinone-squamocin D (**93**) to elucidate the mechanisms of action of the AGEs. The IC_{50} values of quinone-mucocin (**92**) and quinone-squamocin D (**93**) in the inhibition of the mitochondrial NADH-ubiquinone oxidoreductase complex were 3.6 and 1.7 n*M* [298]. These results supported their hypothesis: AGEs are competitive inhibitors at the ubiquinone binding site of complex I based on the structural similarity between the butenolide and the quinone. Arndt et al. synthesized a systematic variation of featured structures, the butenolide and the ether components, to evaluate the critical factors for the interaction of the AGEs with complex I. Their results and data from the smaller substructures indicated that the substructures of the AGEs require the polyether component and the lipophilic side chain for strong binding of the AGEs to complex I [299].

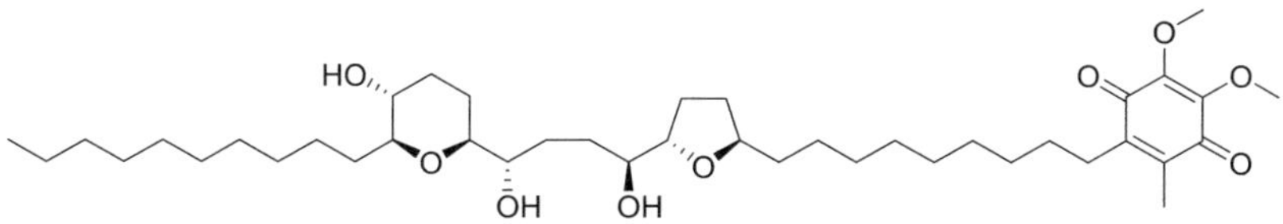

92 (quinone-mucocin)

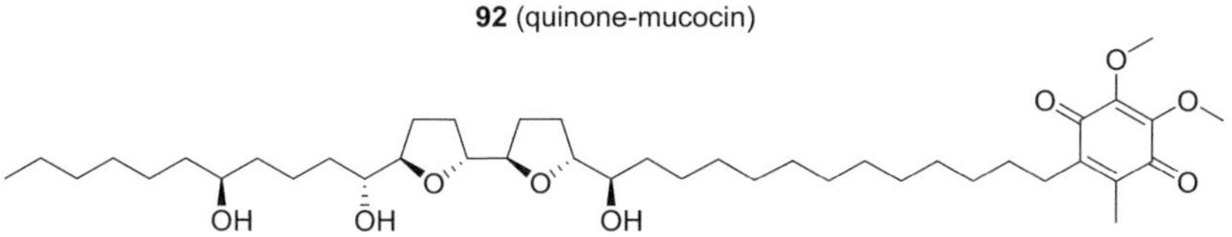

93 (quinone-squamocin D)

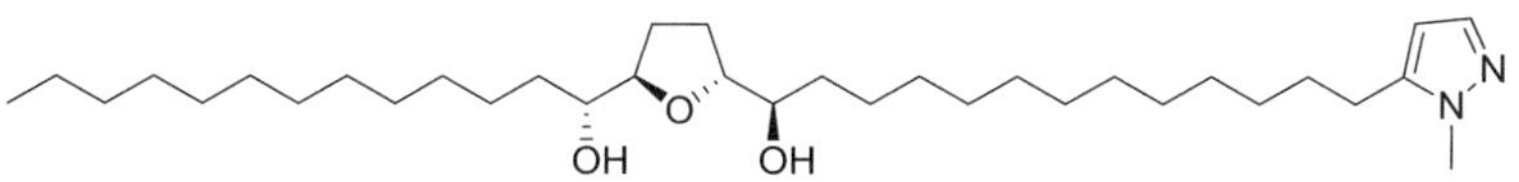

94 (1-methylpyrazol-5-yl derivative of solamin)

In addition, aromatic heterocycles are commonly found as base structures of potent complex I inhibitors. Duval et al. thus replaced the α,β-unsaturated γ-lactone moiety of squamocin (**6**) with benzimidazole via an unusual condensation-oxidative decarboxylation reaction with 1,2-diamines in the presence of acetic acid and oxygen. Although they did not clarify the inhibitory ability of modified squamocin toward complex I, one of the benzimidazole analogues showed cytotoxicity (KB 3-1 cells) with an IC_{50} value of 2.2×10^{-3} μ*M* and induced a 61% accumulation of the G1 phase of the cell cycle at concentrations of 1–5 n*M*, with apoptosis above 10 n*M* [300]. In 2006, this same group of investigators partially synthesized a series of heterocyclic analogues of squamocin (**6**). Their results suggested that the binding of this hybrid inhibitor was responsible for a negative allosteric effect at the level of the first ubiquinone-binding site of mitochondrial complex I [301]. This group also prepared a small assembly of the γ-ketoester derivatives of squamocin (**6**) and screened their biological activities, including their cytotoxicity against KB 3-1 cells, and also evaluated the inhibition of mitochondrial complexes I and III. However, these modified analogues with an open γ-lactone ring did not show any better activity than that of the parent compound, squamocin (**6**) [302].

Except for the adjacent bis-THF and non-adjacent bis-THF AGEs, Kojima et al. prepared a series of α,β-unsaturated-γ-lactone-free, nitrogen-containing heterocyclic analogues of solamin (**53**). The cytotoxic activities of the compounds were investigated against 39 tumor cell lines. One of these, a 1-methylpyrazol-5-yl derivative (**94**) showed a selective increase in cytotoxicity against NCI-H23 cells with potency a 80 times greater than that of solamin (**53**) [303].

7.3 Modification of the THF Ring Moiety of Acetogenins

In 2000, two independent groups proposed replacing the ethylene bridge in the AGE THF rings with normal- and iso-terminal lactone moieties, respectively, based on both the difficulty associated with total syntheses of AGEs and the straightforward means by which their structures can be simplified [304, 305].

Yao and co-workers studied two other series of simplified acetogenin derivatives, AA005 (**95**) and its analogues, which showed potent antitumor activities and significant selectivity between normal cells and cancer cells (see Fig. 42) [306]. Zeng et al. designed and synthesized a series of (4*R*)-hydroxy analogues of AGEs based on the naturally occurring lead compound, bullatacin (**7**). Preliminary screening data showed that the IC_{50} values of (4*R*)-hydroxylated AA005 (**96**) were 1.6×10^{-3} and 8×10^{-2} μg/cm^3 against HT-29 and HCT-8 cells (see Fig. 42). A remarkable enhancement effect was observed for AA005 (**95**) and its (4*R*)-hydroxylated analogues **96** and **97** (see Fig. 42) [307]. The results obtained indicated that both the butenolide and ethylene glycol subunits play essential roles in mediating the cytotoxicity of those compounds against selected tumor cell lines. Recently, additional evidence has indicated that AA005 (**95**) can cause growth inhibition and autophagy of colon cancer cells by depleting ATP, activating AMP-activated protein kinase (AMPK) and inhibiting the mTOR complex 1 (mTORC1) signal pathway. Compound AA005 (**95**) also inhibits cisplatin-triggered up-regulation of mTOR and synergizes with this cancer chemotherapeutic drug in the suppression of proliferation and induction of apoptosis of colon cancer cells [307]. However, the presence of a hydroxy group at C-10 and the absolute configuration of the methyl group on the butenolide moiety are less important for their activities [308].

While Rodier et al. introduced a benzoyl group to adjust the moiety between the ether linkage, analogues **98a–98g** of this type (Fig. 43) displayed interesting cell cycle effects. The analogues were found to be less potent than the cytotoxic agent,

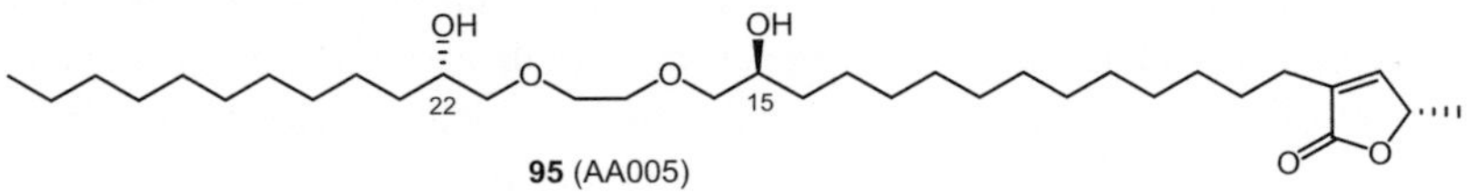

95 (AA005)

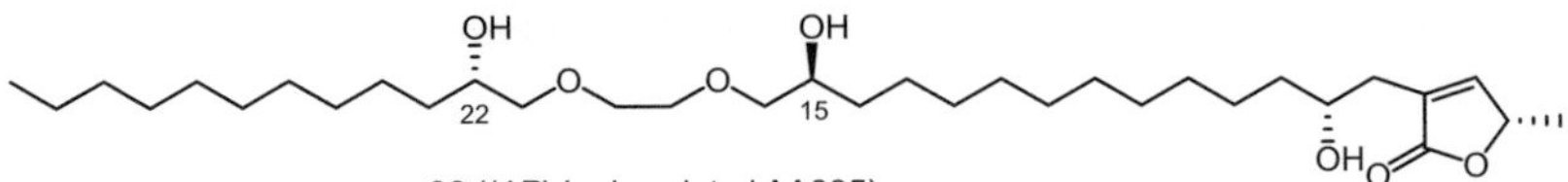

96 ((4*R*)-hydroxylated AA005)

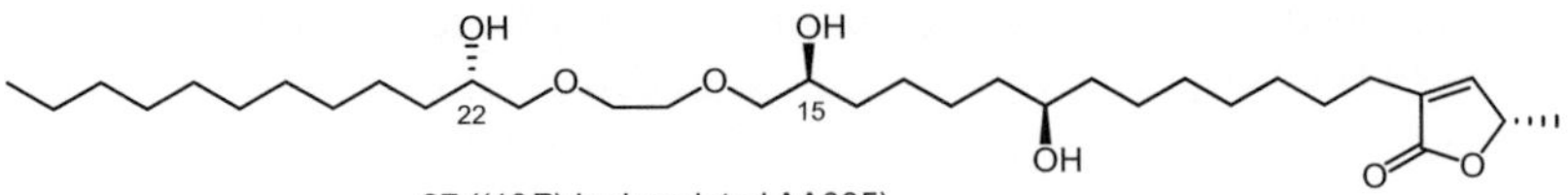

97 ((10*R*)-hydroxylated AA005)

Fig. 42 The structures of Zeng-type simplified mimics **95–97** of AGEs

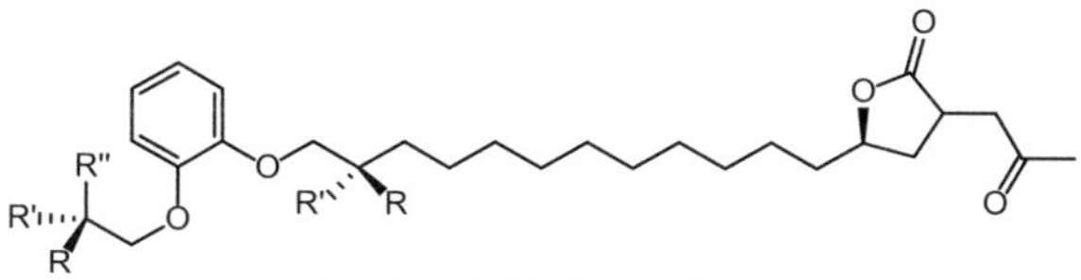

98a: R = OH, R' = H, R" = Ph
98b: R = OH, R' = H, R" = 4-CF3-Ph
98c: R = OH, R' = H, R' = naphthyl
98d: R = OH, R' = H, R' = dodecyl
98e: R = H, R' = OH, R' = dodecyl
98f: R = OH, R' = H, R' = *N*-butylamin
98g: R = OH, R' = H, R' = *N*-acetylamin

Fig. 43 The structures of Rodier-type simplified mimics of AGEs **98a–98g**

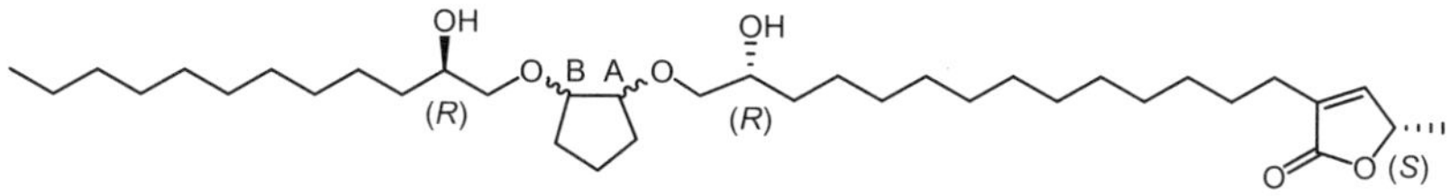

99a A: (*R*), B: (*R*)
99b A: (*R*), B: (*S*)
99c A: (*S*), B: (*R*)
99d A: (*S*), B: (*S*)

Fig. 44 The structures of Fujita-type simplified mimic of AGEs **99a–99d**

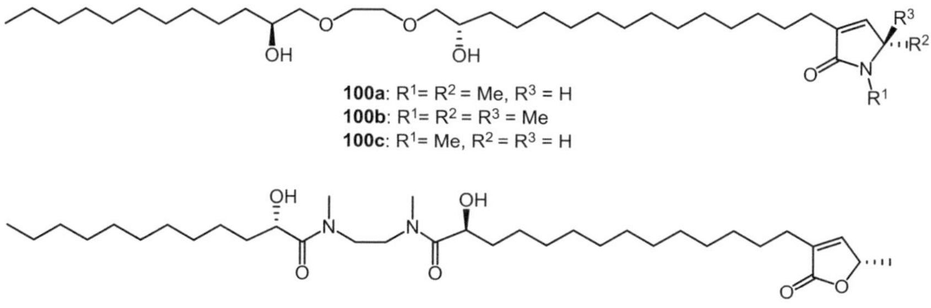

100a: $R^1 = R^2 = Me, R^3 = H$
100b: $R^1 = R^2 = R^3 = Me$
100c: $R^1 = Me, R^2 = R^3 = H$

N,N'-dimethyl-bis-amide mimic **101**

Fig. 45 The structures of modified analogues **100a–100c** and **101** of AA005 (**95**)

bullatacinone (**81**), a compound with the same terminal lactone unit [309]. Fujita et al. replaced the bis-adjacent THF ring by a 1,2-cyclopentanediol bis-ether skeleton to obtain simplified mimics **99a–99d** (see Fig. 44). Based on the evaluation of the inhibitory effects on mitochondrial NADH:ubiquinone oxidoreductase (complex I), compounds containing the 1,2-cyclopentanediol bis-ether motif also showed very potent inhibitory activity at the nanomolar level [310].

For a study of AGE mimics, Liu et al. designed, synthesized, and evaluated a new series of compounds containing a terminal lactam [311]. They found that the N-methylated lactam-containing mimics of AGEs **100a–100c** and **101** (see Fig. 45), the modified analogues of AA005 (**95**), exhibited comparable potencies to that of the parent compound and similar selectivity for the cancer cells represented. It was also revealed that the stereogenic center on the lactam is not essential for cytotoxic activity. Liu et al. synthesized a series of analogues by replacing the acyclic bis-ether functionality of AA005 (**95**) with certain conformationally constrained fragments. Interestingly, most newly synthesized mimetics were found to exhibit potent activities against breast cancer cells and showed selectivity between cancerous and non-cancerous cells. Among those, an *N,N′*-dimethyl bis-amide mimic **101** of AGE exhibited greater potency against MDA-MB-468 cells than did its parent molecule, AA005 (**95**) (see Fig. 42). Xiao et al. indicated that certain bis-amide analogues of AA005 (**95**) make this unique class of anticancer agents simpler and allow more flexibility for their future development [312]. Also, analogues bearing a biphenyl moiety in the hydrocarbon chain part of AA005 (**95**) exhibit more potent antiproliferative activity and preferentially target cancer cells over normal cells [312].

7.4 Replacement of the Hydroxy Group on the Aliphatic Chain

Investigators have studied the effects of the replacement of the hydroxy groups on the aliphatic chain of AGEs. Ye et al. obtained the halogen-substituted AGEs, (4*S*)-chloro-4-deoxygigantetrocin A (**102**) and (4*S*)-18-dichloro-4,18-dideoxyasimilobin (**103**), by treating gigantetrocin A (**47**) with triphenylphosphine and CCl_4. However, the chlorinated products showed decreased bioactivities in the brine shrimp lethality test and when evaluated against selected human tumor cell lines [313]. In contrast, Kojima et al. made C-4-fluorinated solamin (**104**) and evaluated its cytotoxic activities against 39 cancer cell lines (see Fig. 46). They found **104** to show more potent growth inhibitory activity against cancer cell lines than solamin (**53**) [314].

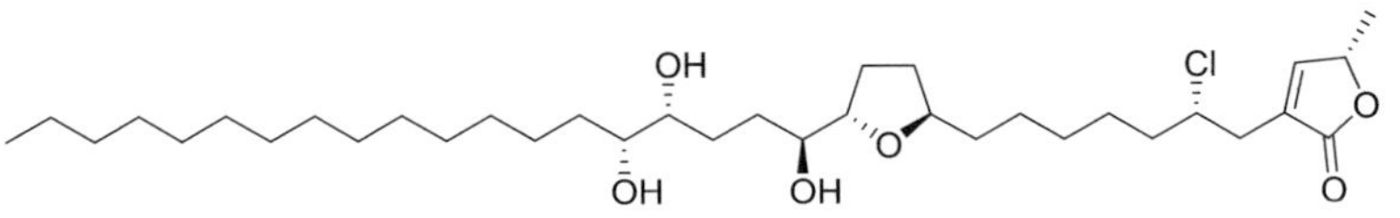

102 ((4*S*)-chloro-4-deoxygigantetrocin A)

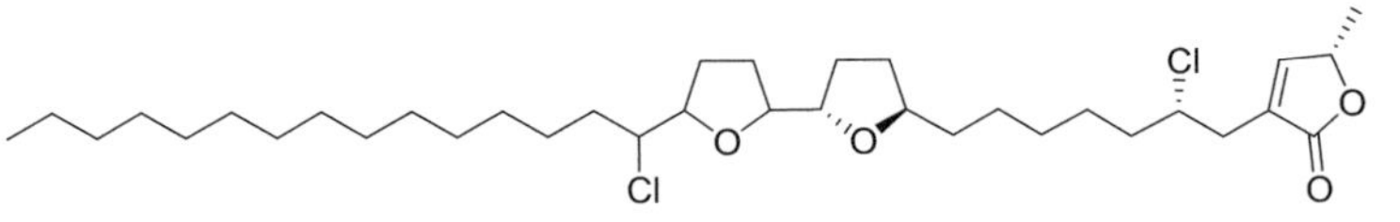

103 ((4*S*),18-dichloro-4,18-dideoxy-asimilobin)

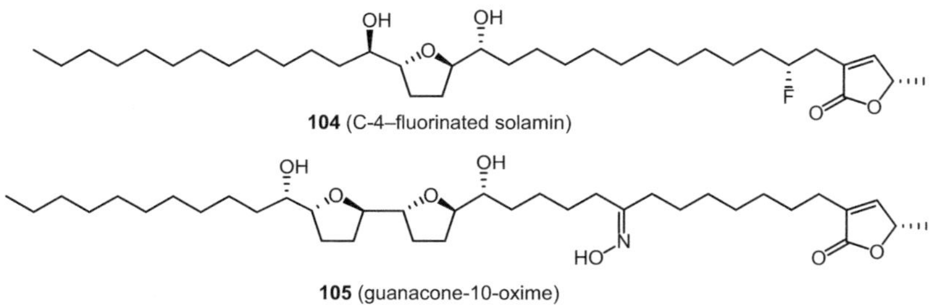

Fig. 46 The structures of C-4–fluorinated solamin (**104**) and guanacone-10-oxime (**105**)

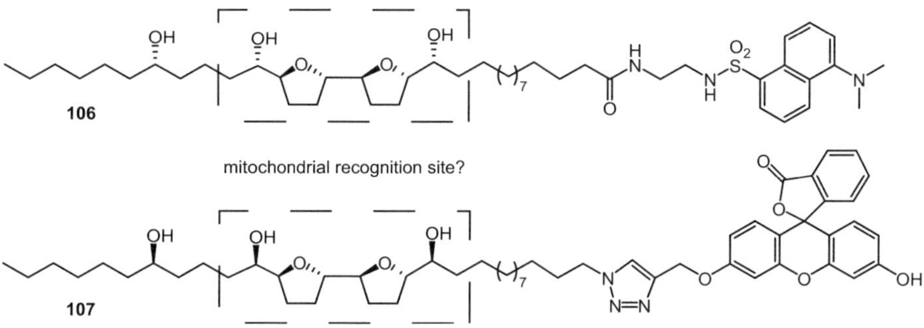

Fig. 47 The structures of fluorescent-hybridized AGEs **106** and **107**

In related work, Gallardo et al. made 10-oximeguanacone (**105**), the first bioactive nitrogenated AGE, which showed potent inhibition toward complex I by the titration of the NADH oxidase and NADH:ubiquinone oxidoreductase activities (see Fig. 46) [315]. Duret et al. partially synthesized amino derivatives from two natural AGEs, rolliniastatin-1 (**28**) and squamocin (**6**). Although it is noteworthy that these amino-AGEs still retain some activity, more studies are required to confirm the potencies of these derivatives as new specific and efficient anticancer agents [316].

A variety of chemical strategies have been applied to investigate biological processes. Recently, fluorescent modifications became powerful tools for visualizing the distribution of bioactive natural products in cells and investigating their targeting. In 2005, Derbre et al. synthesized hybrids consisting of an AGE tail connected to a fluorescent tag (see Fig. 47). Using fluorescence microscopy, two of the fluorescent-hybridized AGEs **106** and **107** were observed initially in Jurkat cell mitochondria, but they diffused into the cytosol of apoptotic cells, supporting the conclusion that squamocin (**6**) passes through the plasma membrane and targets the mitochondria. Indeed, both semi-synthetic fluorescent derivatives were shown to be potent apoptosis inducers that were directed to this organelle. Moreover, they proposed that the lactone moiety seems not to interfere with mitochondrial targeting but apparently influenced the activity [317]. Alexander et al. attached ethyl 7-dimethylaminocoumarin-4-acetate to the diols of (−)-mucocin (**37**) through amide coupling chemistry. Although coumarin-labeled-mucocin can also induce fluorescently coded morphogenic responses, no expected response was found [318]. This result might be due to the occupation of the mitochondrial recognition site by the fluorescent coumarin group. To overcome this above disadvantage, Tanaka et al. labeled the terminal aliphatic chain of solamin (**53**) with the fluorescent groups, 7-nitrobenzo[*c*][1,2,5]oxadiazol-4-yl-amino (NBD-NH-) and 5-dimethylaminonaphthalen-1-yl-sulfonamide (dansyl-NH-), to produce NBD-labeled solamin (**108**) and dansyl-labeled solamin (**109**), in 2007 and 2009 [319, 320]. It was anticipated that these compounds may be used to explore the cellular targeting of AGEs.

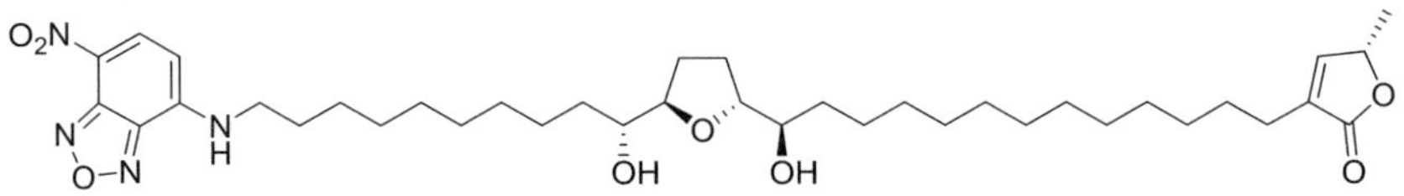

108 ("fluorescent solamin" (NBD-labeled solamin))

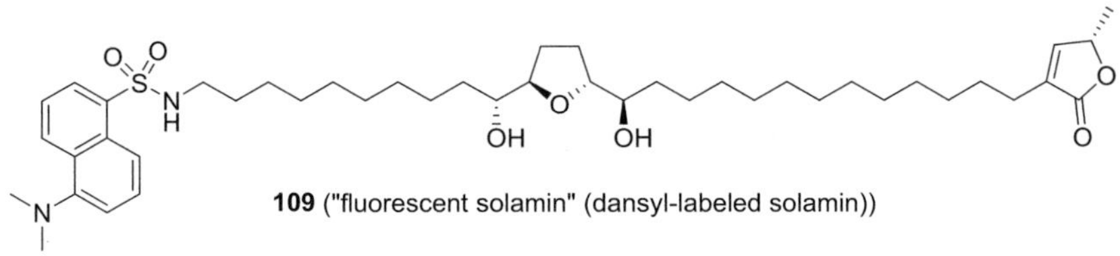

109 ("fluorescent solamin" (dansyl-labeled solamin))

Liu et al. modified the C-10 hydroxy group of some AGE mimics to introduce a label based on the results of the cytotoxicity screening of parallel synthetic analogues. Fluorescent-imaging studies revealed that the AA005-flu derivatives **110** and **111** (AA005 (**95**) connected with fluorescent groups) in normal human cells was significantly different from that in cancer cells (see Fig. 48). AA005-flu accumulated in the mitochondria of the cancer cells. This direct and visible evidence suggests that membrane recognition of AA005 (**95**) is involved in its selective bioactivity [321].

The Yao group recently also developed three representative AA005-like molecules **112–114** via formation of small nitrogen-containing heterocycles or amide bond based on the concept of click chemistry, which was introduced by Sharpless in 2001, describing the tailored chemistry to generate compounds quickly and reliably by joining small units together (see Fig. 49) [322]. These nitrogen-containing analogues exhibited significantly inhibitory activities against several cancer cell lines in low micromolar concentrations.

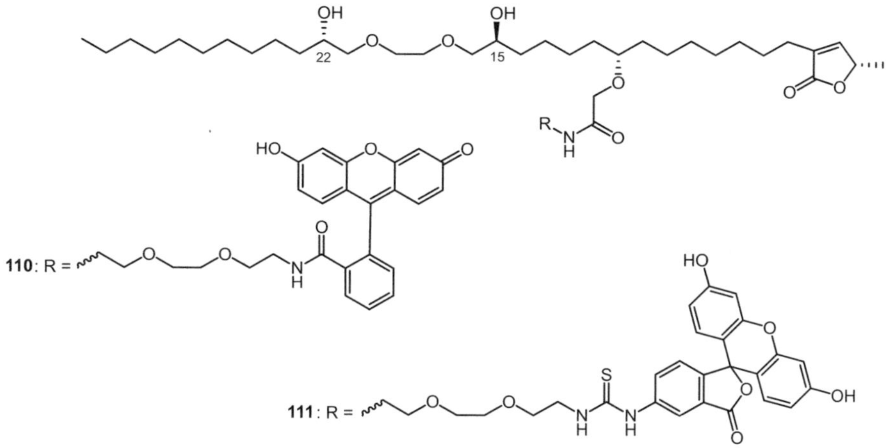

Fig. 48 The structures of AA005-flu derivatives **110** and **111**

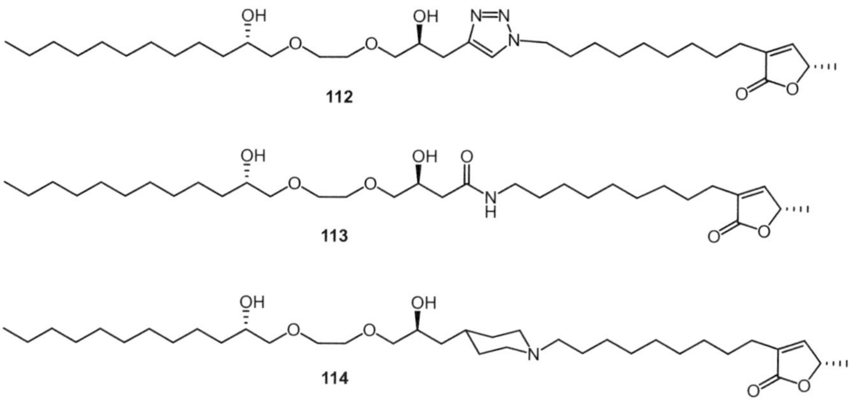

Fig. 49 Three newly proposed AA005-like molecules, **112–114**

7.5 *Mimic Synthesis Study*

Yao's group further designed a series of linear dimeric compounds mimicking naturally occurring AGEs, for example the AGE mimic **115** (see Fig. 50), and evaluated the cytotoxic activities of these analogues toward the growth of cancer cells by the MTT method. Interestingly enough, these compounds showed selective action favoring the human cancer cell lines used. The authors pointed out that with appropriate conformational constraint their assembled moieties of AGEs might be useful to optimize the potential anticancer properties of this class of compounds [323].

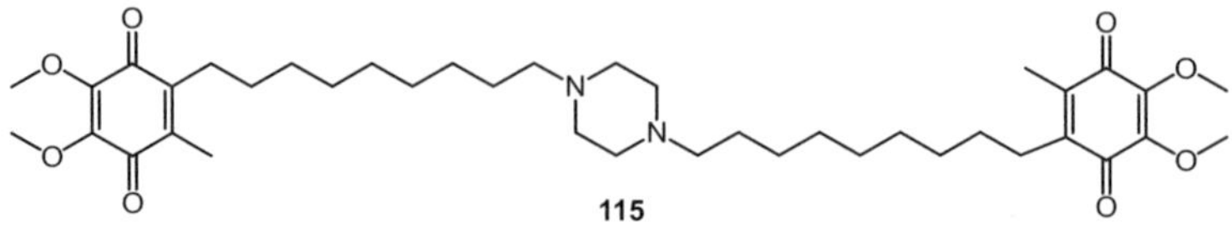

Fig. 50 A new mimicking AGE **115**, the linear dimeric compound with bis-terminal benzoquinone

8 Annonaceous Acetogenin-Containing Products and Their Potential Development

On many websites, paw paw extract-containing products are commonly available; however, there are only a few reports that mention quality control studies for annonaceous acetogenin-containing products [324]. In 1976, McLaughlin et al. found that extracts of the leaves and twigs of the U.S. native paw paw tree, *Asimina triloba*, were active in the cytotoxicity screens of the U.S. National Cancer Institute (NCI). Following their phytochemical study to determine the compounds present [255], McLaughlin et al. used the three most active AGEs, bullatacin (**7**), asimicin (**8**), and trilobacin (**116**) (Fig. 51), to serve as marker compounds for the examination of various extracts of the paw paw tree by LC/MS/MS. In this work, they demonstrated that small twigs collected in the months of May and June were the optimum plant sources for the commercial harvesting of paw paw biomass for extraction. McLaughlin further developed some useful commercial products containing the AGEs, including a head lice shampoo (in 2001), and ointment, lotion, and spray for plant pests, as well as paw paw capsules (in 2003) for human administration. The entire process from the safety and toxicology of the AGEs to the introduction of commercial products was described in McLaughlin's 2008 review [325]. More recently, Cuendet et al. reported the potential of a standardized extract from the twigs of *A. triloba* to mediate a cancer chemopreventative effect in the *N*-methyl-*N*-nitrosourea-induced mammary carcinogenesis model. As McLaughlin et al. did before, they used three potent bioactive AGEs, bullatacin (**7**), asimicin (**8**), and trilobacin (**116**) in their standardized extract. Mammary tumor latency was increased from 55 to 66 days in Sprague-Dawley rats given a diet containing paw paw extract (1250 and 2500 mg/kg diet; based on maximum tolerated dose studies) [326].

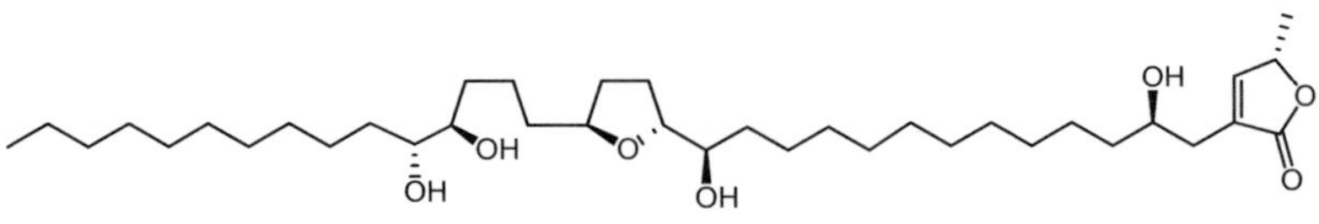

Fig. 51 Structure of trilobacin (**116**)

In Taiwan and elsewhere, plants of the genus *Annona* are important economic crops for their edible fruits. The abundant material obtained from the seeds and the potent cytotoxic effects of the AGEs from this material has stimulated the possible further development of these compounds as pharmaceutical products. In addition to the aforementioned studies, oral gavage (p.o.) animal studies have been performed by MDS Pharma Services-Taiwan for which our group provided the material (unpublished data). An extract powder from *A. muricata*, WYC-AA07, was applied to a xenograft tumor in a SCID mouse model using implanted human MCF-7 breast tumor cells. WYC-AA07 at 10 mg/kg was administered p.o. daily for a total of ten doses. The tumor sizes, body weights and signs of overt animal toxicity after dosing were monitored and recorded for 25 days. This extract caused significant decreases in tumor weights as measured from days 13 to days 25. However, it also caused significant decreases in body weight on days 9, 13, 17, and 21 (unpublished data). Another SCID mouse xenograft experiment was performed on this extract using implanted human HT-29 colon tumor cells. Extract WYC-AA07 at 20 mg/kg p.o. caused death in half of the animals and a significant decrease in body weight on day 8 (unpublished data). Recently, the acute toxicity of squamocin (**6**) was evaluated in nude mice at 20 mg/kg p.o. Most of the animals showed an abnormal movement pattern that caused the mice to move uncontrollably in a lateral direction. However, the mice apparently recovered after the administration of squamocin (**6**) was stopped. Moreover, a Chinese group also investigated the antitumor activities of AGEs in vivo using S180 and HepS xenografts in mice. Their results revealed that three types of AGEs effectively suppressed tumor growth in a dose-dependent fashion. Interesting, their results also matched with those of previous studies: adjacent bis-THF AGEs were more active than nonadjacent bis-THF AGEs, nonadjacent bis-THF AGEs were more active than mono-THF AGEs [327]. Furthermore, the present authors performed a pharmacokinetic study using LC-MS to validate the quantification of AGEs in rats [328].

On the other hand, an ethnopharmacological investigation reported a similar side effect of AGEs in that a neurodegenerative tauopathy endemic to the Caribbean island of Guadeloupe was suspected to be linked to the consumption of annonaceous plants [274], Escobar-Khondiker et al. further found that annonacin induced the retrograde transport of mitochondria to decrease ATP levels, which induces changes in the intracellular distribution of tau in a way that shares characteristics with some neurodegenerative diseases [329]. The demonstrated adverse effects of AGEs on neuronal cells should be of major concern when considering their potential commercial development.

9 Summary and Perspectives

The annonaceous acetogenins (AGEs) are one of the most interesting classes of plant-derived natural products to have been investigated during the last three decades. They exhibit a wide variety of biological activities. Impressively, some of them have comparable cytotoxic potency to the widely used anticancer agent, paclitaxel, against various cancer cells. Two structural features, the hydroxylated tetrahydrofuran (THF) and γ-lactone ring moieties, are considered to be the pharmacophores that block the electron transport system of mitochondrial complex I. Much effort has been dedicated to elucidating the underlying cytotoxic mechanisms of the AGEs and to synthesizing AGE analogues by altering the spacing between the two pharmacophores, or removing either one of these units (ΔLac AGEs or muricatacin (**22**)), or mimicking the THF rings by ether linkages. Although none of the modified AGEs obtained thus far has demonstrated activities comparable to those of the naturally occurring AGEs, studies on the synthesis and mechanisms of action of these compounds have provided solid fundamental knowledge and drug discovery insights. For example, some of the studies on analogues of AGEs have created a series of compounds that are based on completely different skeletons, of which some also show excellent bioactivities against various cancer cells. Not only promising antitumor effects, but also serious side effects of the AGEs have been found, such as neurotoxicity and the generation of symptoms of atypical parkinsonism. These may greatly limit prospects for the drug development of the AGEs. It is recommended that the composition of AGEs and dopaminergic alkaloids in the edible pulps of commercially used annonaceous fruits be evaluated so that their potential for causing neurotoxicity-related side effects can be clarified.

Acknowledgments This work was supported by grants from the Ministry of Science and Technology (National Science Council) of Taiwan NSC100-2320-B-110-001-MY2 awarded to C.C.L., MOST 103-2320-B-037-005-MY2 & 102-2628-B-037-003-MY3 awarded to F.R.C., and NSC 103-2911-I-002-303, NSC101-2632-B039-001-MY3, NSC100-2320-B-110-001-MY2 National Health Research Institutes NHRI-EX102-10241BI, and CMU under the Aim for Top University Plan of the Ministry of Education, Taiwan award to Y.C.W. as well as the Excellence for Cancer Research Center Grant, the Ministry of Health and Welfare, Executive Yuan, Taipei, Taiwan (MOHW104-TDU-B-212-124-003) and the grant for Health and Welfare Surcharge of Tobacco Products.

References

1. Jolad SD, Hoffmann JJ, Schram KH, Cole JR, Tempesta MS, Kreik GR, Bates RB (1982) Uvaricin, a new antitumor agent from *Uvaria acuminata* (Annonaceae). J Org Chem 47:3151
2. Cavé A, Figadère B, Laurens A, Cortes D (1997) Acetogenins from Annonaceae. Prog Chem Org Nat Prod 70:81
3. Hollerhage M, Matusch A, Champy P, Lombes A, Ruberg M, Oertel WH, Hoglinger GU (2009) Natural lipophilic inhibitors of mitochondrial complex I are candidate toxins for sporadic neurodegenerative tau pathologies. Exp Neurol 220:133
4. Londershausen M, Leicht W, Lieb F, Moeschler H, Weiss H (1991) Molecular mode of action of annonins. Pestic Sci 33:427
5. Gu ZM, Fang XP, Miesbauer LR, Smith DL, McLaughlin JL (1993) 30-, 31-, and 32-Hydroxybullatacinones: bioactive terminally hydroxylated annonaceous acetogenins from *Annona bullata*. J Nat Prod 56:870

6. Gu ZM, Fang XP, Rieser MJ, Hui YH, Miesbauer LR, Smith DL, Wood KV, McLaughlin JL (1993) New cytotoxic annonaceous acetogenins: bullatanocin and *cis*- and *trans*-bullatanocinone, from *Annona bullata* (Annonaceae). Tetrahedron 49:747
7. Gu ZM, Fang XP, Zeng L, Wood KV, McLaughlin JL (1993) Bullacin: a new cytotoxic annonaceous acetogenin from *Annona bullata*. Heterocycles 36:2221
8. Zafra-Polo MC, Figadère B, Gallardo T, Tormo JR, Cortes D (1998) Natural acetogenins from Annonaceae, synthesis and mechanisms of action. Phytochemistry 48:1087
9. Duret P, Laurens A, Hocquemiller R, Cortes D, Cavé A (1994) Acetogenins of Annonaceae. 34. Isoacetogenins, artifacts issued from translactonization from annonaceous acetogenins. Heterocycles 39:741
10. Duret P, Fagadère B, Hocquemiller R, Cavé A (1997) Epimerization of annonaceous acetogenins under basic conditions. Tetrahedron Lett 38:8849
11. Colman-Saizarbitoria T, Johnson HA, Alali FQ, Hopp DC, Rogers LL, McLaughlin JL (1998) Annojahnin from *Annona jahnii*: a possible precursor of mono-tetrahydrofuran acetogenins. Phytochemistry 49:1609
12. Colman-Saizarbitoria T, Liu X, Hopp DC, Johnson HA, Alali FQ, Rogers LL, McLaughlin JL (1999) Annodienin and jahnonacin: new bioactive nontetrahydrofuran annonaceous acetogenins from the twigs of *Annona jahnii*. Nat Prod Lett 14:65
13. Gleye C, Raynaud S, Hocquemiller R, Laurens A, Fourneau C, Serani L, Laprevote O, Roblot F, Leboeuf M, Fournet A, de Arias AR, Figadère B, Cavé A (1998) Acetogenins of Annonaceae: part 67. Muricadienin, muridienins and chatenaytrienins, the early precursors of annonaceous acetogenins. Phytochemistry 47:749
14. Wong HF, Brown GD (2002) β-Methoxy-γ-methylene-α,β-unsaturated-γ-butyrolactones from *Artabotrys hexapetalus*. Phytochemistry 59:99
15. Wongsa N, Kanokmedhakul S, Kanokmedhakul K (2011) Cananginones A-I, linear acetogenins from the stem bark of *Cananga latifolia*. Phytochemistry 72:1859
16. Sinz A, Matusch R, Kaempchen T, Fiedler W, Schmidt J, Santisuk T, Wangcharoentrakul S, Chaichana S, Reutrakul V (1998) Novel acetogenins from the leaves of *Dasymaschalon sootepense*. Helv Chim Acta 81:1608
17. Ee GCL, Chuah CH, Sha CK, Goh SH (1996) Disepalin, a new acetogenin from *Disepalum anomalum* (Annonaceae). J Nat Prod 9:141
18. Eparvier V, van Nguyen H, Thoison O, Martin MT, Sévenet T, Guéritte F (2006) Cytotoxic monotetrahydrofuran acetogenins from *Disepalum plagioneurum*. J Nat Prod 69:1289
19. Jiang Z, Chen Y, Chen RY, Yu DQ (1997) Mono-tetrahydrofuran ring acetogenins from *Goniothalamus donnaiensis*. Phytochemistry 46:327
20. Jiang Z, Yu DQ (1997) New type of mono-tetrahydrofuran ring acetogenins from *Goniothalamus donnaiensis*. J Nat Prod 60:122
21. Jiang Z, Yu DQ (1997) Donhexocin, a new non-tetrahydrofuran acetogenin from *Goniothalamus donnaiensis*. Chin Chem Lett 8:233
22. Tantithanaporn S, Wattanapiromsakul C, Itharat A, Keawpradub N (2011) Cytotoxic activity of acetogenins and styryl lactones isolated from *Goniothalamus undulatus* Ridl. root extracts against a lung cancer cell line (COR-L23). Phytomedicine 18:486
23. Li C, Lee D, Graf TN, Phifer SS, Nakanishi Y, Riswan S, Setyowati FM, Saribi AM, Soejarto DD, Farnsworth NR, Falkinham JO III, Kroll DJ, Kinghorn AD, Wani MC, Oberlies NH (2009) Bioactive constituents of the stem bark of *Mitrephora glabra*. J Nat Prod 72:1949
24. Zhang Q, Di YT, He H-P, Li S-L, Hao X-J (2010) Mitregenin, a new annonaceous acetogenin from *Mitrephora maingayi*. Nat Prod Commun 5:1793
25. Sung'hwa F, Mgina CA, Jonker SA, Nkunya MHH, Waibel R, Achenbach H (1999) Constituents of tropical medicinal plants. Part 88. Ophrypetalin and other annonaceous acetogenins from *Ophrypetalum odoratum*. Nat Prod Lett 13:195
26. Panthama N, Kanokmedhakul S, Kanokmedhakul K (2010) Polyacetylenes from the roots of *Polyalthia debilis*. J Nat Prod 73:1366

27. Fevrier A, Ferreira ME, Fournet A, Yaluff G, Inchausti A, Rojas de Arias A, Hocquemiller R, Waechter AI (1999) Acetogenins and other compounds from *Rollinia emarginata* and their antiprotozoal activities. Planta Med 65:47
28. Wang ML, Du J, Zhang PC, Chen RY, Xie FZ, Zhao B, Yu DQ (2002) Saccopetrins A and B, two novel γ-lactones from *Saccopetalum prolificum*. Planta Med 68:719
29. Qin Y, Pan X, Chen R, Yu D (1996) New annonaceous acetogenins from *Uvaria boniana*. Acta Pharm Sin 31:381
30. Zhou GX, Zhou LE, Chen RY, Yu DQ (1999) Calamistrins A and B, two new cytotoxic monotetrahydrofuran annonaceous acetogenins from *Uvaria calamistrata*. J Nat Prod 62:261
31. Raynaud S, Fourneau C, Hocquemiller R, Sévenet T, Hadi HA, Cavé A (1997) Acetogenins of Annonaceae. Part 54. Acetogenins from the bark of *Uvaria pauci-ovulata*. Phytochemistry 46:321
32. Chen Y, Yu DQ (1997) Absolute configurations of tonkinin A and B: two new acetogenins from *Uvaria tonkinensis*. Chin Chem Lett 8:789
33. Colman-Saizarbitoria T, Montilla L, Rodriguez M, Castillo A, Hasegawa M (2009) Xymarginatin: a new acetogenin inhibitor of mitochondrial electron transport from *Xylopia emarginata* Mart., Annonaceae. Rev Bras Farm 19:871
34. Pettit GR, Mukku VJRV, Cragg G, Herald DL, Knight JC, Herald CL, Chapuis JC (2008) Antineoplastic agents. 558. *Ampelocissus* sp. cancer cell growth inhibitory constituents. J Nat Prod 71:130
35. Hopp DC, Conway WD, McLaughlin JL (1999) Using countercurrent chromatography to assist in the purification of new annonaceous acetogenins from *Annona squamosa*. Phytochem Anal 10:339
36. Duret P, Waechter AI, Margraff R, Foucault A, Hocquemiller R, Cavé A (1997) High-speed countercurrent chromatography: a promising method for the separation of the annonaceous acetogenins. J Liq Chromatogr Relat Technol 20:627
37. Liaw CC, Yang YL, Chen M, Chang FR, Chen SL, Wu SH, Wu YC (2008) Mono-tetrahydrofuran annonaceous acetogenins from *Annona squamosa* as cytotoxic agents and calcium ion chelators. J Nat Prod 71:764
38. Gawronski J, Wu YC (1999) A note on the determination of absolute configuration of acetogenins by circular dichroism. Pol J Chem 73:241
39. Duret P, Waechter AI, Figadère B, Hocquemiller R, Cavé A (1998) Determination of absolute configurations of carbinols of annonaceous acetogenins with 2-naphthylmethoxyacetic acid esters. J Org Chem 63:4717–4720
40. Hirayama K, Akashi S, Yuji R, Niitsu U, Fujimoto Y (1993) Structural studies of polyhydroxybis(tetrahydrofuran) acetogenins from *Annona squamosa* using the combination of chemical derivatization and precursor-ion scanning mass spectrometry. Org Mass Spectrom 28:1516
41. Gu ZM, Zhou D, Wu J, Shi G, Zeng L, McLaughlin JL (1997) Screening for annonaceous acetogenins in bioactive plant extracts by liquid chromatography/mass spectrometry. J Nat Prod 60:242
42. Alali FQ, Liu XX, McLaughlin JL (1999) Annonaceous acetogenins: recent progress. J Nat Prod 62:504
43. Chang FR, Chen JL, Lin CY, Chiu HF, Wu MJ, Wu YC (1999) Bioactive acetogenins from the seeds of *Annona atemoya*. Phytochemistry 51:883
44. Gleye C, Laurens A, Hocquemiller R, Laprévote O, Serani L, Cavé A (1997) Cohibins A and B, acetogenins from roots of *Annona muricata*. Phytochemistry 44:1541
45. Gleye C, Raynaud S, Fourneau C, Laurens A, Laprévote O, Serani L, Fournet A, Hocquemiller R (2000) Cohibins C and D, two important metabolites in the biogenesis of acetogenins from *Annona muricata* and *Annona nutans*. J Nat Prod 63:1192
46. Gleye C, Akendengue B, Laurens A, Hocquemiller R (2001) Coronin from roots of *Annona muricata*, a putative intermediate in acetogenin biosynthesis (1). Planta Med 67:570

47. Sahpaz S, Hocquemiller R, Cavé A, Saez J, Cortes D (1997) Diepoxyrollin and diepomuricanin B: two new diepoxyacetogenins from *Rollinia membranacea* seeds. J Nat Prod 60:199
48. ba Ndob IB, Champy P, Gleye C, Lewin G, Akendengue B (2009) Annonaceous acetogenins: precursors from the seeds of *Annona squamosa*. Phytochem Lett 2:72
49. Jiang Z, Chen Y, Chen RY, Yu DQ (1998) Linear acetogenins from *Goniothalamus donnaiensis*. Phytochemistry 49:769
50. Melot A, Fall D, Gleye C, Champy P (2009) Apolar annonaceous acetogenins from the fruit pulp of *Annona muricata*. Molecules 14:43875
51. Chen Y, Jiang Z, Chen RR, Yu DQ (1998) Two linear acetogenins from *Goniothalamus gardneri*. Phytochemistry 49:1317
52. Seidel V, Bailleul F, Waterman PG (1999) Goniothalamusin, a linear acetogenin from *Goniothalamus gardneri*. Phytochemistry 52:1101
53. Liaw CC, Chang FR, Chen SL, Wu CC, Lee KH, Wu YC (2005) Novel cytotoxic monotetrahydrofuranic annonaceous acetogenins from *Annona montana*. Bioorg Med Chem 13:4767
54. Gleye C, Laurens A, Hocquemiller R, Cavé A, Laprévote O, Serani L (1997) Isolation of montecristin, a key metabolite in biogenesis of acetogenins from *Annona muricata* and its structure elucidation by using tandem mass spectrometry. J Org Chem 62:510
55. Yu JG, Gui HQ, Luo XZ, Sun L (1998) Murihexol, a linear acetogenin from *Annona muricata*. Phytochemistry 49:1689
56. Gleye C, Laurens A, Laprevote O, Serani L, Hocquemiller R (1999) Acetogenins of the Annonaceae. Part 81. Isolation and structure elucidation of sabadelin, an acetogenin from roots of *Annona muricata*. Phytochemistry 52:1403
57. Yu JG, Luo XZ, Sun L, Li DY, Huang WH, Liu CY (2005) Chemical constituents from the seeds of *Annona squamosa*. Acta Pharm Sin 40:153
58. Mootoo BS, Ali A, Khan A, Reynolds WF, McLean S (2000) Three novel monotetrahydrofuran annonaceous acetogenins from *Annona montana*. J Nat Prod 63:807
59. Kim DH, Ma ES, Suk KD, Son JK, Lee JS, Woo MH (2001) Annomolin and annocherimolin, new cytotoxic annonaceous acetogenins from *Annona cherimola* seeds. J Nat Prod 64:502
60. Liaw CC, Chang FR, Lin CY, Chou CJ, Chiu HF, Wu MJ, Wu YC (2002) New cytotoxic monotetrahydrofuran annonaceous acetogenins from *Annona muricata*. J Nat Prod 65:470
61. Woo MH, Kim DH, Fotopoulos SS, McLaughlin JL (1999) Annocherin and (2,4)-*cis*- and *trans*-annocherinones, monotetrahydrofuran annonaceous acetogenins with a C-7 carbonyl group from *Annona cherimola* seeds. J Nat Prod 62:1250
62. Liu XX, Alali FQ, Pilarinou E, McLaughlin JL (1999) Two bioactive mono-tetrahydrofuran acetogenins, annoglacins A and B, from *Annona glabra*. Phytochemistry 50:815
63. Liu XX, Pilarinou E, McLaughlin JL (1999) Two novel acetogenins, annoglaxin and 27-hydroxybullatacin, from *Annona glabra*. J Nat Prod 62:848
64. Kim DH, Son JK, Woo MH (2001) Annomocherin, annonacin and annomontacin: a novel and two known bioactive mono-tetrahydrofuran annonaceous acetogenins from *Annona cherimola* seeds. Arch Pharm Res 24:300
65. Son JK, Kim DH, Woo MH (2003) Two new epimeric pairs of acetogenins bearing a carbonyl group from *Annona cherimola* seeds. J Nat Prod 66:1369
66. Alali F, Zeng L, Zhang Y, Ye Q, Hopp DC, Schwedler JT, McLaughlin JL (1997) 4-Deoxyannomontacin and (2,4-*cis* and *trans*)-annomontacinone, new bioactive mono-tetrahydrofuran annonaceous acetogenins from *Goniothalamus giganteus*. Bioorg Med Chem 5:549
67. Kim GS, Zeng L, Alali F, Rogers LL, Wu FE, McLaughlin JL, Sastrodihardjo S (1998) Two new mono-tetrahydrofuran ring acetogenins, annomuricin E and muricapentocin, from the leaves of *Annona muricata*. J Nat Prod 61:432
68. Liaw CC, Chang FR, Wu CC, Chen SL, Bastow KF, Hayashi K, Nozaki H, Lee KH, Wu YC (2004) Nine new cytotoxic monotetrahydrofuranic annonaceous acetogenins from *Annona montana*. Planta Med 70:948

69. Hopp DC, Alali FQ, Gu ZM, McLaughlin JL (1998) Mono-THF ring acetogenins from *Annona squamosa*. Phytochemistry 47:803
70. Woo MH, Kim DH, McLaughlin JL (1999) Asitrilobins A and B: cytotoxic mono-THF Annonaceous acetogenins from the seeds of *Asimina triloba*. Phytochemistry 50:1033
71. Woo MH, Chung SO, Kim DH (2000) Asitrilobins C and D: two new cytotoxic mono-tetrahydrofuran Annonaceous acetogenins from *Asimina triloba* seeds. Bioorg Med Chem 8:285
72. Kim EJ, Tian F, Woo MH (2000) Asitrocin, (2,4)-*cis*- and *trans*-asitrocinones: novel bioactive mono-tetrahydrofuran acetogenins from *Asimina triloba* seeds. J Nat Prod 63:1503
73. Zhou GX, Chen RY, Zhang YJ, Yu DQ (2000) New annonaceous acetogenins from the roots of *Uvaria calamistrata*. J Nat Prod 63:1201
74. Meneses Da Silva EL, Roblot F, Laprévote O, Serani L, Cavé A (1997) Coriaheptocins A and B, the first heptahydroxylated acetogenins, isolated from the roots of *Annona coriacea*. J Nat Prod 60:162
75. Sun L, Zhu J, Yu J, Yu D, Li D, Zhou L (2003) Chemical constituents from the seeds of *Annona glabra*. Acta Pharm Sin 38:32
76. Liu XX, Alali FQ, Pilarinou E, McLaughlin JL (1998) Glacins A and B: two novel bioactive mono-tetrahydrofuran acetogenins from *Annona glabra*. J Nat Prod 61:620
77. Waechter AI, Szlosek M, Hocquemiller R, Laurens A, Cavé A (1998) Acetogenins of Annonaceae. 65. Glaucabellin and glaucaflorin, two acetogenins from *Annona glauca*. Phytochemistry 48:141
78. Alali FQ, Zhang Y, Rogers L, McLaughlin JL (1998) Mono-tetrahydrofuran acetogenins from *Goniothalamus giganteus*. Phytochemistry 49:761
79. Chang FR, Chen JL, Chiu HF, Wu MJ, Wu YC (1998) Acetogenins from seeds of *Annona reticulata*. Phytochemistry 47:1057
80. Chávez D, Acevedo LA, Mata R (1998) Jimenezin, a novel annonaceous acetogenin from the seeds of *Rollinia mucosa* containing adjacent tetrahydrofuran-tetrahydropyran ring systems. J Nat Prod 61:419
81. Liaw CC, Chang FR, Wu YC, Wang HK, Nakanishi Y, Bastow KF, Lee KH (2004) Montacin and *cis*-montacin, two new cytotoxic monotetrahydrofuran annonaceous acetogenins from *Annona montana*. J Nat Prod 67:1804
82. Wang LQ, Nakamura N, Meselhy MR, Hattori M, Zhao WM, Cheng KF, Yang RZ, Qin GW (2000) Four mono-tetrahydrofuran ring acetogenins, montanacins B-E, from *Annona montana*. Chem Pharm Bull 48:1109
83. Wang LQ, Min BS, Li Y, Nakamura N, Qin GW, Li CJ, Hattori M (2002) Annonaceous acetogenins from the leaves of *Annona montana*. Bioorg Med Chem 10:561
84. Wang LQ, Li Y, Min BS, Nakamura N, Qin GW, Li CJ, Hattori M (2001) Cytotoxic mono-tetrahydrofuran ring acetogenins from leaves of *Annona montana*. Planta Med 67:847
85. Hopp DC, Zeng L, Gu ZM, Kozlowski JF, McLaughlin JL (1997) Novel mono-tetrahydrofuran ring acetogenins, from the bark of *Annona squamosa*, showing cytotoxic selectivities for the human pancreatic carcinoma cell line, PACA-2. J Nat Prod 60:581
86. Chang FR, Wu YC (2001) Novel cytotoxic annonaceous acetogenins from *Annona muricata*. J Nat Prod 64:925
87. Kim GS, Lu Z, Alali F, Rogers LL, Wu F-E, Sastrodihardjo S, McLaughlin JL (1998) Muricoreacin and murihexocin C, monotetrahydrofuran acetogenins, from the leaves of *Annona muricata*. Phytochemistry 49:565
88. Gleye C, Duret P, Laurens A, Hocquemiller R, Cavé A (1998) *cis*-Monotetrahydrofuran acetogenins from the roots of *Annona muricata*. J Nat Prod 61:576
89. Queiroz EF, Roblot F, Laprevote O, de Paulo M, Hocquemiller R (2003) Two unusual acetogenins from the roots of *Annona salzmanii*. J Nat Prod 66:755
90. Liaw CC, Chang FR, Chen YY, Chiu HF, Wu MJ, Wu YC (1999) New annonaceous acetogenins from *Rollinia mucosa*. J Nat Prod 62:1613

91. Liaw CC, Chang FR, Wu MJ, Wu YC (2003) A novel constituent from *Rollinia mucosa*, rollicosin, and a new approach to develop annonaceous acetogenins as potential antitumor agents. J Nat Prod 66:279
92. Alvarez Colom O, Neske A, Chahboune N, Zafra-Polo MC, Bardon A (2009) Tucupentol, a novel mono-tetrahydrofuranic acetogenin from *Annona montana*, as a potent inhibitor of mitochondrial complex I. Chem Biodivers 6:335
93. Alali FQ, Rogers L, Zhang Y, McLaughlin JL (1999) Goniotriocin and (2,4-*cis*- and -*trans*)-xylomaticinones, bioactive annonaceous acetogenins from *Goniothalamus giganteus*. J Nat Prod 62:31
94. Chang FR, Liaw CC, Lin CY, Chou CJ, Chiu HF, Wu YC (2003) New adjacent bis-tetrahydrofuran annonaceous acetogenins from *Annona muricata*. Planta Med 69:241
95. Duret P, Hocquemiller R, Cavé A (1997) Acetogenins from Annonaceae. Part 55. Annonisin, a bis-tetrahydrofuran acetogenin from *Annona atemoya* seeds. Phytochemistry 45:1423
96. Chen Y, Chen JW, Wang Y, Xu SS, Li X (2012) Six cytotoxic annonaceous acetogenins from *Annona squamosa* seeds. Food Chem 135:960
97. Chen Y, Chen JW, Li X (2011) Cytotoxic bistetrahydrofuran annonaceous acetogenins from the seeds of *Annona squamosa*. J Nat Prod 74:2477
98. Hopp DC, Alali FQ, Gu ZM, McLaughlin JL (1998) Three new bioactive bis-adjacent THF-ring acetogenins from the bark of *Annona squamosa*. Bioorg Med Chem 6:569
99. Kim EJ, Suh KM, Kim DH, Jung EJ, Seo CS, Son JK, Woo MH, McLaughlin JL (2005) Asimitrin and 4-hydroxytrilobin, new bioactive annonaceous acetogenins from the seeds of *Asimina triloba* possessing a bis-tetrahydrofuran ring. J Nat Prod 68:194
100. Duret P, Hocquemiller R, Cavé A (1998) Acetogenins from Annonaceae. Part 66. Bulladecin and atemotetrolin, two bis-tetrahydrofuran acetogenins from *Annona atemoya* seeds. Phytochemistry 48:499
101. Queiroz EF, Roblot F, Figadère B, Laurens A, Duret P, Hocquemiller R, Cavé A, Serani L, Laprévote O, Cotte-Laffitte J, Quéro AM (1998) Three new bistetrahydrofuran acetogenins from the seeds of *Annona spinescens*. J Nat Prod 61:34
102. Fall D, Duval Romain A, Gleye C, Laurens A, Hocquemiller R (2004) Chamuvarinin, an acetogenin bearing a tetrahydropyran ring from the roots of *Uvaria chamae*. J Nat Prod 67:1041
103. Rodrigues dos Santos LA, Boaventura MAD, Pimenta LPS (2006) Cornifolin, a new bis-tetrahydrofuran annonaceous acetogenin from *Annona cornifolia*. Biochem Syst Ecol 34:78
104. dos Santos Lima LAR, Pimenta LPS, Boaventura MAD (2009) Two new adjacent bis-tetrahydrofuran annonaceous acetogenins from seeds of *Annona cornifolia*. Planta Med 75:80
105. Liu XX, Alali FQ, Hopp DC, Rogers LL, Pilarinou E, McLaughlin JL (1998) Glabracins A and B, two new acetogenins from *Annona glabra*. Bioorg Med Chem 6:959
106. Chahboune N, Barrachina I, Royo I, Romero V, Saez J, Tormo JR, De Pedro N, Estornell E, Zafra-Polo MC, Pelaez F, Cortes D (2006) Guanaconetins, new antitumoral acetogenins, mitochondrial complex I and tumor cell growth inhibitors. Bioorg Med Chem 14:1089
107. Fall D, Pimentel L, Champy P, Gleye C, Laurens A, Hocquemiller R (2006) A new adjacent bis-tetrahydrofuran annonaceous acetogenin from the seeds of *Uvaria chamae*. Planta Med 72:938
108. Chávez D, Mata R (1999) Chemical studies on Mexican plants used in traditional medicine. 36. Purpuracenin: a new cytotoxic adjacent bis-tetrahydrofuran annonaceous acetogenin from the seeds of *Annona purpurea*. Phytochemistry 50:823
109. Chávez D, Mata R (1998) Purpurediolin and purpurenin, two new cytotoxic adjacent bis-tetrahydrofuran annonaceous acetogenins from the seeds of *Annona purpurea*. J Nat Prod 61:580

110. Gu ZM, Zhou D, Lewis NJ, Wu J, Shi G, McLaughlin JL (1997) Isolation of new bioactive annonaceous acetogenins from *Rollinia mucosa* guided by liquid chromatography/mass spectrometry. Bioorg Med Chem 5:1911
111. Shi G, MacDougal JM, McLaughlin JL (1997) Bioactive annonaceous acetogenins from *Rollinia mucosa*. Phytochemistry 45:719
112. Araya H, Sahai M, Singh S, Singh AK, Yoshida M, Hara N, Fujimoto Y (2002) Squamocin-O1 and squamocin-O2, new adjacent bis-tetrahydrofuran acetogenins from the seeds of *Annona squamosa*. Phytochemistry 61:999
113. He K, Zhao GX, Shi G, Zeng L, Chao JF, McLaughlin JL (1997) Additional bioactive annonaceous acetogenins from *Asimina triloba* (Annonaceae). Bioorg Med Chem 5:501
114. Barrachina I, Neske A, Granell S, Bermejo A, Chahboune N, El Aouad N, Alvarez O, Bardon A, Zafra-Polo MC (2004) Tucumanin, a β-hydroxy-γ-lactone bistetrahydrofuranic acetogenin from *Annona cherimola*, is a potent inhibitor of mitochondrial complex I. Planta Med 70:866–868
115. Souza MMC, Bevilaqua CML, Morais SM, Costa CTC, Silva ARA, Braz-Filho R (2008) Anthelmintic acetogenin from *Annona squamosa* L. seeds. An Acad Bras Cienc 80:271
116. Chen CY, Chang FR, Chiu HF, Wu MJ, Wu YC (1999) Aromin-A, an annonaceous acetogenin from *Annona cherimola*. Phytochemistry 51:429
117. Alali FQ, Zhang Y, Rogers L, McLaughlin JL (1997) (2,4-*cis* and *trans*)-Gigantecinone and 4-deoxygigantecin, bioactive nonadjacent bis-tetrahydrofuran annonaceous acetogenins, from *Goniothalamus giganteus*. J Nat Prod 60:929
118. Yang HJ, Li X, Zhang N, Chen JW, Wang MY (2009) Two new cytotoxic acetogenins from *Annona squamosa*. J Asian Nat Prod Res 11:250
119. Yang HJ, Zhang N, Li X, He L, Chen J (2009) New nonadjacent bis-THF ring acetogenins from the seeds of *Annona squamosa*. Fitoterapia 80:1771
120. Alfonso D, Colman-Saizarbitoria T, Zhao GX, Shi G, Ye Q, Schwedler JT, McLaughlin JL (1996) Aromin and aromicin, two new bioactive annonaceous acetogenins, possessing an unusual bis-THF ring structure, from *Xylopia aromatica* (Annonaceae). Tetrahedron 52:4215
121. Xie HH, Wei XY, Wang JD, Liu MF, Yang RZ (2003) A new cytotoxic acetogenin from the seeds of *Annona squamosa*. Chin Chem Lett 14:588
122. Rieser MJ, Kozlowski JF, Wood KV, McLaughlin JL (1991) Muricatacin: a simple biologically active acetogenin derivative from the seeds of *Annona muricata* (Annonaceae). Tetrahedron Lett 32:1137
123. Myint SH, Laurens A, Hocquemiller R, Cavé A, Davoust D, Cortes D (1990) Murisolin: a new cytotoxic mono-tetrahydrofuran-γ-lactone from *Annona muricata*. Heterocycles 31:861
124. Hattori Y, Kimura Y, Moroda A, Konno H, Abe M, Miyoshi H, Goto T, Makabe H (2006) Synthesis of murisolin, (15*R*,16*R*,19*R*,20*S*)-murisolin A, and (15*R*,16*R*, 19*S*,20*S*)-16,19-*cis*-murisolin and their inhibitory action with bovine heart mitochondrial complex I. Chem Asian J 1:894
125. Ye Q, Zeng L, Zhang Y, Zhao GX, McLaughlin JL, Woo MH, Evert DR (1995) Longicin and goniothalamicinone: novel bioactive monotetrahydrofuran acetogenins from *Asimina longifolia*. J Nat Prod 58:1398
126. Hanessian S, Giroux S, Buffat M (2005) Total synthesis and structural confirmation of (+)-longicin. Org Lett 7:3989
127. Goksel H, Stark CBW (2006) Total synthesis of *cis*-solamin: exploiting the RuO_4-catalyzed oxidative cyclization of dienes. Org Lett 8:3433
128. Zhao H, Gorman JST, Pagenkopf BL (2006) Advances in Lewis acid controlled carbon-carbon bond-forming reactions enable a concise and convergent total synthesis of bullatacin. Org Lett 8:4379
129. Pettit GR, Cragg GM, Polonsky J, Herald DL, Goswami A, Smith CR, Moretti C, Schmidt JM, Weisleder D (1987) Isolation and structure of rolliniastatin 1 from the South American tree *Rollinia mucosa*. Can J Chem 65:1433

130. Keum G, Hwang CH, Kang SB, Kim Y, Lee E (2005) Stereoselective syntheses of rolliniastatin 1, rollimembrin, and membranacin. J Am Chem Soc 127:10396
131. Gonzalez MC, Tormo JR, Bermejo A, Zafra-Polo MC, Estornell E, Cortes D (1997) Rollimembrin, a novel acetogenin inhibitor of mammalian mitochondrial complex I. Bioorg Med Chem Lett 7:1113
132. Yang SW, Shen YC, Chen CH (1996) Steroids and triterpenoids of *Antodia cinnamomea*–a fungus parasitic on *Cinnamomum micranthum*. Phytochemistry 41:1389
133. Nattrass GL, Diez E, McLachlan MM, Dixon DJ, Ley SV (2005) The total synthesis of the annonaceous acetogenin 10-hydroxyasimicin. Angew Chem Int Ed 44:580–584
134. Saez J, Sahpaz S, Villaescusa L, Hocquemiller R, Cavé A, Cortes D (1993) Acetogenins of the Annonaceae. 18. Rioclarin and membranacin, two new bis-tetrahydrofuran acetogenins of the seeds of *Rollinia membranacea*. J Nat Prod 56:351
135. Rupprecht JK, Chang CJ, Cassady JM, McLaughlin JL, Mikolajczak KL, Weisleder D (1986) Asimicin, a new cytotoxic and pesticidal acetogenin from the paw paw, *Asimina triloba* (Annonaceae). Heterocycles 24:1197
136. Marshall JA, Sabatini JJ (2006) An outside-in approach to adjacent bistetrahydrofuran annonaceous acetogenins with C_2 core symmetry. Total synthesis of asimicin and a C32 analogue. Org Lett 8:3557
137. Ye Q, He K, Oberlies NH, Zeng L, Shi G, Evert D, McLaughlin JL (1996) Longimicins A-D: novel bioactive acetogenins from *Asimina longifolia* (Annonaceae) and structure-activity relationships of asimicin type of annonaceous acetogenins. J Med Chem 39:1790
138. Tominaga H, Maezaki N, Yanai M, Kojima N, Urabe D, Ueki R, Tanaka T (2006) First total synthesis of longimicin D. Eur J Org Chem 2006:1422
139. Shi G, Kozlowski JF, Schwedler JT, Wood KV, MacDougal JM, McLaughlin JL (1996) Muconin and mucoxin: additional nonclassical bioactive acetogenins from *Rollinia mucosa*. J Org Chem 61:7988
140. Narayan RS, Borhan B (2006) Synthesis of the proposed structure of mucoxin via regio- and stereoselective tetrahydrofuran ring-forming strategies. J Org Chem 71:1416
141. Shi G, Zheng L, Gu ZM, MacDougal JM, McLaughlin JL (1995) Absolute stereochemistries of sylvaticin and 12,15-*cis*-sylvaticin, bioactive C-20,23-*cis* nonadjacent bistetrahydrofuran annonaceous acetogenins, from *Rollinia mucosa*. Heterocycles 41:1785
142. Donohoe TJ, Harris RM, Burrows J, Parker J (2006) Total synthesis of (+)-*cis*-sylvaticin: double oxidative cyclization reactions catalyzed by osmium. J Am Chem Soc 128:13704
143. Alkofahi A, Rupprecht JK, Liu YM, Chang CJ, Smith DL, McLaughlin JL (1990) Gigantecin: a novel antimitotic and cytotoxic acetogenin, with nonadjacent tetrahydrofuran rings, from *Goniothalamus giganteus* (Annonaceae). Experientia 46:539
144. Hoye TR, Eklov BM, Jeon J, Khoroosi M (2006) Sequencing of three-component olefin metatheses: total synthesis of either (+)-gigantecin or (+)-14-deoxy-9-oxygigantecin. Org Lett 8:3383
145. Bandur NG, Bruckner D, Hoffmann RW, Koert U (2006) Total synthesis of jimenezin via an intramolecular allylboration. Org Lett 8:3829
146. Shi G, Alfonso D, Fatope MO, Zeng L, Gu ZM, Zhao GX, He K, MacDougal JM, McLaughlin JL (1995) Mucocin: A new annonaceous acetogenin bearing a tetrahydropyran ring. J Am Chem Soc 117:10409
147. Crimmins MT, Zhang Y, Diaz FA (2006) Total synthesis of (–)-mucocin. Org Lett 8:2369
148. Alali FQ, Rogers L, Zhang Y, McLaughlin JL (1998) Unusual bioactive annonaceous acetogenins from *Goniothalamus giganteus*. Tetrahedron 54:5833
149. Strand D, Rein T (2004) Total synthesis of pyranicin. Org Lett 7:199
150. Strand D, Rein T (2005) Synthesis of pyragonicin. Org Lett 7:2779
151. Takahashi S, Ogawa N, Koshino H, Nakata T (2005) Total synthesis of the proposed structure for pyragonicin. Org Lett 7:2783
152. Quinn KJ, Isaacs AK, DeChristopher BA, Szklarz SC, Arvary RA (2005) Asymmetric total synthesis of rollicosin. Org Lett 7:1243

153. Makabe H, Kimura Y, Higuchi M, Konno H, Murai M, Miyoshi H (2006) Synthesis of (4*R*,15*R*,16*R*,21*S*)- and (4*R*,15*S*,16*S*,21*S*)-rollicosin, squamostolide, and their inhibitory action with bovine heart mitochondrial complex I. Bioorg Med Chem 14:3119
154. Harcken C, Bruckner R (2001) Elucidation of the stereostructure of the annonaceous acetogenin (+)-montecristin through total synthesis. New J Chem 25:40
155. Solladie G, Hanquet G, Izzo I, Crumbie R (1999) Asymmetric synthesis of *syn* and *anti* 1,2-diols from diethyl oxalate using the stereoselective sulfoxide directed reduction of 1,2-diketone derivatives. Tetrahedron Lett 40:3071
156. Baylon C, Prestat G, Heck MP, Mioskowski C (2000) Synthesis of (–)-(4*R*,5*R*)-muricatacin using a regio- and stereospecific ring-opening of a vinyl epoxide. Tetrahedron Lett 41:3833
157. Konno H, Hiura N, Yanaru M (2002) Syntheses of (4*R*,5*S*)- and (4*S*,5*R*)-muricatacins, and (4*S*,5*R*)-aza-muricatacin, unnatural analogues of the annonaceous acetogenin. Heterocycles 57:1793
158. Chandrasekhar M, Chandra KL, Singh VK (2002) An efficient strategy for the synthesis of 5-hydroxyalkylbutan-4-olides from D-mannitol: total synthesis of (–)-muricatacin. Tetrahedron Lett 43:2773
159. Bernard AM, Frongia A, Piras PP, Secci F (2003) Unexpected stereochemistry in the lithium salt catalyzed ring expansion of nonracemic oxaspiropentanes. Formal syntheses of (−)-(4*R*,5*R*)-muricatacin and the pheromone (*R*)-japonilure. Org Lett 5:2923
160. Chen Y, Yu DQ (1996) Tonkinelin, a novel annonaceous acetogenin from *Uvaria tonkinesis.* Planta Med 62:512
161. Hattori T, Konno H, Abe M, Miyoshi H, Goto T, Makabe H (2007) Synthesis, determination of the absolute configuration of tonkinelin, and inhibitory action with bovine heart mitochondrial complex I. Bioorg Med Chem 15:3026
162. Hu TS, Wu YL, Wu Y (2000) The first total synthesis of annonacin, the most typical monotetrahydrofuran annonaceous acetogenins. Org Lett 2:887
163. Hu TS, Yu Q, Wu YL, Wu Y (2001) Enantioselective syntheses of monotetrahydrofuran annonaceous acetogenins tonkinecin and annonacin starting from carbohydrates. J Org Chem 66:853
164. Yu Q, Yao ZJ, Chen XG, Wu YL (1999) Total synthesis of (10*R*)- and (10*S*)-corossolin: Determination of the stereochemistry at C-10 of the natural corossolin and the differential toxicity toward cancer cells caused by the configuration at C-10. J Org Chem 64:2440
165. Wang F, Kawamura A, Mootoo DR (2008) Synthesis and antitumor activity of C-9 epimers of the tetrahydrofuran containing acetogenin 4-deoxyannoreticuin. Bioorg Med Chem 16:8413
166. Yu Q, Wu Y, Ding H, Wu YL (1999) The first total synthesis of 4-deoxyannomontacin. J Chem Soc Perkin Trans 1:1183
167. Orru RVA, Groenendaal B, Heyst JV, Hunting M, Wesseling C, Schmitz RF, Mayer SF, Faber K (2003) Biomimetic approach to the stereoselective synthesis of acetogenins. Pure Appl Chem 75:259
168. Wang ZM, Tian SK, Shi M (2000) Total synthesis of gigantetrocin A. Chirality 12:581
169. Wang ZM, Tian SK, Shi M (1999) A facile route to the total synthesis of gigantetrocin A. Tetrahedron Asymmetry 10:667
170. Marshall JA, Jiang H (1998) Total synthesis of the *threo, trans, threo*-monotetrahydrofuran annonaceous acetogenin longifolicin. Tetrahedron Lett 39:1493
171. Maezaki N, Kojima N, Sakamoto A, Iwata C, Tanaka T (2001) First total synthesis of mosin B. Org Lett 3:429
172. Maezaki N, Kojima N, Sakamoto A, Tominaga H, Iwata C, Tanaka T, Monden M, Damdinsuren B, Nakamori S (2003) Total synthesis of the antitumor acetogenin mosin B: desymmetrization approach to the stereodivergent synthesis of *threo/trans/erythro*-type acetogenins. Chem Eur J 9:389
173. Dixon DJ, Ley SV, Reynolds DJ (2002) The total synthesis of the annonaceous acetogenin, muricatetrocin C. Chem Eur J 8:1621

174. Dixon DJ, Ley SV, Reynolds DJ (2000) The total synthesis of the annonaceous acetogenin, muricatetrocin C. Angew Chem Int Ed 39:3622
175. Maezaki N, Tominaga H, Kojima N, Yanai M, Urabe D, Ueki R, Tanaka T, Yamori T (2005) Total synthesis of murisolins and evaluation of tumor-growth inhibitory activity. Chem Eur J 11:6237
176. Hanessian S, Grillo TA (1998) Stereocontrolled total synthesis of an annonacin A-type acetogenin: pseudoannonacin A? J Org Chem 63:1049
177. Makabe H, Miyawaki A, Takahashi R, Hattori Y, Konno H, Abe M, Miyoshi H (2004) Synthesis of two possible diastereomers of reticulatain-1. Tetrahedron Lett 45:973
178. Kuriyama W, Ishigami K, Kitahara T (1999) Synthesis of solamin. Heterocycles 50:981
179. Cecil ARL, Brown RCD (2002) Synthesis of *cis*-solamin using a permanganate-mediated oxidative cyclization. Org Lett 4:3715
180. Makabe H, Hattori Y, Tanaka A, Oritani T (2002) Total synthesis of *cis*-solamin. Org Lett 4:1083
181. Makabe H, Hattori Y, Kimura Y, Konno H, Abe M, Miyoshi H, Tanaka A, Oritani T (2004) Total synthesis of *cis*-solamin and its inhibitory action with bovine heart mitochondrial complex I. Tetrahedron 60:10651–10657
182. Prestat G, Baylon C, Heck MP, Grasa GA, Nolan SP, Mioskowski C (2004) New strategy for the construction of a monotetrahydrofuran ring in annonaceous acetogenin based on a ruthenium ring-closing metathesis: application to the synthesis of solamin. J Org Chem 69:5770
183. Konno H, Okuno Y, Makabe H, Nosaka K, Onishi A, Abe Y, Sugimoto A, Akaji K (2008) Total synthesis of *cis*-solamin A, a mono-tetrahydrofuran acetogenin isolated from *Annona muricata*. Tetrahedron Lett 49:782–785
184. Konno H, Makabe H, Hattori Y, Nosaka K, Akaji K (2010) Synthesis of solamin type mono-THF acetogenins using cross-metathesis. Tetrahedron 66:7946
185. Wang M, Chen Y, Lou L, Tang W, Wang X, Shen J (2005) Synthesis of pyrrolidine analogues of solamin. Tetrahedron Lett 46:5309
186. Hu TS, Yu Q, Lin Q, Wu YL, Wu Y (1999) The first synthesis of tonkinecin, an annonaceous acetogenin with a C-5 carbinol center. Org Lett 1:399
187. Avedissian H, Sinha SC, Yazbak A, Sinha A, Neogi P, Sinha SC, Keinan E (2000) Total synthesis of asimicin and bullatacin. J Org Chem 65:6035
188. Mertz E, Tinsley JM, Roush WR (2005) [3 + 2]-Annulation reactions of chiral allylsilanes and chiral aldehydes. Studies on the synthesis of bis-tetrahydrofuran substructures of annonaceous acetogenins. J Org Chem 70:8035
189. Marshall JA, Sabatini JJ, Valeriote F (2007) ABC synthesis and antitumor activity of a series of annonaceous acetogenin analogs with a *threo*, *trans*, *threo*, *trans*, *threo*-bis-tetrahydrofuran core unit. Bioorg Med Chem Lett 17:2434
190. Wang ZM, Tian SK, Shi M (1999) The synthesis of asimilobin and the correction of its tetrahydrofuran segment's configuration. Tetrahedron Lett 40:977
191. Wang ZM, Tian SK, Shi M (2000) A facile route to bulladecin-type acetogenins – total synthesis of asimilobin and correction of the configuration of its tetrahydrofuran segment. Eur J Org Chem 2000:349
192. Marshall JA, Jiang H (1999) Total synthesis of the cytotoxic *threo*, *trans*, *threo*, *trans*, *threo* annonaceous acetogenin asimin and its C-10 epimer: unambiguous confirmation of absolute stereochemistry. J Nat Prod 62:1123
193. Mohapatra DK, Nayak S, Mohapatra S, Chorghade MS, Gurjar MK (2007) Double intramolecular oxymercuration: the first stereoselective synthesis of the C10-34 fragment of asimitrin. Tetrahedron Lett 48:5197
194. Marshall JA, Hinkle KW (1998) Total synthesis of the cytotoxic annonaceous acetogenin (30*S*)-bullanin. Tetrahedron Lett 39:1303

195. Tinsley JM, Mertz E, Chong PY, Rarig RAF, Roush WR (2005) Synthesis of (+)-bullatacin via the highly diastereoselective [3+2] annulation reaction of a racemic aldehyde and a nonracemic allylsilane. Org Lett 7:4245
196. Marshall JA, Piettre A, Paige MA, Valeriote F (2003) Total synthesis and structure confirmation of the annonaceous acetogenins (30*S*)-hydroxybullatacin, uvarigrandin A, and 5(*R*)-uvarigrandin A (narumicin I?). J Org Chem 68:1780
197. Huh CW, Roush WR (2008) Highly stereoselective and modular syntheses of 10-hydroxytrilobacin and three diastereomers via stereodivergent (3+2)-annulation reactions. Org Lett 10:3371
198. He YT, Xue S, Hu TS, Yao ZJ (2005) An iterative acetylene–epoxide coupling strategy for the total synthesis of longimicin C. Tetrahedron Lett 46:5393
199. Head GD, Whittingham WG, Brown RCD (2004) Synthesis of membranacin. Synlett 8:1437
200. Ruan Z, Dabideen D, Blumenstein M, Mootoo DR (2000) A modular synthesis of the bis-tetrahydrofuran core of rolliniastatin from pyranoside precursors. Tetrahedron 56:9203
201. Emde U, Koert U (2000) Total syntheses of squamocin A and squamocin D, bi-tetrahydrofuran acetogenins from Annonaceae. Eur J Org Chem 2000:1889
202. Emde U, Koert U (1999) Total syntheses of squamocin A and squamocin D. Tetrahedron Lett 40:5979
203. Sinha SC, Keinan E (1999) Total synthesis of squamotacin. J Org Chem 64:7067
204. Das S, Li LS, Abraham S, Chen Z, Sinha SC (2005) A bidirectional approach to the synthesis of a complete library of adjacent-bis-THF annonaceous acetogenins. J Org Chem 70:5922
205. Ruan ZM, Mootoo DR (1999) A novel desymmetrization reaction of an acetogenin precursor: a formal synthesis of trilobacin and asimicin. Tetrahedron Lett 40:49
206. Marshall JA, Jiang H (1999) Total synthesis of the cytotoxic *threo*, *trans*, *erythro*, *cis*, *threo* annonaceous acetogenin trilobin. J Org Chem 64:971
207. Sinha A, Sinha SC, Sinha SC, Keinan E (1999) Total synthesis of trilobin. J Org Chem 64:2381
208. Burke SD, Jiang L (2001) Formal synthesis of uvaricin via palladium-mediated double cyclization. Org Lett 3:1953
209. Yazbak A, Sinha SC, Keinan E (1998) Total synthesis of uvaricin. J Org Chem 63:5863
210. Marshall JA, Piettre A, Paige MA, Valeriote F (2003) A modular synthesis of annonaceous acetogenins. J Org Chem 68:1771
211. Gu Z-M, Zeng L, Schwedler JT, Wood KV, McLaughlin JL (1995) New bioactive adjacent bis-THF annonaceous acetogenins from *Annona bullata*. Phytochemistry 40:467
212. Pan X, Yu D (1997) Studies on new cytotoxic annonaceous acetogenins from *Uvaria grandiflora* and absolute configurations. Acta Pharmacol Sin 32:286
213. Zhu L, Mootoo DR (2003) Synthesis of nonadjacently linked tetrahydrofurans: an iodoetherification and olefin metathesis approach. Org Lett 5:3475
214. Zhu L, Mootoo DR (2004) Total synthesis of the nonadjacently linked bis-tetrahydrofuran acetogenin bullatanocin (squamostatin C). J Org Chem 69:3154
215. Makabe H, Tanaka A, Oritan T (1998) Total synthesis of (+)-4-deoxygigantecin. Tetrahedron 54:6329
216. Crimmins MT, She J (2004) Enantioselective total synthesis of (+)-gigantecin: exploiting the asymmetric glycolate aldol reaction. J Am Chem Soc 126:12790
217. Marshall JA, Jiang H (1998) Total synthesis of the nonadjacent bis-tetrahydrofuran annonaceous acetogenin squamostatin-D. J Org Chem 63:7066
218. Donohoe TJ, Harris RM, Williams O, Hargaden GC, Burrows J, Parker J (2009) Concise syntheses of the natural products (+)-sylvaticin and (+)-*cis*-sylvaticin. J Am Chem Soc 131:12854
219. Quinn KJ, Smith AG, Cammarano CM (2007) Convergent total synthesis of squamostolide. Tetrahedron 63:4881
220. Makabe H, Tanaka A, Oritani T (1997) Total synthesis of (+)-4-deoxygigantecin. Tetrahedron Lett 38:4247

221. Donohoe TJ, Williams O, Churchill GH (2008) Hydride shift generated oxonium ions: evidence for mechanism and intramolecular trapping experiments to form trans THF derivatives. Angew Chem Int Ed 47:2869
222. Takahashi S, Maeda K, Hirota S, Nakata T (1999) Total synthesis of a new cytotoxic acetogenin, jimenezin, and the revised structure. Org Lett 1:2025
223. Takahashi S, Maeda K, Hirota S, Nakata T (2000) Total synthesis of antitumor acetogenin, jimenzein, and the revised structure. Tennen Yuki Kagobutsu Toronkai Koen Yoshishu 129:769
224. Schaus SE, Brånalt J, Jacobsen EN (1998) Total synthesis of muconin by efficient assembly of chiral building blocks. J Org Chem 63:4876
225. Yang WQ, Kitahara T (1999) Convergent synthesis of (+)-muconin. Tetrahedron Lett 40:7827
226. Yang WQ, Kitahara T (2000) Total synthesis of a nonclassical bioactive acetogenin, (+)-muconin. Tetrahedron 56:1451
227. Takahashi S, Kubota A, Nakata T (2002) Total synthesis of muconin. Tetrahedron Lett 43:8661
228. Takahashi S, Kubota A, Nakata T (2003) Stereoselective total synthesis of muconin. Tetrahedron 59:1627
229. Yoshimitsu T, Makino T, Nagaoka H (2004) Total synthesis of (+)-muconin. J Org Chem 69:1993
230. Crisóstomo FRP, Carrillo R, León LG, Matrín T, Padón JM, Martín VS (2006) Molecular simplification in bioactive molecules: formal synthesis of (+)-muconin. J Org Chem 71:2339
231. Takahashi S, Hongo Y, Tsukagoshi Y, Koshino H (2008) Structural determination of montanacin D by total synthesis. Org Lett 10:4223
232. Neogi P, Doundoulakis T, Yazbak A, Sinha SC, Sinha SC, Keinan E (1998) Total synthesis of mucocin. J Am Chem Soc 120:11279
233. Takahashi S, Nakata T (1998) Total synthesis of an anticancer agent, mucocin. Tennen Yuki Kagobutsu Toronkai Koen Yoshishu, 40th, p. 637
234. Takahashi S, Nakata T (1999) Total synthesis of an anticancer agent, mucocin. 2. A novel approach to a gamma-hydroxy butenolide derivative and completion of total synthesis. Tetrahedron Lett 40:727
235. Takahashi S, Nakata T (1999) Total synthesis of an anticancer agent, mucocin. 1. Stereoselective synthesis of the left-half segment. Tetrahedron Lett 40:723
236. Takahashi S, Nakata T (2002) Total synthesis of an antitumor agent, mucocin, based on the "chiron approach". J Org Chem 67:5739
237. Bäurle S, Hoppen S, Koert U (1999) Total synthesis of (−)-mucocin. Angew Chem Int Ed 38:1263
238. Hoppen S, Bäurle S, Koert U (2000) A convergent total synthesis of (−)-mucocin: an acetogenin from Annonaceae. Chem Eur J 6:2382
239. Evans PA, Cui J, Gharpure SJ, Polosukhin A, Zhang HR (2003) Enantioselective total synthesis of the potent antitumor agent (−)-mucocin using a temporary silicon-tethered ring-closing metathesis cross-coupling reaction. J Am Chem Soc 125:14702
240. Zhu L, Mootoo DR (2005) Synthesis of the non-classical acetogenin mucocin: a modular approach based on olefinic coupling reactions. Org Biomol Chem 3:2750
241. Takahashi S, Hongo Y, Ogawa N, Koshino H, Nakata T (2006) Convergent synthesis of pyragonicin. J Org Chem 71:6305
242. Takahashi S, Kubota A, Nakata T (2006) Total synthesis of a cytotoxic acetogenin, pyranicin. Org Lett 5:1353
243. Strand D, Norrby PO, Rein T (2006) Divergence en route to nonclassical annonaceous acetogenins, synthesis of pyranicin and pyragonicin. J Org Chem 71:1879
244. Griggs ND, Phillips AJ (2008) A concise and modular synthesis of pyranicin. Org Lett 10:4955

245. Sinha SC, Sinha A, Sinha SC, Keinan E (1997) Tandem oxidative cyclization with rhenium oxide. Total synthesis of 17,18-bisepi-goniocin. J Am Chem Soc 119:12014
246. Owens GS, Arias J, Abu-Omar MM (2000) Rhenium oxo complexes in catalytic oxidations. Catal Today 55:317
247. Sinha SC, Sinha A, Sinha SC, Keinan E (1998) Goniocin and its heptaepimer, cyclogoniodenin T. Unique biosynthetic implications. J Am Chem Soc 120:4017
248. Makabe H, Higuchi M, Konno H, Murai M, Miyoshi H (2005) Synthesis of (4*R*,15*R*,16*R*,21*S*)- and (4*R*,15*S*,16*S*,21*S*)-rollicosin. Tetrahedron Lett 46:4671
249. Wolvetang EJ, Johnson KL, Krauer K, Ralph SJ, Linnane AW (1994) Mitochondrial respiratory chain inhibitors induce apoptosis. FEBS Lett 339:40
250. Chih HW, Chiu HF, Tang KS, Chang FR, Wu YC (2001) Bullatacin, a potent antitumor annonaceous acetogenin, inhibits proliferation of human hepatocarcinoma cell line 2.2.15 by apoptosis induction. Life Sci 69:1321
251. Cartagena E, Colom OA, Neske A, Valdez JC, Bardon A (2007) Effects of plant lactones on the production of biofilm of *Pseudomonas aeruginosa*. Chem Pharm Bull 55:22
252. Morton JF (1987) Fruits of warm climates. Media, Inc., Greensboro, NC, p 65
253. Rupprecht JK, Hui YH, McLaughlin JL (1990) Annonaceous acetogenins: a review. J Nat Prod 53:237
254. Hui YH, Rupprecht JK, Liu YM, Anderson JE, Smith DL, Chang CJ, McLaughlin JL (1989) Bullatacin and bullatacinone: two highly potent bioactive acetogenins from *Annona bullata*. J Nat Prod 52:463
255. Ratnayake S, Rupprecht JK, Potter WM, McLaughlin JL (1992) Evaluation of various parts of the paw paw tree, *Asimina triloba* (Annonaceae), as commercial sources of the pesticidal annonaceous acetogenins. J Econ Entomol 85:2353
256. He K, Zeng L, Ye Q, Shi G, Oberlies NH, Zhao GX, Njoku CJ, McLaughlin JL (1997) Comparative structure-activity relationship evaluations of annonaceous acetogenins for pesticidal activity. Pestic Sci 49:372
257. Alali FQ, Kaakeh W, Bennett GW, McLaughlin JL (1998) Annonaceous acetogenins as natural pesticides: potent toxicity against insecticide-susceptible and -resistant German cockroaches (Dictyoptera: Blattellidae). J Econ Entomol 91:641
258. Ohsawa K, Atsuzawa S, Mitsui T, Yamamoto I (1991) Isolation and insecticidal activity of three acetogenins from seeds of pond apple, *Annona glabra* L. Nippon Noyaku Gakkaishi 16:93
259. Guadano A, Gutierrez C, de La Pena E, Cortes D, Gonzalez-Coloma A (2000) Insecticidal and mutagenic evaluation of two annonaceous acetogenins. J Nat Prod 63:773
260. Lewis MA, Arnason JT, Philogene BJR, Rupprecht JK, McLaughlin JL (1993) Inhibition of respiration at site I by asimicin, an insecticidal acetogenin of the pawpaw, *Asimina triloba* (Annonaceae). Pestic Biochem Physiol 45:15
261. Friedrich T, Ohnishi T, Forche E, Kunze B, Jansen R, Trowitzsch W, Hofle G, Reichenbach H, Weiss H (1994) Two binding sites for naturally occurring inhibitors in mitochondrial and bacterial NADH:ubiquinone oxidoreductase (complex I). Biochem Soc Trans 22:226
262. Hollingworth RM, Ahammadsahib KI, Gadelhak G, McLaughlin JL (1994) New inhibitors of complex I of the mitochondrial electron transport chain with activity as pesticides. Biochem Soc Trans 22:230
263. Vila-Nova NS, Morais SM, Falcao MJC, Machado Lyeghyna KAA, Bevilaqua CMLL, Costa IRS, de Brasil NVGPS, Andrade HF (2011) Leishmanicidal activity and cytotoxicity of compounds from two Annonaceae species cultivated in Northeastern Brazil. Rev Soc Bras Med Trop 44:567
264. Ahammadsahib KI, Hollingworth RM, McGovren JP, Hui YH, McLaughlin JL (1993) Mode of action of bullatacin: a potent antitumor and pesticidal annonaceous acetogenin. Life Sci 53:1113

265. Degli Esposti M, Ghelli A, Ratta M, Cortes D, Estornell E (1994) Natural substances (acetogenins) from the family Annonaceae are powerful inhibitors of mitochondrial NADH dehydrogenase (complex I). Biochem J 301:161
266. Oberlies NH, Jones JL, Corbett TH, Fotopoulos SS, McLaughlin JL (1995) Tumor cell growth inhibition by several annonaceous acetogenins in an in vitro disk diffusion assay. Cancer Lett 96:55
267. Oberlies NH, Croy VL, Harrison ML, McLaughlin JL (1997) The annonaceous acetogenin bullatacin is cytotoxic against multidrug-resistant human mammary adenocarcinoma cells. Cancer Lett 115:73
268. Shimada H, Grutzner JB, Kozlowski JF, McLaughlin JL (1998) Membrane conformations and their relation to cytotoxicity of asimicin and its analogs. Biochemistry 37:854
269. Kuwabara K, Takada M, Iwata J, Tatsumoto K, Sakamoto K, Iwamura H, Miyoshi H (2000) Design syntheses and mitochondrial complex I inhibitory activity of novel acetogenin mimics. Eur J Biochem 267:2538
270. Yuan SSF, Chang HL, Chen HW, Yeh YT, Kao YH, Lin KH, Wu YC, Su JH (2003) Annonacin, a mono-tetrahydrofuran acetogenin, arrests cancer cells at the G1 phase and causes cytotoxicity in a Bax- and caspase-3-related pathway. Life Sci 72:2853
271. Lu MC, Yang SH, Hwang SL, Lu YJ, Lin YH, Wang SR, Wu YC, Lin SR (2006) Induction of G2/M phase arrest by squamocin in chronic myeloid leukemia (K562) cells. Life Sci 78:2378
272. Caparros-Lefebvre D, Elbaz A (1999) Possible relation of atypical parkinsonism in the French West Indies with consumption of tropical plants: a case-control study. Lancet 354:281
273. Chaudhuri KR, Hu MTM, Brooks DJ (2000) Atypical parkinsonism in Afro-Caribbean and Indian origin immigrants to the UK. Mov Disord 15:18
274. Caparros-Lefebvre D, Steele J (2005) Atypical parkinsonism on Guadeloupe, comparison with the parkinsonism-dementia complex of Guam, and environmental toxic hypotheses. Environ Toxicol Pharmacol 19:407
275. Lannuzel A, Michel PP, Caparros-Lefebvre D, Abaul J, Hocquemiller R, Ruberg M (2002) Toxicity of Annonaceae for dopaminergic neurons: potential role in atypical parkinsonism in Guadeloupe. Mov Disord 17:84
276. Lannuzel A, Michel PP, Hoglinger GU, Champy P, Jousset A, Medja F, Lombes A, Darios F, Gleye C, Laurens A, Hocquemiller R, Hirsch EC, Ruberg M (2003) The mitochondrial complex I inhibitor annonacin is toxic to mesencephalic dopaminergic neurons by impairment of energy metabolism. Neuroscience 121:287
277. Champy P, Hoglinger Gunter U, Feger J, Gleye C, Hocquemiller R, Laurens A, Guerineau V, Laprevote O, Medja F, Lombes A, Michel PP, Lannuzel A, Hirsch Etienne C, Ruberg M (2004) Annonacin, a lipophilic inhibitor of mitochondrial complex I, induces nigral and striatal neurodegeneration in rats: possible relevance for atypical parkinsonism in Guadeloupe. J Neurochem 88:63
278. Angibaud G, Gaultier C, Rascol O (2004) Atypical parkinsonism and Annonaceae consumption in New Caledonia. Mov Disord 19:603
279. Potts LF, Luzzio FA, Smith SC, Hetman M, Champy P, Litvan I (2012) Annonacin in *Asimina triloba* fruit: implication for neurotoxicity. NeuroToxicology 33:53
280. Wu TY, Yang IH, Tsai YT, Wang JY, Shiurba R, Hsieh TJ, Chang FR, Chang WC (2012) Isodesacetyluvaricin, an annonaceous acetogenin, specifically inhibits gene expression of cyclooxygenase-2. J Nat Prod 75:572
281. Parellada EA, Ramos AN, Ferrero M, Cartagena E, Bardon A, Valdez JC, Neske A (2011) Squamocin mode of action to stimulate biofilm formation of *Pseudomonas plecoglossicida* J26, a PAHs degrading bacterium. Int Biodeterior Biodegrad 65:1066
282. Parellada EA, Ramos AN, Ferrero M, Cartagena E, Bardon A, Neske A (2012) Effect of the annonaceous acetogenins, squamocin and laherradurin, on the degradation of naphthalene mediated by *Pseudomonas plecoglossicida* J26. Int Biodeterior Biodegrad 72:82

283. Bombasaro JA, Blessing LDT, Diaz S, Neske A, Suvire FD, Enriz RD, Rodriguez AM (2011) Theoretical and experimental study of the interactions of annonaceous acetogenins with artificial lipid bilayers. J Mol Struct 1003:87
284. Marsh D (1996) Lateral pressure in membranes. Biochim Biophys Acta 1286:183
285. Zafra-Polo MC, Gonzalez MC, Estornell E, Sahpaz S, Cortes D (1996) Acetogenins from Annonaceae, inhibitors of mitochondrial complex I. Phytochemistry 42:253
286. Sasaki S, Naito H, Maruta K, Kawahara E, Maeda M (1994) Novel calcium ionophores: supramolecular complexation by the hydroxylated-bistetrahydrofuran skeleton of potent antitumor annonaceous acetogenins. Tetrahedron Lett 35:3337
287. Araya H, Fujimoto Y, Hirayama K (1994) Structural elucidation of tetrahydrofuranic acetogenins by means of precursor-ion scanning method. J Synth Org Chem Jpn 52:765
288. Sasaki S, Maruta K, Naito H, Sugihara H, Hiratani K, Maeda M (1995) New calcium-selective electrodes based on annonaceous acetogenins and their analogs with neighboring bistetrahydrofuran. Tetrahedron Lett 36:5571
289. Sasaki S, Maruta K, Naito H, Maemura R, Kawahara E, Maeda M (1998) Novel acyclic ligands for metal cations based on the adjacent bistetrahydrofuran as analogs of natural annonaceous acetogenins. Tetrahedron 54:2401
290. Peyrat JF, Mahuteau J, Figadère B, Cavé A (1997) NMR Studies of Ca^{2+} complexes of annonaceous acetogenins. J Org Chem 62:4811
291. Wu SN, Chiang HT, Chang FR, Liaw CC, Wu YC (2003) Stimulatory effects of squamocin, an annonaceous acetogenin, on $Ca(^{2+})$-activated K^+ current in cultured smooth muscle cells of human coronary artery. Chem Res Toxicol 16:15
292. Liaw CC, Liao WY, Chen CS, Jao SC, Wu YC, Shen CN, Wu SH (2011) The calcium-chelating capability of tetrahydrofuranic moieties modulates the cytotoxicity of annonaceous acetogenins. Angew Chem Int Ed 50:7885
293. Miyoshi H, Ohshima M, Shimada H, Akagi T, Iwamura H, McLaughlin JL (1998) Essential structural factors of annonaceous acetogenins as potent inhibitors of mitochondrial complex I. Biochim Biophys Acta 1365:443–452
294. Miyoshi H (2005) Inhibitors of mitochondrial respiratory enzymes. J Pestic Sci 30:127
295. Takada M, Kuwabara K, Nakato H, Tanaka A, Iwamura H, Miyoshi H (2000) Definition of crucial structural factors of acetogenins, potent inhibitors of mitochondrial complex I. Biochim Biophys Acta 1460:302
296. Murai M, Ichimaru N, Abe M, Nishioka T, Miyoshi H (2006) Synthesis of photolabile Δlac-acetogenin for photoaffinity labeling of mitochondrial complex I. J Pestic Sci 31:156
297. Ichimaru N, Murai M, Abe M, Hamada T, Yamada Y, Makino S, Nishioka T, Makabe H, Makino A, Kobayashi T, Miyoshi H (2005) Synthesis and inhibition mechanism of Δlac-acetogenins, a novel type of inhibitor of bovine heart mitochondrial complex I. Biochemistry 44:816
298. Hoppen S, Emde U, Friedrich T, Grubert L, Koert U (2000) Natural-product hybrids: design, synthesis, and biological evaluation of quinone-annonaceous acetogenins. Angew Chem Int Ed 39:2099
299. Arndt S, Emde U, Baurle S, Friedrich T, Grubert L, Koert U (2001) Quinone-annonaceous acetogenins: synthesis and complex I inhibition studies of a new class of natural product hybrids. Chem Eur J 7:993
300. Duval R, Lewin G, Hocquemiller R (2003) Semisynthesis of heterocyclic analogues of squamocin, a cytotoxic annonaceous acetogenin, by an unusual oxidative decarboxylation reaction. Bioorg Med Chem 11:3439
301. Duval RA, Lewin G, Peris E, Chahboune N, Garofano A, Droese S, Cortes D, Brandt U, Hocquemiller R (2006) Heterocyclic analogues of squamocin as inhibitors of mitochondrial complex I. On the role of the terminal lactone of annonaceous acetogenins. Biochemistry 45:2721
302. Duval RA, Poupon E, Romero V, Peris E, Lewin G, Cortes D, Brandt U, Hocquemiller R (2006) Analogues of cytotoxic squamocin using reliable reactions: new insights into the

reactivity and role of the α,β-unsaturated lactone of the annonaceous acetogenins. Tetrahedron 62:6248
303. Kojima N, Fushimi T, Maezaki N, Tanaka T, Yamori T (2008) Synthesis of hybrid acetogenins, α,β-unsaturated-γ-lactone-free nitrogen-containing heterocyclic analogues, and their cytotoxicity against human cancer cell lines. Bioorg Med Chem Lett 18:1637
304. Zeng BB, Wu Y, Yu Q, Wu YL, Li Y, Chen XG (2000) Enantiopure simple analogues of annonaceous acetogenins with remarkable selective cytotoxicity towards tumor cell lines. Angew Chem Int Ed 39:1934
305. Rodier S, Le Huerou Y, Renoux B, Doyon J, Renard P, Pierre A, Gesson JP, Gree R (2000) Synthesis and cytotoxic activity of acetogenin analogues. Bioorg Med Chem Lett 10:1373
306. Yao ZJ, Wu HP, Wu YL (2000) Polyether mimics of naturally occurring cytotoxic annonaceous acetogenins. J Med Chem 43:2484
307. Liu YQ, Cheng X, Guo LX, Mao C, Chen YJ, Liu HX, Xiao QC, Jiang S, Yao ZJ, Zhou GB (2012) Identification of an annonaceous acetogenin mimetic, AA005, as an AMPK activator and autophagy inducer in colon cancer cells. PLoS One 7:e47049
308. Zeng BB, Wu Y, Jiang S, Yu Q, Yao ZJ, Liu ZH, Li HY, Li Y, Chen XG, Wu YL (2003) Studies on mimicry of naturally occurring annonaceous acetogenins: non-THF analogues leading to remarkable selective cytotoxicity against human tumor cells. Chem Eur J 9:282
309. Rodier S, Le Huerou Y, Renoux B, Doyon J, Renard P, Pierre A, Gesson JP, Gree R (2001) New cytotoxic analogues of annonaceous acetogenins. Anticancer Drug Des 16:109
310. Fujita D, Ichimaru N, Abe M, Murai M, Hamada T, Nishioka T, Miyoshi H (2005) Synthesis of non-THF analogs of acetogenin toward simplified mimics. Tetrahedron Lett 46:5775
311. Liu HX, Huang GR, Zhang HM, Wu JR, Yao ZJ (2007) annonaceous acetogenin mimics bearing a terminal lactam and their cytotoxicity against cancer cells. Bioorg Med Chem Lett 17:3426
312. Xiao Q, Liu Y, Qiu Y, Zhou G, Mao C, Li Z, Yao ZJ, Jiang S (2011) Potent antitumor mimetics of annonaceous acetogenins embedded with an aromatic moiety in the left hydrocarbon chain part. J Med Chem 54:525
313. Ye Q, Shi G, He K, McLaughlin JL (1996) Chlorinated annonaceous acetogenins and their bioactivities. J Nat Prod 59:994
314. Kojima N, Hayashi H, Suzuki S, Tominaga H, Maezaki N, Tanaka T, Yamori T (2008) Synthesis of C4-fluorinated solamins and their growth inhibitory activity against human cancer cell lines. Bioorg Med Chem Lett 18:6451
315. Gallardo T, Saez J, Granados H, Tormo JR, Velez ID, Brun N, Torres B, Cortes D (1998) 10-Oximeguanacone, the first nitrogenated acetogenin derivative found to be a potent inhibitor of mitochondrial complex I. J Nat Prod 61:1001
316. Duret P, Hocquemiller R, Gantier JC, Figadère B (1999) Semisynthesis and cytotoxicity of amino acetogenins and derivatives. Bioorg Med Chem 7:1821
317. Derbre S, Roue G, Poupon E, Susin Santos A, Hocquemiller R (2005) Annonaceous acetogenins: the hydroxyl groups and THF rings are crucial structural elements for targeting the mitochondria, demonstration with the synthesis of fluorescent squamocin analogues. ChemBioChem 6:979
318. Alexander MD, Burkart MD, Leonard MS, Portonovo P, Liang B, Ding X, Joullie MM, Gulledge BM, Aggen JB, Chamberlin AR, Sandler J, Fenical W, Cui J, Gharpure SJ, Polosukhin A, Zhang HR, Evans PA, Richardson AD, Harper MK, Ireland CM, Vong BG, Brady TP, Theodorakis EA, La Clair JJ (2006) A central strategy for converting natural products into fluorescent probes. ChemBioChem 7:409
319. Maezaki N, Urabe D, Yano M, Tominaga H, Morioka T, Kojima N, Tanaka T (2007) Synthesis of fluorescent solamin for visualization of cell distribution. Heterocycles 73:159
320. Kojima N, Morioka T, Yano M, Suga Y, Maezaki N, Tanaka T (2009) Convergent synthesis of fluorescence labeled solamin. Heterocycles 79:387

321. Liu HX, Huang GR, Zhang HM, Jiang S, Wu JR, Yao ZJ (2007) A structure-activity guided strategy for fluorescent labeling of annonaceous acetogenin mimetics and their application in cell biology. ChemBioChem 8:172
322. Mao C, Han B, Wang LS, Wang S, Yao ZJ (2011) Modular assembly of cytotoxic acetogenin mimetics by click linkage with nitrogen functionalities. MedChemComm 2:918
323. Xiao Q, Liu Y, Qiu Y, Yao Z, Zhou G, Yao ZJ, Jiang S (2011) Design, synthesis of symmetrical bivalent mimetics of annonaceous acetogenins and their cytotoxicities. Bioorg Med Chem Lett 21:3613
324. Le Ven J, Schmitz-Afonso I, Lewin G, Brunelle A, Touboul D, Champy P (2014) Identification of the environmental neurotoxins annonaceous acetogenins in an *Annona cherimola* Mill. alcoholic beverage using HPLC-ESI-LTQ-orbitrap. J Agric Food Chem 62:8696
325. McLaughlin JL (2008) Paw paw and cancer: annonaceous acetogenins from discovery to commercial products. J Nat Prod 71:1311
326. Cuendet M, Oteham CP, Moon RC, Keller WJ, Peaden PA, Pezzuto JM (2008) Dietary administration of *Asimina triloba* (paw paw) extract increases tumor latency in *N*-methyl-*N*-nitrosourea-treated rats. Pharm Biol 46:3
327. Chen Y, Chen JW, Xu SS, Wang Y, Li X, Cai BC, Fan NB (2012) Antitumor activity of annonaceous acetogenins in HepS and S180 xenografts bearing mice. Bioorg Med Chem Lett 22:2717
328. Chen Y, Chen JW, Liu SI, Xu CL, Xu HQ, Cai BC, Li X, Ju WZ (2012) Determination of bullatacin in rat plasma by liquid chromatography–mass spectrometry. J Chromatogr B 897:94
329. Escobar-Khondiker M, Hoellerhage M, Muriel MP, Champy P, Bach A, Depienne C, Respondek G, Yamada ES, Lannuzel A, Yagi T, Hirsch EC, Oertel WH, Jacob R, Michel PP, Ruberg M, Hoeglinger GU (2007) Annonacin, a natural mitochondrial complex I inhibitor, causes tau pathology in cultured neurons. J Neurosci 27:7827

Chih-Chuang Liaw received his B.S. (1991) and M.S. (1994) degrees in Marine Resources at National Sun Yat-sen University, Kaohsiung, Taiwan and a Ph.D. in pharmaceutical science from Kaohsiung Medical University (KMU, 2004), Taiwan. At KMU, under the guidance of Prof. Yang-Chang Wu, he studied bioactive annonaceous acetogenins from plants of the genus *Annona* collected in Taiwan. In 2005, he worked as a postdoctoral research associate at the Graduate Institute of Natural Products, KMU, and elucidated the structures of several Taiwanese herbal constituents. From 2006 to 2010, he was an Assistant Professor of the Graduate Institute of Pharmceutical Chemistry, China Medical University, Taichung, Taiwan, focusing on the biological principles of medicinal ferns and fungi. In 2010, he moved to his current position as Associate Professor in the Department of Marine Biotechnology and Resources, National Sun Yat-sen University. His current research relates to the discovery of novel secondary metabolites from marine microbes and their biomedical applications.

Jing-Ru Liou was born in Taoyuan, Taiwan, in 1985. She received a B.S. degree, majoring in Pharmacy and Dental Hygiene, from Kaohsiung Medical University (KMU), Kaohsiung, Taiwan in 2009, and a M.S. degree (2011) from the Fauclty of Pharmacy and Graduate Institute of Natural Products (GINP), KMU. Currently at KMU, under the guidance of Professors Yang-Chang Wu and Fang-Rong Chang, she is a Ph.D. student. Her present work is focused on pharmaceutical analysis, inclusive of the discovery of bioactive natural products from *Annona* species.

Tung-Ying Wu received his B.S. degree (2001) in Medicinal Chemistry from Chia Nan University of Pharmacy and Science, and a M.S. degree (2005) from the Institute of Pharmaceutical Science of Taipei Medical University (TMU), Taiwan. During 2006–2008, he was an Assistant Research Fellow at the Institute of Pharmaceutics, Development Center for Biotechnology (DCB), Taipei, Taiwan. Currently, he is a Ph.D. student working under Prof. Fang-Rong Chang, at the Faculty of Pharmacy and Graduate Institute of Natural Products (GINP), Kaohsiung Medical University (KMU), Taiwan. His present research deals with the investigation of Taiwanese medicinal mushrooms and herbs.

Fang-Rong Chang was born in Tainan, Taiwan, in 1966. He received a B.S. degree majoring in Nutrition from Chung Shan Medical University (CSMU) in 1988, and M.S. and Ph.D. degrees in Pharmacognosy from the College of Pharmacy, Kaohsiung Medical University (KMU), Taiwan in 1991 and 1995. In 1998 and 2001, he performed postdoctoral research with Prof. K.-H. Lee at the University of North Carolina, and in 2004 he performed further postdoctoral work at Okayama University of Science, Okayama, Japan. During 1998–2005, he was an Assistant and Associate Professor in Pharmacognosy, Faculty of Pharmacy and the Graduate Institute of Natural Products (GINP), KMU, becoming a Full Professor in 2005. During 2005–2006, he was the R&D leader of College of Pharmacy, KMU, and during 2006–2012, Director of GINP. Presently, he is Professor of Pharmacognosy of GINP and the Vice Dean of Global Affairs of KMU. In 1998, he won the Outstanding Academic Alumnus, CSMU, Taiwan. In 2004 and 2009, he received a Best Young Researcher Award and an Outstanding Research Award, both of KMU. Up to the present, Prof Chang has published more than 250 research articles, and has more than 30 patents issued or filed. He has performed more than 90 invited lectures.

Yang-Chang Wu was born in Chiayi, Taiwan, in 1951. He received his B.S. degree in Pharmacy from Kaohsiung Medical University (KMU), Kaohsiung, Taiwan in 1975. He completed his M.S. (1982) and Ph.D. (1986) degrees in Pharmacognosy from the College of Pharmacy, KMU. He then joined the group of Prof. Yoshimasa Hirata from 1986 to 1987 at Meijo University, Nagoya, Japan as a postdoctoral researcher. He later joined the laboratory of Prof. Kuo-Hsiung Lee for further postdoctoral research at the University of North Carolina (UNC), Chapel Hill, North Carolina, USA. In 1986, he was promoted to Associate Professor at the College of Pharmacy at KMU, and was promoted to Professor in 1990, and became Director of the Graduate Institute of Natural Products (GINP) in 1992 at KMU. He was Dean of the Office of Research and Development at KMU, from 2006 to 2009. Then, he was selected as Chair Professor and Vice-President of the Graduate Institute of Integrated Medicine and the College of Chinese Medicine at China Medical University (CMU), Taiwan from 2010 to 2012. In 2012, he was appointed also as Dean of the School of Pharmacy, at CMU, Taiwan. The professional expertise of Prof. Wu concerns translational research on Chinese herbal medicines, functional foods, and new drug development. In 2007, he was cited by the Wang Ming-Ning Foundation for outstanding merit and high scholastic achievement in medical and pharmaceutical research. In 2009, he gained an outstanding research award from the National Science Council, Taiwan. Prof. Wu also received an outstanding medical and pharmaceutical technology prize in 2010 awarded by the TienTe Lee Biomedical Foundation, Taiwan. Prof. Wu serves as editorial board member of six journals, and he has published more than 430 research articles and has authored several book chapters. He has been named as co-inventor of more than 30 patents.

Erratum to: Dimeric Sesquiterpenoids

Shang-Gao Liao and Jian-Min Yue

Erratum to :
Chapter "Dimeric Sesquiterpenoids" in: A.D. Kinghorn, H. Falk, S. Gibbons, J. Kobayashi (eds.), *Progress in the Chemistry of Organic Natural Products, Vol. 101,* DOI 10.1007/978-3-319-22692-7_1

The affiliation of J.-M. Yue was incorrect. The correct affiliation is given below:

State Key Laboratory of Drug Research, Shanghai Institute of Materia Medica, Chinese Academy of Sciences, Zhangjiang Hi-Tech Park, Shanghai 201203, People's Republic of China
e-mail: jmyue@simm.ac.cn

The updated original online version for this chapter can be found at DOI 10.1007/978-3-319-22692-7_1

S.-G. Liao
School of Pharmacy, Guizhou Medical University, Zhangjiang Hi-Tech Park, Shanghai 201203, People's Republic of China

Engineering Research Center for the Development and Application of Ethnic Medicine and TCM (Ministry of Education), School of Pharmacy, Guizhou Medical University, Guizhou 550004, People's Republic of China
e-mail: lshangg@163.com

J.-M. Yue (✉)
State Key Laboratory of Drug Research, Shanghai Institute of Materia Medica, Chinese Academy of Sciences, Zhangjiang Hi-Tech Park, Shanghai 201203, People's Republic of China
e-mail: jmyue@simm.ac.cn

A.D. Kinghorn, H. Falk, S. Gibbons, J. Kobayashi (eds.), *Progress in the Chemistry of Organic Natural Products, Vol. 101*, DOI 10.1007/978-3-319-22692-7_3

MIX
Papier aus verantwortungsvollen Quellen
Paper from responsible sources
FSC® C105338

If you have any concerns about our products,
you can contact us on
ProductSafety@springernature.com

In case Publisher is established outside the EU,
the EU authorized representative is:
Springer Nature Customer Service Center GmbH
Europaplatz 3, 69115 Heidelberg, Germany

Printed by Libri Plureos GmbH
in Hamburg, Germany